Musculoskeletal Ultrasound

CLINICS IN DIAGNOSTIC ULTRASOUND VOLUME 30

EDITORIAL BOARD

Kenneth J. W. Taylor, M.D., Ph.D., Chairman

Diane Babcock, M.D.

Joshua A. Copel, M.D.

Barry B. Goldberg, M.D.

John C. Hobbins, M.D.

Joseph A. Kisslo, M.D.

George R. Leopold, M.D.

Christopher R. B. Merritt, M.D.

P. N. T. Wells, Ph.D., D.Sc.

Marvin C. Ziskin, M.D.

Volumes Already Published

Vol. 1 Diagnostic Ultrasound in Gastrointestinal Disease, Kenneth J. W. Taylor, Guest Editor

Vol. 2 Genitourinary Ultrasonography, Arthur T. Rosenfield, Guest Editor

Vol. 3 Diagnostic Ultrasound in Obstetrics, John C. Hobbins, Guest Editor

Vol. 4 Two-Dimensional Echocardiography, Joseph A. Kisslo, Guest Editor

Vol. 5 New Techniques and Instrumentation in Ultrasonography, P. N. T. Wells and Marvin C. Ziskin, Guest Editors

Vol. 6 Ultrasound in Cancer, Barry B. Goldberg, Guest Editor

Vol. 7 Ultrasound in Emergency Medicine, Kenneth J. W. Taylor and Gregory N. Viscomi, Guest Editors

Vol. 8 Diagnostic Ultrasound in Pediatrics, Jack O. Haller and Arnold Shkolnik, Guest Editors

Vol. 9 Case Studies in Ultrasound, Harris J. Finberg, Guest Editor

Vol. 10 Real-Time Ultrasonography, Fred Winsberg and Peter L. Cooperberg, Guest Editors

Vol. 11 Ultrasound in Inflammatory Disease, Anton E. A. Joseph and David O. Cosgrove, Guest Editors

Vol. 12 Ultrasound in Breast and Endocrine Disease, George R. Leopold, Guest Editor

Vol. 13 Vascular and Doppler Ultrasound, C. Carl Jaffe, Guest Editor

Vol. 14 Coordinated Diagnostic Imaging, Joseph F. Simeone, Guest Editor
Vol. 15 Gynecologic Ultrasound, William B. Steel and William J. Cochrane, Guest Editors
Vol. 16 Biological Effects of Ultrasound, Wesley L. Nyborg and Marvin C. Ziskin, Guest Editors
Vol. 17 Basic Doppler Echocardiography, Joseph A. Kisslo, David Adams, and Daniel B. Mark, Guest Editors
Vol. 18 Genitourinary Ultrasound, Hedvig Hricak, Guest Editor
Vol. 19 Ultrasound in Perinatology, Nabil F. Maklad, Guest Editor
Vol. 20 Controversies in Ultrasound, John P. McGahan, Guest Editor
Vol. 21 Interventional Ultrasound, Eric vanSonnenberg, Guest Editor
Vol. 22 Intraoperative and Endoscopic Ultrasonography, Matthew D. Rifkin, Guest Editor
Vol. 23 Gastrointestinal Ultrasonography, Alfred B. Kurtz and Barry B. Goldberg, Guest Editors
Vol. 24 Neonatal and Pediatric Ultrasonography, Diane S. Babcock, Guest Editor
Vol. 25 Diagnosis and Therapy of Fetal Anomalies, John C. Hobbins and Beryl R. Benacerraf, Guest Editors
Vol. 26 Duplex Doppler Ultrasound, Kenneth J. W. Taylor and D. Eugene Strandness, Jr., Guest Editors
Vol. 27 Doppler Color Imaging, Christopher R. B. Merritt, Guest Editor
Vol. 28 Advances in Ultrasound Techniques and Instrumentation, Peter N. T. Wells, Guest Editor
Vol. 29 Ultrasound in Gastroenterology, Paul A. Dubbins and A. E. A. Joseph, Guest Editors

Musculoskeletal Ultrasound

Edited by

Bruno D. Fornage, M.D.

Professor and Chief, Section of Ultrasound
Department of Diagnostic Radiology
The University of Texas M. D. Anderson Cancer Center
Houston, Texas

Churchill Livingstone
New York, Edinburgh, London, Melbourne, Tokyo

Library of Congress Cataloging-in-Publication Data

Musculoskeletal ultrasound / edited by Bruno D. Fornage.
    p. cm. — (Clinics in diagnostic ultrasound ; v. 30)
  Includes bibliographical references and index.
  ISBN 0-443-08909-4
  1. Musculoskeletal system—Ultrasonic imaging. I. Fornage,
Bruno. II. Series .
  [DNLM: 1. Musculoskeletal Diseases—ultrasonography.
2. Extremities—ultrasonography. W1 CL831BC v. 30 1995 / WE 141
M9859 1995]
RC925.7.M88 1995
616.7'07543—dc20
DNLM/DLC
for Library of Congress   94-44679
                                                                           CIP

© Churchill Livingstone Inc. 1995

All rights reserved. No part of this publication may be reproduced, stored in a retrieval system, or transmitted in any form or by any means, electronic, mechanical, photocopying, recording, or otherwise, without prior permission of the publisher (Churchill Livingstone, 650 Avenue of the Americas, New York, NY 10011).

Distributed in the United Kingdom by Churchill Livingstone, Robert Stevenson House, 1–3 Baxter's Place, Leith Walk, Edinburgh EH1 3AF, and by associated companies, branches, and representatives throughout the world.

Accurate indications, adverse reactions, and dosage schedules for drugs are provided in this book, but it is possible that they may change. The reader is urged to review the package information data of the manufacturers of the medications mentioned.

The Publishers have made every effort to trace the copyright holders for borrowed material. If they have inadvertently overlooked any, they will be pleased to make the necessary arrangements at the first opportunity.

Acquisitions Editor: *Miranda Bromage*
Production Editor: *Elizabeth A. Bowman-Schulman*
Production Supervisor: *Laura Mosberg Cohen*

Printed in the United States of America

First published in 1995   7 6 5 4 3 2 1

Contributors

Ronald S. Adler, M.D., Ph.D.
Associate Professor, Department of Radiology, University of Michigan Medical School; Director of Musculoskeletal Imaging, University of Michigan Medical Center, Ann Arbor, Michigan

Lori L. Barr, M.D.
Assistant Professor of Radiology and Pediatrics, Department of Radiology, University of Cincinnati College of Medicine; Staff Radiologist, Department of Radiology, Children's Hospital Medical Center, Cincinnati, Ohio

Germain Beauregard, M.D.
Clinical Assistant Professor, Department of Radiology, Université de Montréal Faculty of Medicine; Head, Musculoskeletal Division, Department of Radiology, Hôpital Sacré-Coeur, Montreal, Canada

J. Antonio Bouffard, M.D.
Senior Staff Radiologist, Department of Diagnostic Radiology, Henry Ford Hospital, Detroit, Michigan

Rethy K. Chhem, M.D.
Assistant Professor, Department of Radiology, McGill University Faculty of Medicine; Head, Musculoskeletal Division, Department of Radiology, The Montréal General Hospital, Montréal, Quebec, Canada

Luca De Flaviis, M.D.
Adjunct Professor, Institute of Radiology, University of Pavia, Milan, Italy

Asma Q. Fischer, M.D.
Associate Clinical Professor, Department of Pediatrics, Medical College of Georgia; Pediatric Neurologist and Neurosonologist, University Hospital, Augusta, Georgia

Bruno D. Fornage, M.D.
Professor and Chief, Section of Ultrasound, Department of Diagnostic Radiology, The University of Texas M. D. Anderson Cancer Center, Houston, Texas

Gretchen A. W. Gooding, M.D.
Professor and Vice Chairman, Department of Radiology, University of California, San Francisco, School of Medicine; Chief, Radiology Service, Department of Veterans Affairs Medical Center, San Francisco, California

Moshe Graif, M.D.
Associate Professor, Department of Diagnostic Radiology, Tel-Aviv University Sackler School of Medicine; Chairman, Department of Diagnostic Radiology, Tel-Aviv Sourasky Medical Center, Tel-Aviv, Israel

H. Theodore Harcke, M.D.
Professor, Departments of Radiology and Pediatrics, Jefferson Medical College of Thomas Jefferson University, Philadelphia, Pennsylvania; Chairman, Department of Medical Imaging, Alfred I. duPont Institute, Wilmington, Delaware

Laurence A. Mack, M.D.
Professor, Department of Radiology, University of Washington School of Medicine; Director of Ultrasound, Department of Radiology, University of Washington Medical Center, Seattle, Washington

Frederick A. Matsen III, M.D.
Professor and Chairman, Department of Orthopaedics, University of Washington School of Medicine; Chief, Division of Shoulder and Elbow Surgery, Department of Orthopaedics, University Hospital, Seattle, Washington

Maurizio Giulio Musso, M.D.
Assistant Surgeon, Department of Hand Surgery, Orthopedic and Trauma Center, Milan, Italy

Alex Powell, M.D.
Resident, Department of Diagnostic Radiology, Henry Ford Hospital, Detroit, Michigan

Glenn M. Strome, M.D.
Resident, Department of Diagnostic Radiology, Henry Ford Hospital, Detroit, Michigan

Marnix van Holsbeeck, M.D.
Director, Sections of Emergency Radiology and Skeletal Radiology, Department of Diagnostic Radiology, Henry Ford Hospital, Detroit, Michigan

Preface

Sonography's unique real-time capability, which permits examination during movement and allows guidance of biopsy needles, combined with the exquisite resolution of state-of-the-art high-frequency transducers and advances in color Doppler imaging makes sonography a powerful tool, in expert hands, for diagnosing abnormalities of the soft tissues, from the surface of the skin to the surface of the bones. Musculoskeletal sonography has been underused in the United States because of the availability of magnetic resonance imaging. However, sonography can often provide similar diagnostic information for only a fraction of the cost of MRI, and in this era of cost containment, sonography should be—as it is in Europe and other parts of the world—the first-line examination technique for many pathologic conditions of the soft tissues.

In this issue of the Clinics in Diagnostic Ultrasound, the most popular as well as the newest applications of musculoskeletal sonography are covered. Chapters on general topics are followed by chapters on specific anatomic segments of the extremities.

I offer my sincere thanks to the contributors, world-renowned experts in their fields, who have shared their outstanding insights, valuable time, and professional commitment to achieve our common goal of promoting this growing field of diagnostic ultrasound. I also am deeply indebted to Melissa G. Burkett of the Department of Scientific Publications at The University of Texas M. D. Anderson Cancer Center; her exceptional editorial and organizational skills were essential to the completion of this project.

It is my hope that the material in this issue of the Clinics in Diagnostic Ultrasound will encourage readers to champion the use of musculoskeletal sonography.

Bruno D. Fornage, M.D.

Contents

1	**Muscular Trauma** Bruno D. Fornage	1
2	**Neuromuscular Diseases** Asma Q. Fischer	11
3	**Soft-Tissue Masses** Bruno D. Fornage	21
4	**Synovial Diseases** Rethy K. Chhem and Germain Beauregard	43
5	**Bone and Articular Cartilage** Ronald S. Adler	59
6	**Peripheral Nerves** Moshe Graif	73
7	**Skin and Subcutaneous Tissues** Bruno D. Fornage	85
8	**Foreign Bodies** Gretchen A. W. Gooding	99
9	**Rotator Cuff** Laurence A. Mack and Frederick A. Matsen III	113
10	**Elbow** Lori L. Barr	135
11	**Hand and Wrist** Luca De Flaviis and Maurizio Giulio Musso	151

12	**Hip in Infants and Children** H. Theodore Harcke	179
13	**Knee** Glenn M. Strome, J. Antonio Bouffard, and Marnix van Holsbeeck	201
14	**Ankle and Foot** Marnix van Holsbeeck and Alex Powell	221
	Index	239

Color Plates

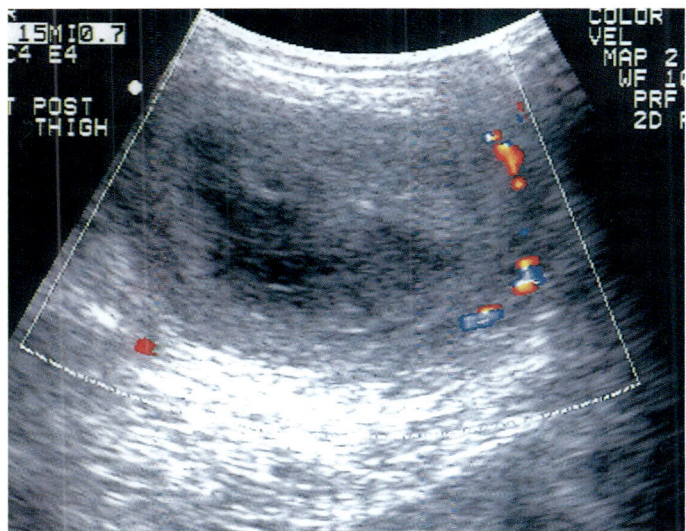

Plate 3-1

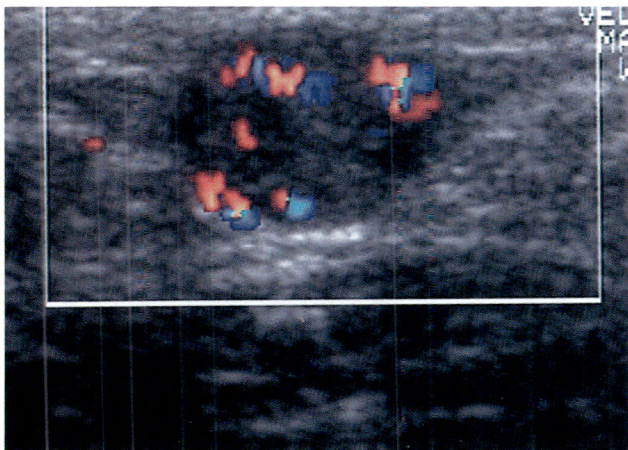

Plate 3-2

Plate 3-1 Abscess in the soft tissues of the posterior thigh. Sonogram shows a complex mass with thick irregular wall. Note the presence of color Doppler signals in the wall.

Plate 3-2 Subcutaneous angioma. Color Doppler study shows marked vascularity inside the tumor.

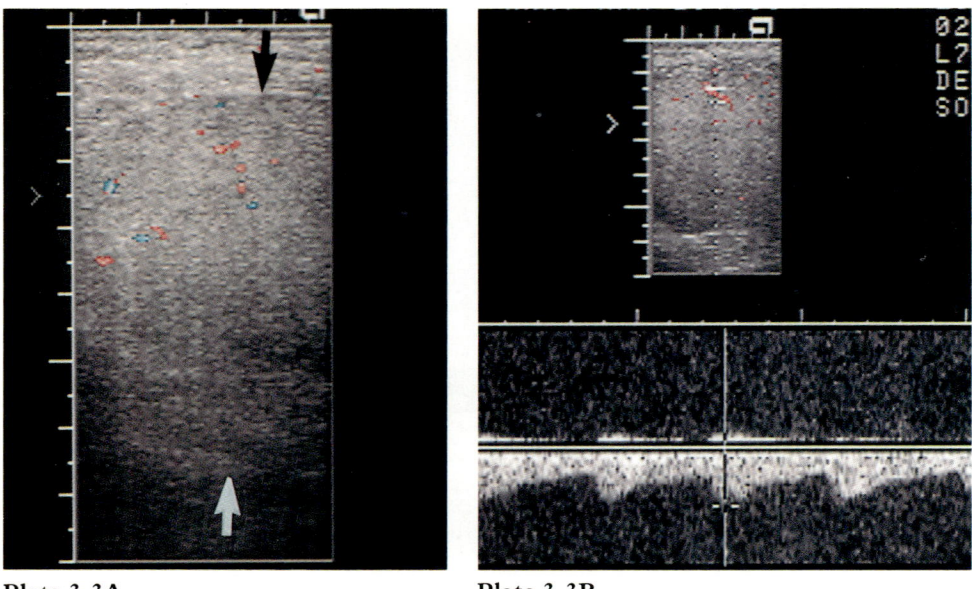

Plate 3-3A Plate 3-3B

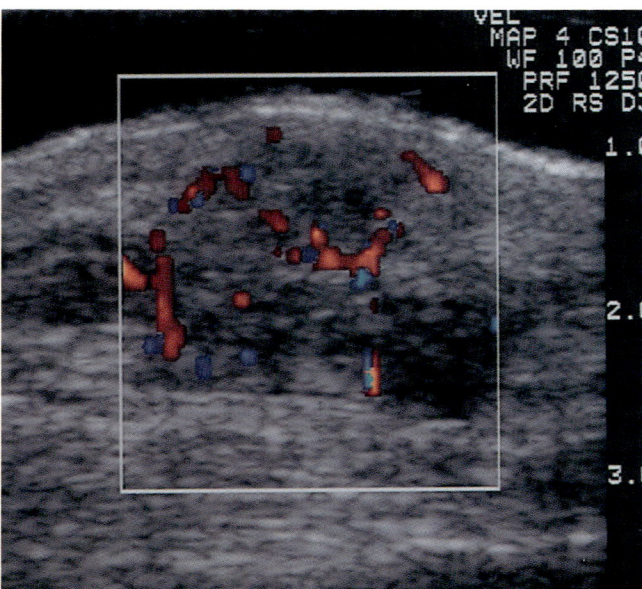

Plate 3-4

Plate 3-3 Myxoid liposarcoma of the thigh. **(A)** Color Doppler scan shows a large, diffusely echogenic tumor (*arrows*) containing several scattered color Doppler signals. **(B)** Spectral analysis reveals low-resistance blood flow.

Plate 3-4 Recurrent malignant fibrous histiocytoma in the arm. Color Doppler study shows chaotic hypervascularity inside the tumor, suggestive of malignancy.

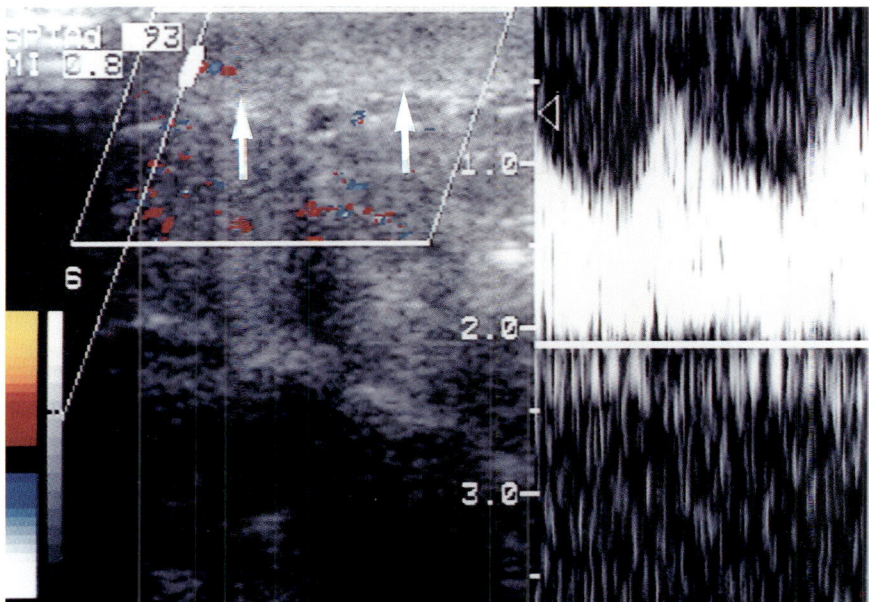

Plate 7-1

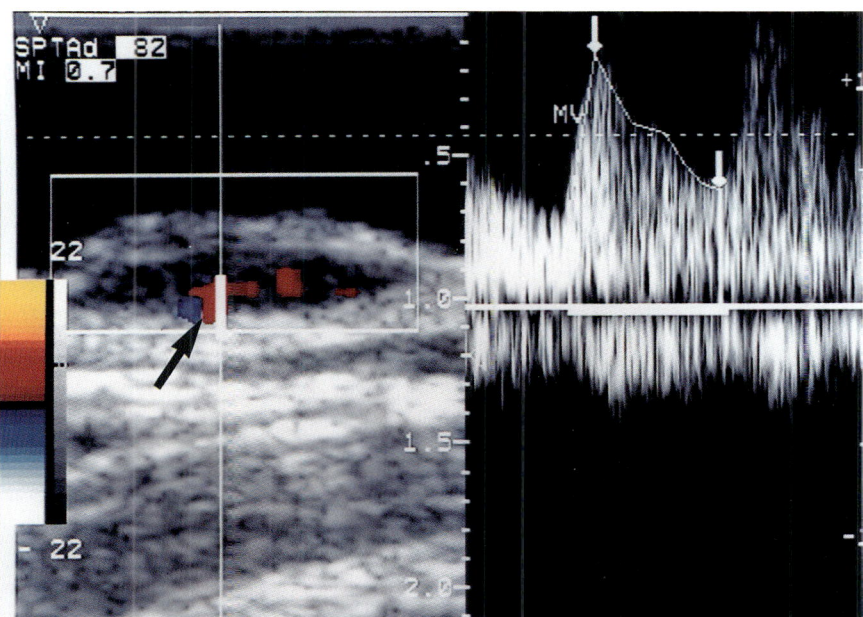

Plate 7-2

Plate 7-1 Postradiation skin edema. Color Doppler sonogram shows marked thickening of the skin (*arrows*) with decreased echogenicity and the presence of Doppler signals in the dermis. Spectral analysis (right) shows a low-resistance flow pattern.

Plate 7-2 Primary malignant melanoma of the skin. Color Doppler sonogram shows hypervascularity at the base of the tumor (*arrow*). Spectral analysis (right) shows a low-resistance flow pattern with a resistivity index of 0.53.

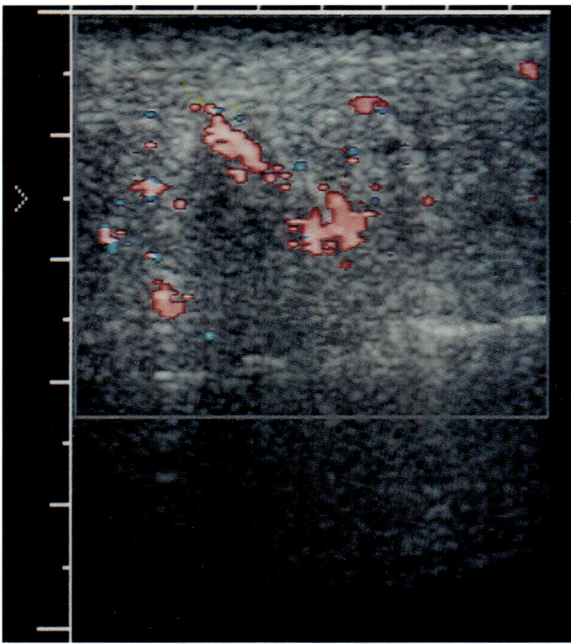

Plate 7-3

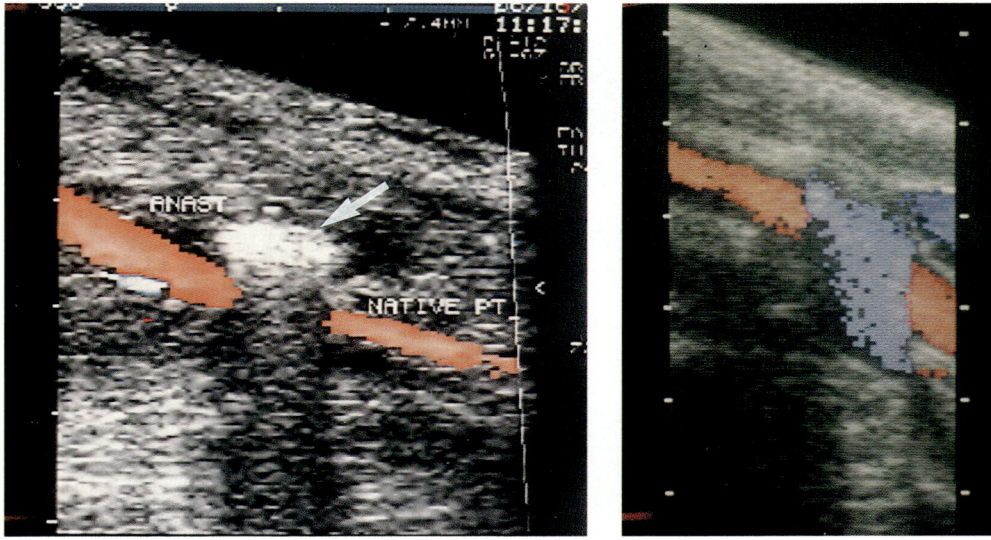

Plate 8-1   Plate 8-2

Plate 7-3 A woman presenting with a sore mass in the anterior abdominal wall. Color Doppler sonogram shows increased vascularity in the diffusely echogenic subcutaneous fat. The interface with the skin is no longer visible. Sonography confirms the diagnosis of cellulitis and rules out the presence of an abscess.

Plate 8-1 Longitudinal color Doppler sonogram shows in situ saphenous graft, partially obscured by a brightly echogenic overlying metallic clip (*arrow*).

Plate 8-2 Color Doppler sonogram of the common carotid artery shows artifactual flow reversal (in blue), which represents a mirror artifact phenomenon and misregistration of actual flow associated with the metal clamp.

1
Muscular Trauma

Bruno D. Fornage

The most common source of pathologic conditions involving muscles is trauma. The muscles most often affected are those of the extremities. Other common sites of muscular trauma include the neck, spine, and thoracic and abdominal walls. Because of the increased importance of athletic activities in social life, evaluation of trauma-related muscular conditions has become an everyday problem for physicians. Accurate cross-sectional imaging of muscles is important in initial assessment and follow-up of injuries, and close cooperation between radiologists, sports medicine physicians, and orthopedic surgeons has been a key factor in developing the application of high-resolution sonography for muscular trauma.

This chapter focuses on trauma-related injuries to muscles. Traumatic lesions of tendons are discussed in the chapters in which the relevant tendons are covered; for example, Achilles tendon injuries are discussed in the chapter on the ankle (Ch. 14).

TECHNICAL CONSIDERATIONS

Because of their wider field of view and higher resolution in the near-field, linear-array electronic transducers provide the best images of superficial structures. Large and deep muscles are best examined using 5-MHz and occasionally 3.5-MHz transducers, whereas the examination of superficial muscles or lesions in the distal extremities requires probes of higher frequency.[1] The use of a stand-off pad allows visualization of the skin and also palpation during real-time sonography for correlation of images and palpation findings.[2] Muscles should be examined at rest and during contraction. In some cases, it is helpful to scan the contralateral region to provide a normal anatomy reference. A combination of longitudinal and transverse scans is mandatory for three-dimensional localization of lesions and calculation of the volume of hematomas. When surgery is to be performed, preoperative sonographically guided skin marking of the projected location of a nonpalpable lesion is recommended to spare normal muscle as much as possible. In selected cases, intraoperative sonography can be used for localization of a lesion.[3]

With high-sensitivity color Doppler equipment, blood vessels can be identified in normal muscles, and increased blood flow during exercise can be demonstrated. The role of color Doppler imaging in muscle trauma has yet to be evaluated.

MECHANISMS OF TRAUMA

Muscular trauma can be categorized as external or direct, causing mostly contusions and—if severe—ruptures, and internal or indirect, causing strains and ruptures. Traumatic foreign bodies are discussed in Chapter 8.

Direct Trauma

Direct external trauma to a muscle can be caused by virtually any blunt object. During athletic activities, causes of direct trauma include collisions between two players and between a player and a stationary object such as a pole or even the ground. The muscles of the lower extremities are the most exposed. In a direct trauma, the involved muscle is crushed against the underlying bone(s). The resulting lesions can range from a bruise to a fast-expanding intramuscular hematoma requiring emergency surgery. When the muscle is under tension at the time of the trauma, contusion can result in a frank muscular rupture.

Any condition that causes ischemia of the muscle fibers, including prolonged compression or extensive direct trauma, can result in rhabdomyolysis, in which the integrity of the cell membranes is sufficiently altered to allow the release of cell contents into the extracellular fluid.

Indirect Trauma

Indirect (or internal) trauma does not involve an external agent; it is common in athletic activities. Indirect trauma can affect top-level athletes who exceed their physiologic limitations as well as nonathletes who put a normal-intensity stress onto a weakened or untrained muscle.

In indirect trauma, the muscle is stressed beyond its limits through excessive passive stretching or, more often, through sudden hypertension during contraction. Such hypertension can result from the sudden blockade of a normal movement (e.g., missing the ball and instead kicking the ground in soccer) or from performing a movement in a misaligned position. Different groups of muscles are involved depending on the sport. For example, the most frequently injured muscles in soccer are the quadriceps—in particular the rectus femoris muscle—and the hamstrings, whereas the gastrocnemius medialis muscle is often injured in high jumpers, long jumpers, and volleyball, basketball, and tennis players.[3]

NORMAL ANATOMY AND SONOGRAPHIC APPEARANCE

Skeletal muscles consist of muscle fibers grouped in fascicles separated from one another by septa of fibroadipose tissue, the perimysium; the whole muscle is wrapped in the epimysium. Blood supply is rich, and dramatic variations in blood flow can occur within the muscles according to the level of exercise.

The echo pattern is similar in all skeletal muscles. On longitudinal scans, the perimysium appears as oblique, parallel, echogenic striae against a markedly hypoechoic background representing the bulk of the muscle fibers (Fig. 1-1).[4,5] On transverse scans, the perimysium appears in cross-section as finely dotted echoes and short lines scattered throughout the hypoechoic background. Occasionally, large intramuscular echogenic septa can be visualized, resulting in a reticular pattern on transverse scans.[1] The intermuscular fasciae are brightly echogenic. During muscle contraction, real-time sonography demonstrates the changes in muscle shape and the change in orientation of the echogenic striae when the muscle is scanned longitudinally. Doppler studies can demonstrate intramuscular blood flow, even in ves-

MUSCULAR TRAUMA

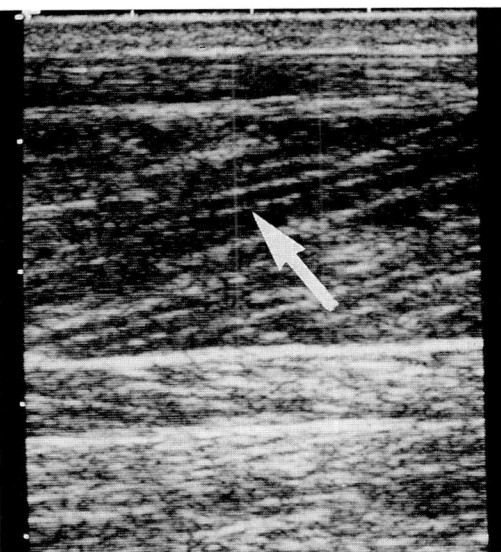

Fig. 1-1 Normal ultrasound anatomy of skeletal muscle. Longitudinal sonogram of the tibialis anterior muscle shows the typical arrangement of parallel, oblique, echogenic striae (*arrow*) against a hypoechoic background.

sels that are too small to be seen on B-mode scans.

SONOGRAPHIC FINDINGS IN TRAUMA

In muscular trauma, questions that sonography can help answer include: Is there a lesion? In which muscle or compartment is the lesion located? Is the muscle ruptured? What is the extent of the rupture and how large is the hematoma? How does the lesion change over time with conservative treatment?

Acute Lesions

STRAINS

Strains usually result from overstretching or overuse of muscles, resulting in stiffness and soreness. Because of the absence of a macroscopically detectable lesion, sonographic examination of strained muscles demonstrates no abnormalities.

CONTUSIONS

Minor contusions involve the crush of a few muscle fibers, with minor hemorrhaging and infiltration of blood between the fibers. In the few studies of the sonographic appearances of soft-tissue hematomas in humans, sonographic findings have varied significantly with the type and age of the hematoma.[6-14] At an early stage, a hemorrhagic suffusion appears as a diffuse, ill-defined area of increased echogenicity in the muscle. However, a recent intramuscular hematoma may be poorly defined with little change in echogenicity, so the enlargement of the injured muscle compared with the contralateral muscle is the sole sign of an intramuscular hematoma.[6] In such a case, it is recommended that the examination be repeated a few days later, by which time changes in the echogenicity of the hematoma should facilitate its detection. Subsequently, the hematoma becomes better defined, with an overall decrease in echogenicity (Fig. 1-2). It should be kept in mind that high-frequency transducers are better able to demonstrate fine, low-level internal echoes in a hematoma than are low-frequency probes. A clotted hematoma examined a few days after an injury is already markedly hypoechoic and can mimic a fluid-filled collection; at this point, any attempt at aspiration will fail. The hematoma then liquefies and is gradually resorbed (Fig. 1-3). At this late stage, it is not uncommon to observe echogenic fibrinous strands or septa in the resolving collection. Special reference must be made to muscular hematomas of the calf, which can mimic thrombophlebitis[7] and cause posterior compartment syndrome.[8]

MUSCULAR RUPTURES

Sonographic findings in muscular ruptures include the discontinuity of muscle fibers

4 MUSCULOSKELETAL ULTRASOUND

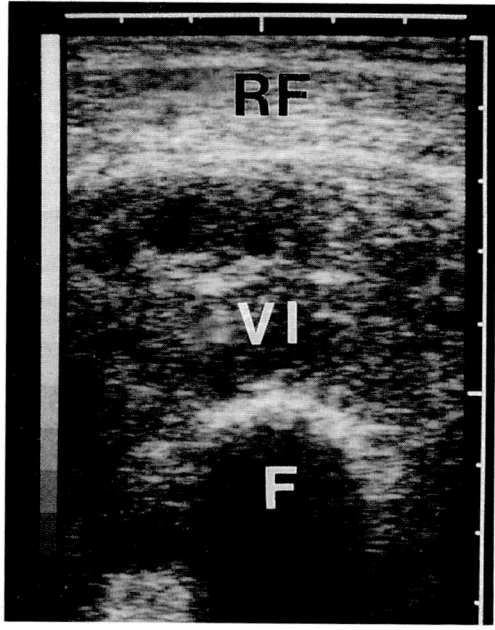

Fig. 1-2 Contusion of the anterior thigh with large hematoma of the vastus intermedius muscle. Transverse sonogram shows the enlarged vastus intermedius muscle, with a diffusely nonhomogeneous echotexture and overall decrease in echogenicity. F, femur; RF, rectus femoris; VI, vastus intermedius.

(direct sign) and the presence of an associated hematoma (indirect sign).[12]

Complete Ruptures

Complete ruptures are usually diagnosed by history and physical examination. In a complete rupture, sonograms show the retracted, echogenic muscle fragment surrounded by the hypoechoic hematoma, which is known as the *clapper-in-the-bell* sign (Fig. 1-4). However, identifying the torn fragments of the ruptured muscle can be difficult when the hematoma is small or already organized. Real-time examination during contraction better demonstrates the gap between the ruptured muscle fragments[12] (Fig. 1-5).

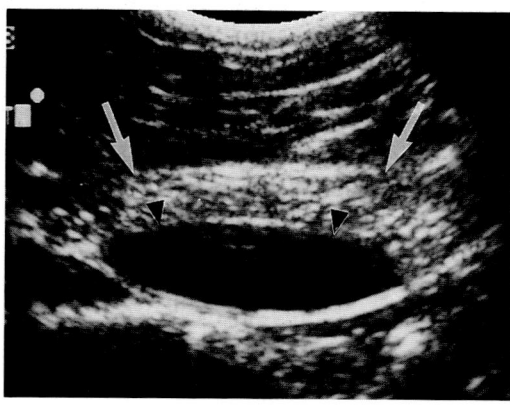

Fig. 1-3 Circumscribed hematoma in the sheath of the rectus abdominis muscle. Transverse sonogram of the anterior abdominal wall shows a liquefied anechoic hematoma (*arrowheads*) at the posterior aspect of the rectus abdominis muscle (*arrows*).

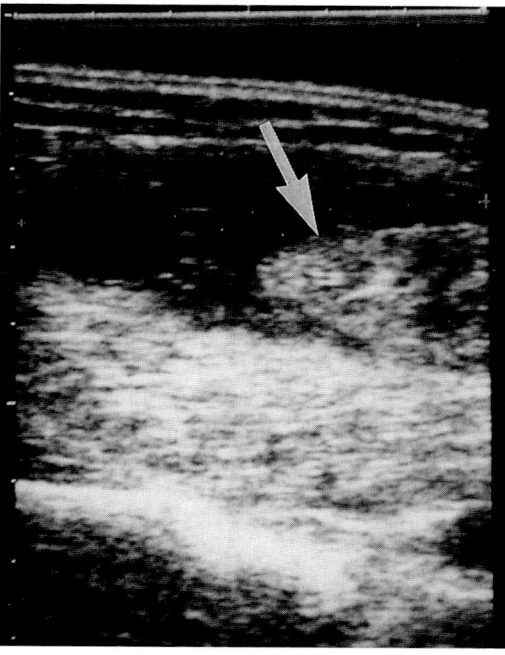

Fig. 1-4 Rupture of the long head of the biceps femoris muscle. Longitudinal sonogram of the posterior thigh shows the clapper-in-the-bell sign with the retracted, torn muscle (*arrow*) surrounded by the anechoic hematoma.

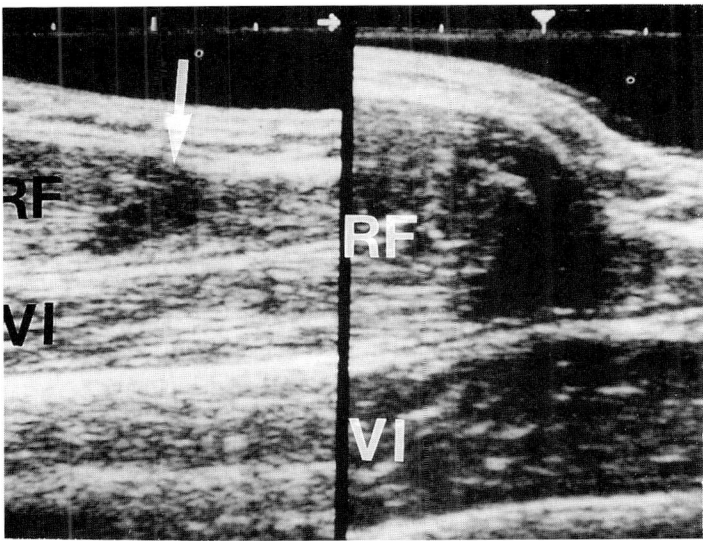

Fig. 1-5 Complete rupture of the lower rectus femoris muscle in a soccer player. **(Left)** Longitudinal sonogram of the anterior thigh at rest shows discontinuity of the perimysium/fibers with a hematoma filling the gap between the torn fragments (*arrow*). **(Right)** Longitudinal sonogram obtained during contraction of the quadriceps shows the round deformity of the proximal fragment; the distal fragment remains stationary. RF, rectus femoris; VI, vastus intermedius.

Partial Ruptures

Partial ruptures affect a limited number of fascicles. When a partial rupture is located superficially, the gap can be felt on palpation; however, when it lies more deeply in the muscle, only imaging techniques can demonstrate it. On sonograms, the discontinuity of muscle fibers is inferred from the discontinuity of the echogenic perimysial striae. The area of rupture shows jagged margins, and a surrounding hyperechoic halo is sometimes present (Fig. 1-6). When the associated hematoma is markedly hypoechoic, even a minute tear can be detected as a focal area of disruption of the smooth fibrillar echotexture of the muscle.[4] In fact, the sonographer should search for any focal change in the otherwise smooth architecture of a skeletal muscle; in the context of a recent trauma, such a change would indicate a partial tear. Examination during contraction is useful because contraction can enhance both the size and the contrast of the focal area of

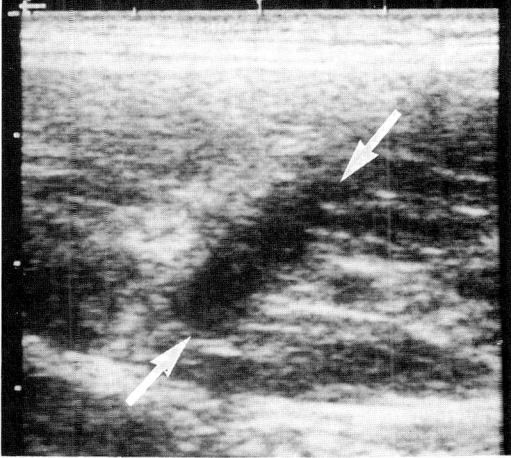

Fig. 1-6 Partial rupture of the rectus femoris muscle. Longitudinal sonogram of the upper portion of the rectus femoris muscle shows the partial discontinuity of the perimysium/fibers (*arrows*), with a small hematoma and a subtle echogenic rim around the rupture.

rupture. A clapper-in-the-bell sign can be present with partial ruptures and is pathognomonic of a rupture. A combination of longitudinal and transverse sonograms is required to determine the exact location and size of a partial rupture.

The extent of the rupture and the volume of the associated hematoma are crucial in deciding whether to perform surgical repair of a partial rupture. However, the extent of a partial tear may be difficult to determine in the acute phase or in the presence of a large associated hematoma. The volume of a hematoma can be derived from its largest diameters using the simplified volume formula for the ellipsoid: $V = 1/2 \ (D1 \times D2 \times D3)$, where D1, D2, and D3 are the three greatest diameters of the hematoma. Large hematomas are drained surgically, while the resorption of small ones can be monitored sonographically. Hematomas of small to moderate volume can also be aspirated under sonographic guidance, to relieve pain and perhaps accelerate the healing process.

A peculiar type of partial rupture is the partial detachment of muscle fibers from an adjacent fascia, the typical example being the detachment of the distal fibers of the gastrocnemius medialis muscle from the aponeurosis common to the gastrocnemii and soleus muscles, at the origin of the Achilles tendon. The detached distal portion of the muscle appears round and surrounded by a thin hemorrhagic collection, which extends superiorly along the aponeurosis (Fig. 1-7).

Sonography is ideal for the follow-up of minor muscular ruptures, which usually do not require surgical repair. To ensure reproducibility, it is best to have the follow-up examinations done by the same operator, using the same equipment, settings, and measuring technique, with the findings sketched on paper. In minor uncomplicated ruptures, the echotexture of the muscle returns to normal within a few weeks. Sonographic documen-

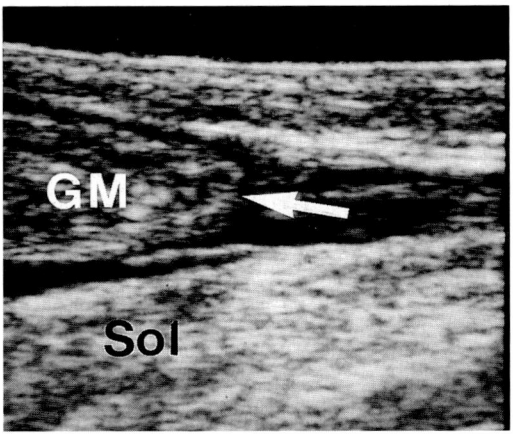

Fig. 1-7 Detachment of the lower end of the gastrocnemius medialis muscle from the common aponeurosis of the triceps surae. Longitudinal sonogram shows the round, retracted lower portion of the gastrocnemius medialis muscle (*arrow*) surrounded by a small hematoma containing a few echogenic fibrinous strands. GM, gastrocnemius medialis; Sol, soleus.

tation of the complete return to normal of the muscular echotexture is required before the patient resumes any full-strength occupational or athletic activity involving the injured muscle.

RHABDOMYOLYSIS

Traumatic causes of rhabdomyolysis include prolonged muscular compression, extensive direct trauma (e.g., crush injuries), and burns. Few reports on the sonographic appearances of rhabdomyolysis are found in the literature.[15-18] On sonograms obtained in the acute phase, the smooth architecture of the muscle is replaced with nonhomogeneous patchy areas of mixed echogenicity (Fig. 1-8). Later, the areas may appear more hypoechoic[15,16] and become fluid-filled collections.[17] Recently, magnetic resonance imaging has been reported to be more accurate than sonography and computed tomography (CT).[18]

MUSCULAR TRAUMA

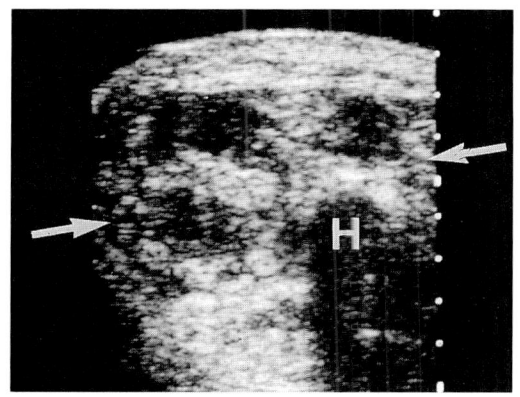

Fig. 1-8 Rhabdomyolysis of the triceps brachii muscle. Transverse sonogram of the posterior arm shows a diffusely nonhomogeneous enlarged muscle (*arrows*) with patchy areas of decreased and increased echogenicity. H, humerus.

Chronic Lesions

Chronic muscular lesions of traumatic origin include fibrous scars, cystic hematomas, myositis ossificans, and hernias.

FIBROUS SCARS

In most cases, a minor rupture will heal without sequelae. In some cases of neglected or recurrent ruptures, however, the presence of a granuloma impairs the regeneration of muscle fibers and results in the formation of a permanent fibrous scar. Sonographically, a focal, stellate, echogenic area is seen, often adhering to the epimysium[12] (Fig. 1-9). The abnormality does not change in shape during contraction of the muscle. The presence of a scar predisposes the muscle to recurrent rupture. In such a case, the preexisting chronic lesion may limit the accuracy of sonography in the detection and diagnosis of a new, acute traumatic change.

In the case of a chronic complete rupture, a thick, rope-like scar can develop between the two retracted torn fragments, resulting in a digastric muscle.

CYSTIC HEMATOMAS

Rarely, a hematoma does not resolve and turns into a cystic collection. The presence of such a fluid-filled collection prevents adequate healing and predisposes the muscle to recurrent rupture.

MYOSITIS OSSIFICANS

Significant hemorrhage from direct muscular trauma can result in myositis ossificans, the formation of heterotopic non-neoplastic bone in or adjacent to muscle and close to bone. It must be kept in mind that a minor but repeated trauma can have a similar consequence, with formation of an osteoma in a location that is typical for the occupational or athletic activity involved, e.g., the deltoid muscle in rifle-shooting, the brachialis muscle in fencing, the adductor muscles in horseback riding (cavalryman's osteoma), or the soleus muscle in ballet.

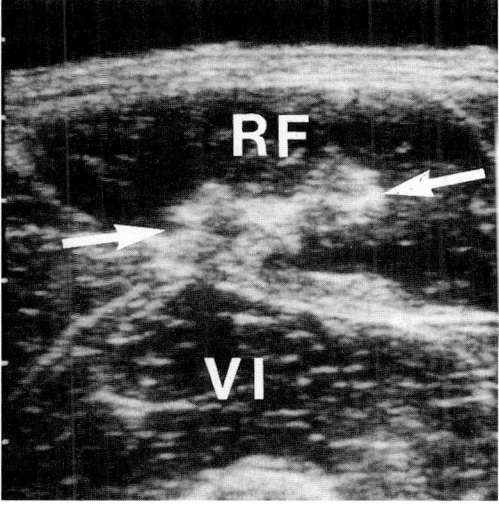

Fig. 1-9 Fibrous scar in the rectus femoris muscle. Transverse sonogram of the thigh shows a stellate hyperechoic area (*arrows*) abutting the epimysium of the rectus femoris muscle. RF, rectus femoris; VI, vastus intermedius.

8 MUSCULOSKELETAL ULTRASOUND

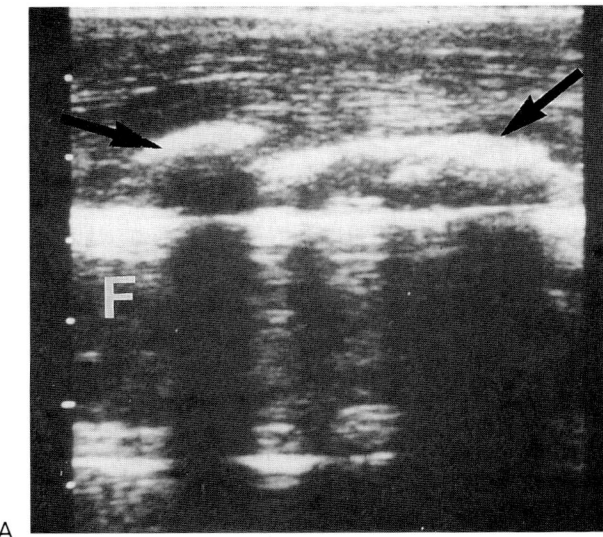

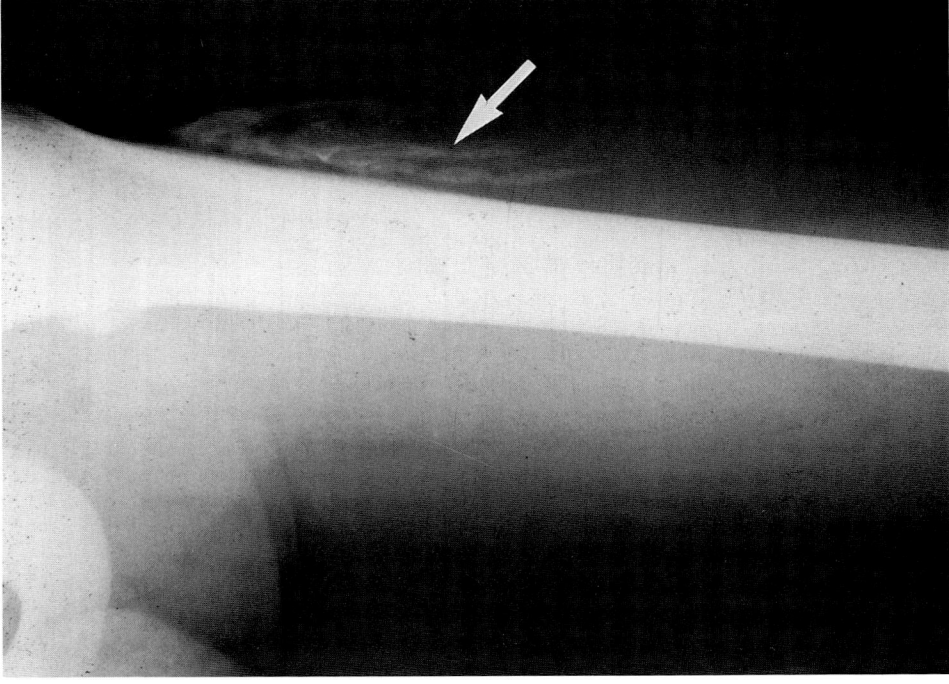

Fig. 1-10 Myositis ossificans complicating a post-traumatic hematoma in the vastus intermedius muscle. **(A)** Longitudinal sonogram obtained 4 weeks after injury shows large calcifications (*arrows*) at the site of the hematoma with shadowing partially obscuring the underlying femur (F). **(B)** Low-kilovoltage lateral radiograph of the thigh shows an ill-defined, faint calcific density (*arrow*) oriented parallel to the femur diaphysis.

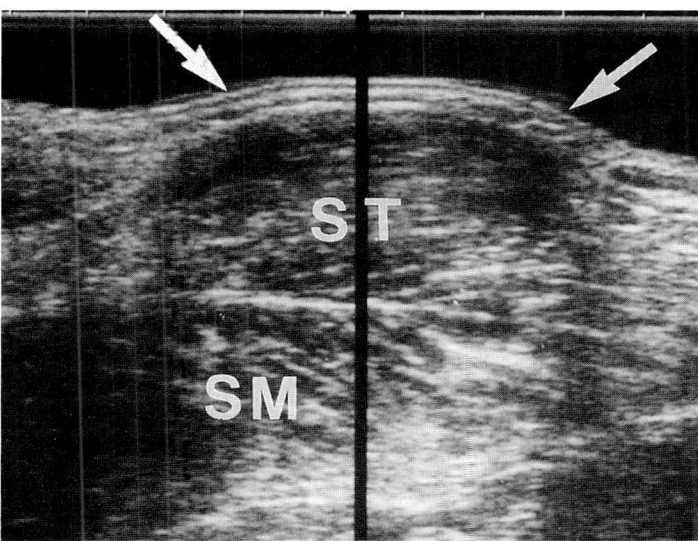

Fig. 1-11 Muscular hernia of the semitendinosus muscle. Montage of two contiguous longitudinal sonograms of the posterior thigh obtained during contraction confirms that the palpable mass (*arrows*) is made of normal muscular tissue bulging through a weakened aponeurosis. SM, semimembranosus; ST, semitendinosus.

One of the most frequent locations for traumatic myositis ossificans is the quadriceps muscle. In a series of 60 rugby players who had a hematoma in the quadriceps, myositis ossificans was found in 10 (17 percent).[19] In another study, myositis ossificans developed in 13 of 18 (72 percent) young military recruits who had sustained a severe trauma to the quadriceps muscle.[20] The relationship between the severity of the trauma and the subsequent development of myositis ossificans is not clear.[19,20] Massages of the injured area at the acute phase are thought to be a contributing factor.

In the early stage of myositis ossificans, lamellar calcifications appear on sonograms as sheets of echogenic material.[21] As calcifications become coarser, they appear as large hyperechoic foci with typical acoustic shadowing, usually aligned parallel to the adjacent diaphysis[3,22] (Fig. 1-10). This differs from the typical thin, rim-like pattern of calcific deposits usually seen in nontraumatic myositis ossificans (see Ch. 3). The calcifications can usually be confirmed on plain radiographs after the fourth week. In areas of complex osseous anatomy, CT may be needed to document the ossification.

MUSCULAR HERNIAS

The bulge of a herniated muscle through a weakened or ruptured aponeurosis or fascia is readily seen by sonography when a stand-off pad is used. Real-time sonography during contraction of the involved muscle confirms that the palpable mass is actually made of normal muscle tissue[3] (Fig. 1-11). The rectus femoris muscle, the hamstrings, and the tibialis anterior muscle are most often involved.

CONCLUSION

Because of its unrivaled cost-effectiveness in imaging soft tissues and high negative predictive value, sonography should be used first (after physical examination and, when a bone lesion may be associated, plain radiography) to screen patients for the presence of trauma-related muscular lesions. A negative sonographic examination performed by a well-trained operator using state-of-the-art equipment rules out a traumatic lesion (except for strains) and avoids unnecessary exploratory surgery. In complete muscular ruptures, the role of sonography is limited because the diagnosis is readily made on the basis of physical examination alone. In contrast, sonography is instrumental in detecting a small partial rupture, although assessing the extent of the rupture is not always easy, particularly if the injury is recent. In chronic lesions, sonography readily confirms the presence of a focal abnormality, which may require further imaging if clinically indicated.

REFERENCES

1. Fornage BD: Ultrasonography of Muscles and Tendons. Examination Technique and Atlas of Normal Anatomy of the Extremities. Springer-Verlag, New York, 1988
2. Fornage BD, Touche DH, Rifkin MD: Small parts real-time sonography: a new "waterpath." J Ultrasound Med 3:355, 1984
3. Fornage BD: Echographie des Membres [Sonography of the Extremities]. Vigot, Paris, 1991
4. Fornage BD: Musculoskeletal evaluation. p. 1. In Mittelstaedt CA (ed): General Ultrasound. Churchill Livingstone, New York, 1992
5. Dock W, Grabenwoger F, Happak W et al: [Sonography of the skeletal muscles using high-frequency ultrasound probes]. ROFO 152:47, 1990
6. Aspelin P, Pettersson H, Sigurjonsson S, Nilsson IM: Ultrasonographic examinations of muscle hematomas in hemophiliacs. Acta Radiol 25:513, 1984
7. Giyanani VL, Grozinger KT, Gerlock AJ et al: Calf hematoma mimicking thrombophlebitis: sonographic and computed tomographic appearance. Radiology 154:779, 1985
8. Auerbach DN, Bowen AD III: Sonography of leg in posterior compartment syndrome. AJR 136:407, 1981
9. Kaftori JK, Rosenberger A, Pollack S, Fish JH: Rectus sheath hematoma: ultrasonographic diagnosis. AJR 128:283, 1977
10. Kumari S, Fulco JD, Karayalcin G, Lipton R: Gray scale ultrasound: evaluation of iliopsoas hematomas in hemophiliacs. AJR 133:103, 1979
11. Wicks JD, Silver TM, Bree RL: Gray scale features of hematomas: an ultrasonic spectrum. AJR 131:977, 1978
12. Fornage BD, Touche DH, Segal P, Rifkin MD: Ultrasonography in the evaluation of muscular trauma. J Ultrasound Med 2:549, 1983
13. Giannini S, Lipparini M, Della VS et al: Ultrasonography in pathological conditions of muscles, tendons and joints. Ital J Orthop Traumatol 13:253, 1987
14. Aspelin P, Ekberg O, Thorsson O et al: Ultrasound examination of soft tissue injury of the lower limb in athletes. Am J Sports Med 20:601, 1992
15. Fornage BD, Nérot C: Sonographic diagnosis of rhabdomyolysis. J Clin Ultrasound 14:389, 1986
16. Vukanovic S, Hauser H, Curati WL: Myonecrosis induced by drug overdose: pathogenesis, clinical aspects, and radiological manifestations. Eur J Radiol 3:314, 1983
17. Kaplan GN: Ultrasonic appearance of rhabdomyolysis. AJR 134:375, 1980
18. Lamminen AE, Hekali PE, Tiula E et al: Acute rhabdomyolysis: evaluation with magnetic resonance imaging compared with computed tomography and ultrasonography. Br J Radiol 62:326, 1989
19. Jackson DW, Feagin JA: Quadriceps contusions in young athletes. Relation of severity of injury to treatment and prognosis. J Bone Joint Surg 55A:95, 1973
20. Rothwell AG: Quadriceps hematoma. A prospective clinical study. Clin Orthop 171:97, 1982
21. Peck RJ, Metreweli C: Early myositis ossificans: a new echographic sign. Clin Radiol 39:586, 1988
22. Kramer FL, Kurtz AB, Rubin C, Goldberg BB: Ultrasound appearance of myositis ossificans. Skeletal Radiol 4:19, 1979

2
Neuromuscular Diseases
Asma Q. Fischer

TECHNIQUE

Real-time ultrasound has been used successfully to evaluate normal and diseased muscles in children and adults.[1-6]

Muscle sonography should be performed using a standardized procedure, to minimize variation between tests and between patients. The standard method for performing muscle sonography has been well described in the literature.[7,8]

Since muscle depth and characteristics change with age, several healthy subjects in the age groups of newborn to 12 months, 12 months to 5 years, and 5 years to adult should be examined sonographically to establish age-matched norms for the individual ultrasound laboratory. Using high-resolution real-time ultrasound equipment, the depth-gain compensation, overall gain, preprocessing, postprocessing, and persistence settings can be optimized for patient age and the thickness of the muscle being examined. The optimized combinations for each age group are then stored in the equipment's memory for use in sonographic examination of patients with neuromuscular diseases.

Muscle sonography is ideally performed with the patient supine and in a comfortable position. The muscle groups to be examined are best imaged if they are relaxed. Children can usually be reassured of the benign nature of the test by sliding the transducer on the examiner's or the parent's arm. Sonography should be performed in transverse and longitudinal planes for all four limbs, with hard copies being obtained at approximately the midsection of each limb. This enables delineation of anatomy, identification of pathology, and measurement of the absolute and relative thicknesses of muscle and subcutaneous fat. Minor instrumental adjustments may be needed while scanning patients with neuromuscular diseases, but major departures from the age-standardized settings can obscure subtle findings.

Most patients with neuromuscular diseases can be evaluated by general sampling of the muscles using the above method; however, in cases of focal disease—such as a localized peripheral neuropathy—it is advisable to scan the muscles innervated by the affected peripheral nerve. A similar unaffected muscle in the opposite or same limb should be scanned for comparison.

12 MUSCULOSKELETAL ULTRASOUND

Table 2-1. Descriptive Terms Used in Interpretation of Sonograms in Neuromuscular Diseases

Sonographic Variable	Standard Descriptive Terms
Echogenicity	
Intensity	Euechoic, hyperechoic, hypoechoic
Distribution	Homogeneous, heterogeneous
Visibility of bony interfaces	Normal, diminished, indeterminate, absent
Visibility of bone shadow	Normal, diminished, indeterminate, absent
Visibility of fascial interfaces	Normal, diminished, indeterminate, absent

Using a standardized method for interpretation of sonograms provides reliability among interpreters. Five major characteristics are assessed and described using standard terms (Table 2-1), and the thicknesses of subcutaneous fat and muscle are measured.

In healthy subjects, fascial planes and bony interfaces appear hyperechoic; the bones cast crisp shadows. Normal muscle parenchyma has a low- to mid-level echogenicity and is described as euechoic. The exact signal intensity that corresponds with the term euechoic differs slightly between age groups (Figs. 2-1 and 2-2).[7,8]

MYOPATHIES

Myopathic diseases are those in which almost all the skeletal muscles are involved in the disease process; therefore, sampling the

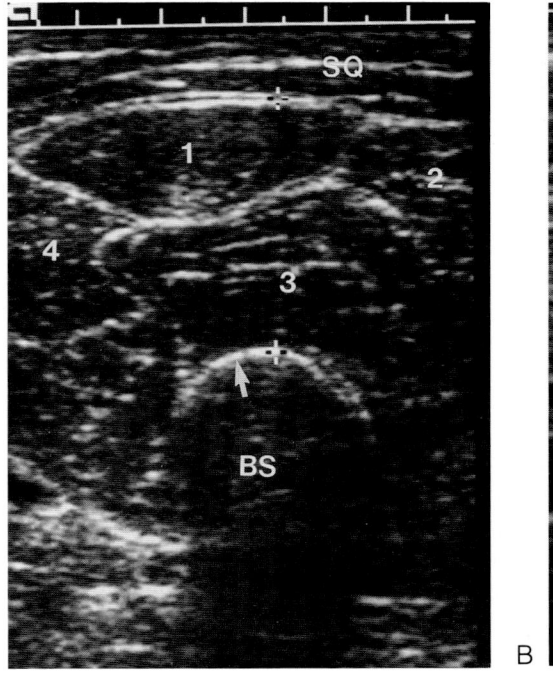
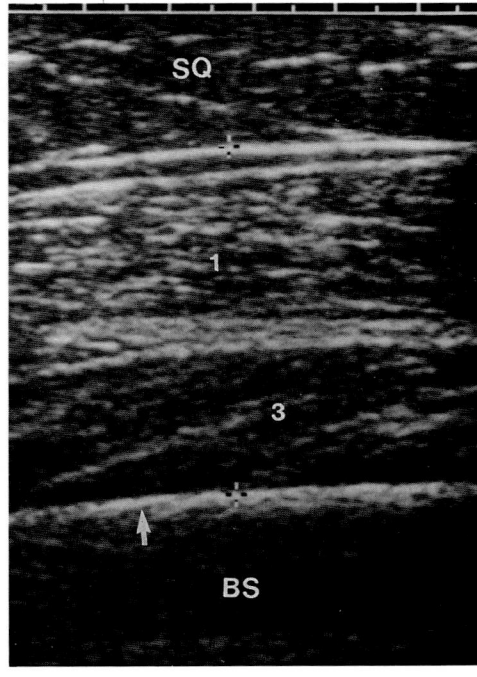

Fig. 2-1 (A) Transverse and **(B)** longitudinal sonograms of the left midthigh in a healthy adult. 1, rectus femoris; 2, vastus lateralis; 3, vastus intermedius; 4, vastus medialis; area between + cursors, muscle thickness; *arrow,* bone interface; BS, bone shadow; SQ, subcutaneous fat.

NEUROMUSCULAR DISEASES

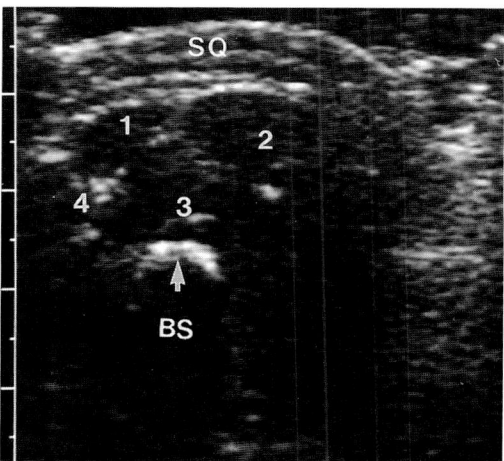

Fig. 2-2 Transverse sonogram of the left mid-thigh in an infant with no muscle disease. 1, rectus femoris; 2, vastus lateralis 3, vastus intermedius; 4, vastus medialis; *arrow*, bone interface; BS, bone shadow; SQ, subcutaneous tissue or fat. (From Fischer et al,[9] with permission.)

midarm and midthigh is sufficient. However, if the patient has an inflammatory myopathy, it is advisable to scan the clinically weak muscles and then compare their appearance to that of the clinically unaffected muscles since the distribution of pathology is patchy in such diseases. Table 2-2 outlines the general sonographic findings in patients with myopathies and compares them to the findings in neuropathies. The hallmarks of myopathies are increased background echogenicity and decreased distinction of myofascial planes, bone edge interfaces, and shadows cast by the bones.[9,10]

Progressive Muscular Dystrophies

In the progressive muscular dystrophies, which include Duchenne-type muscular dystrophy, limb-girdle muscular dystrophy, and Emery-Dreyfuss dystrophy, the degree of abnormality on sonograms often reflects the stage of the disease (Figure 2-3); thus the greater the abnormal collection of fibrous tissue, the degeneration of muscle, and the replacement of muscle with other tissue, the more abnormal the sonograms will appear. The muscle is mildly affected initially, and an increase in the muscle parenchymal echogenicity without major loss of anatomic landmarks is noted. As the disease ravages the muscle architecture, these changes are seen on the sonograms as a marked increase in granular echogenicity in the muscle parenchyma, reminiscent of a snowstorm (Fig. 2-3). This blizzard-like echogenic haze obliterates the usual landmarks in the worst cases and causes diminution of the landmarks in mild cases.

MYOTONIC DYSTROPHIES

The sonographic findings in myotonic dystrophies are varied. Some reports indicate abnormally increased echogenicity, but at other times the sonograms may be normal. Therefore, a normal sonogram for a patient suspected of having a myotonic dystrophy cannot rule out the disease.[11]

Table 2-2. Sonographic Characteristics of Myopathies and Neuropathies

Sonographic Characteristic	Myopathies	Neuropathies
Echogenicity	Increased	Increased
Echotexture	Homogeneous	Heterogeneous
Site of abnormalities	Entire muscle	Part of muscle
Visibility of bony interfaces	Decreased +++	Decreased ++
Visibility of bone shadow	Diminished	Relatively preserved
Affected muscles	Proximal or all	Distal or focal
Muscle thickness	Increased or unchanged	Decreased

Table 2-3. Sonographic Characteristics of Progressive Muscular Dystrophies and Inflammatory Myopathies

Sonographic Characteristic	Progressive Muscular Dystrophies	Inflammatory Myopathies
Echogenicity	Increased	Increased
Echotexture	Homogeneous	Homogeneous
Visibility of bony interfaces	Diminished	Preserved
Visibility of bone shadow	Diminished	Mostly preserved
Affected muscles	All	Some (patchy pattern)
Muscle thickness	Increased or unchanged	Decreased or unchanged

CONGENITAL MUSCULAR DYSTROPHIES

Patients with Fukuyama's or occidental type of congenital muscular dystrophies present in early infancy with hypotonia.[12,13] Sonographic evaluation of these patients reveals increased echogenicity in the quadriceps femoris muscle; this increase may be generalized or may appear selectively in the vastus muscles, sparing the rectus femoris muscle.[14]

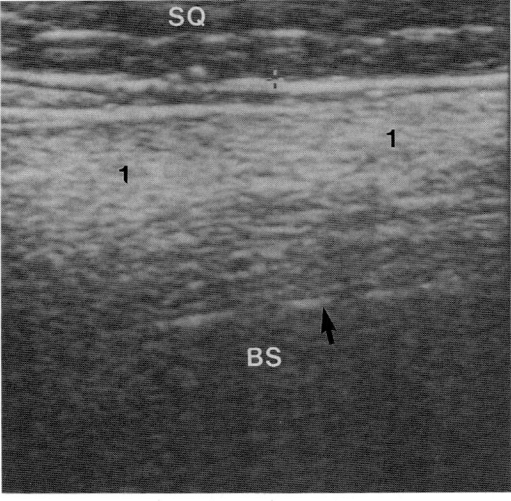

Fig. 2-3 Longitudinal sonogram of the midthigh in a child with advanced Duchenne-type muscular dystrophy. + Cursor, muscle–subcutaneous fat interface; SQ, subcutaneous fat; 1, hyperechoic muscle parenchyma (obliterating the myofascial interfaces); arrow, diminished bone interface; BS, bone shadow (no longer crisp).

Congenital Myopathies

In congenital myopathies, such as nemaline myopathy, central core disease, and myotubular myopathy, sonograms may appear normal early on. In the advanced disease stage, a mild-to-moderate increase in echogenicity may become noticeable.[11]

Metabolic Myopathies

Patients with glycogen storage disease and Kearns-Sayre disease usually have normal sonograms. Mitochondrial myopathies have been reported to cause increased echogenicity of the muscles, but the sonographic findings vary.[10]

Inflammatory Myopathies

Polymyositis and dermatomyositis are characterized by patchy muscle involvement. The individual muscles affected by inflammatory myopathies have increased echogenicity (Table 2-3). A patchy distribution of hyperechoic muscles in the extremities is seen on sonograms, along with a relatively decreased visibility of bone edge interfaces and myofascial planes (Fig. 2-4A). This pattern of patchy echogenicity alters as the disease progresses or regresses, affecting new muscles or burning out in previously affected muscles.[14,15] These abnormal sonographic findings have been reported to correlate well with biopsy findings.[16] In the late stages of inflammatory myopathies, the increase in echogenicity that previously may

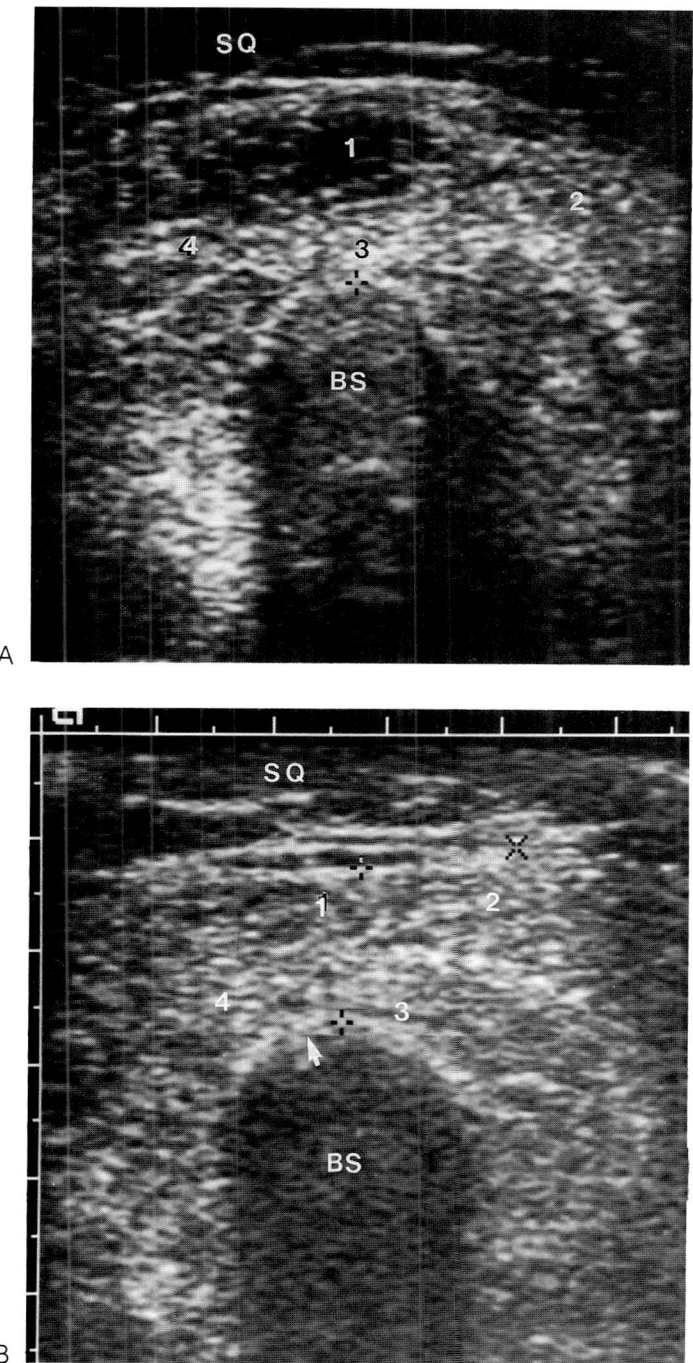

Fig. 2-4 Transverse sonograms of the midthigh in a child with polymyositis. **(A)** Sonogram at a stage of moderate involvement. 1, rectus femoris (normal echogenicity or relatively hypoechoic); 2, 3, and 4, hyperechoic vastus lateralis, intermedius, and medialis muscles; + *cursor*, bone interface; BS bone shadow (no longer crisp); SQ, subcutaneous fat. **(B)** Sonogram at a more advanced stage of disease with severe weakness. Area between the + *cursors* represents muscle thickness. All four elements of the quadriceps (1, 2, 3, 4) are hyperechoic, with loss of distinction of the myofascial planes. *Arrow,* bone interface (diminished in clarity); BS, bone shadow (present but not crisp); SQ, subcutaneous fat.

have been seen only in isolated muscles may be present in all muscles (Fig. 2-4B). With regression of disease, the muscles usually do not return to a completely normal state. Instead, they retain a mildly increased echogenicity or reflect signs of denervation if the disease has been severe and of long duration.[14]

Sonography is particularly useful in biopsy site selection. Ideally, the biopsy specimen should be taken from a diseased muscle site, but since involvement is patchy, site selection can be difficult. Visualization of the diseased muscle by ultrasound gives optimum site selection for biopsy material.[17]

NEUROPATHIES

Table 2-2 outlines the general sonographic appearances of neuropathic conditions.

Peripheral and Focal Neuropathies

Effective imaging of the muscles in a patient with a peripheral neuropathy requires knowledge of the muscles supplied by the abnormal nerve. In the path of the affected peripheral nerve, signs of denervation will be manifested as areas of low echogenicity alternating with areas of normal muscle (Table 2-4) (Fig. 2-5).[9]

Hereditary Sensorimotor Neuropathy

Sonography of the distal muscles in patients with hereditary sensorimotor neuropathy reveals an increase in muscle echogenicity and a loss of muscle mass as a result of atrophy. The distribution of sonographic abnormalities in patients with this disorder is proximal in some cases; the lower extremities may show greater involvement than the upper extremities. In some cases of type 1 hereditary sensorimotor neuropathy, the muscles may appear sonographically normal.[10]

Spinal Muscular Atrophy

The classic pattern of findings in early but established floppy infant syndrome with spinal muscular atrophy is a heterogeneous increase in muscle parenchymal echogenicity (Table 2-4). A pattern of heterogeneous hyperechogenicity is found within the muscle parenchyma of each muscle, giving it a "moth-eaten" appearance; i.e., areas of increased echogenicity with islands of hypoechoic or normal echogenicity are seen within one muscle. The changes in echogenicity of the muscle parenchyma may be associated with apparent decreased echogenicity of bone edge interfaces and indistinguishable myofascial interfaces. The acoustic shadowing associated with bones remains discernible until late in the disease process. The moth-eaten appearance of the muscle parenchyma is the earliest diagnostic change detected by ultrasound (Fig. 2-6).[9,18]

Table 2-4. Sonographic Characteristics of Focal/Peripheral Neuropathies and Spinal Muscular Atrophy

Sonographic Characteristic	Focal/Peripheral Neuropathies	Spinal Muscular Atrophy
Echogenicity	Mild increase	Moderate increase
Echotexture	Heterogeneous	Heterogeneous
Distribution of lesions	Focal	Multiple muscles
Atrophy	Focal	General (late finding)

NEUROMUSCULAR DISEASES 17

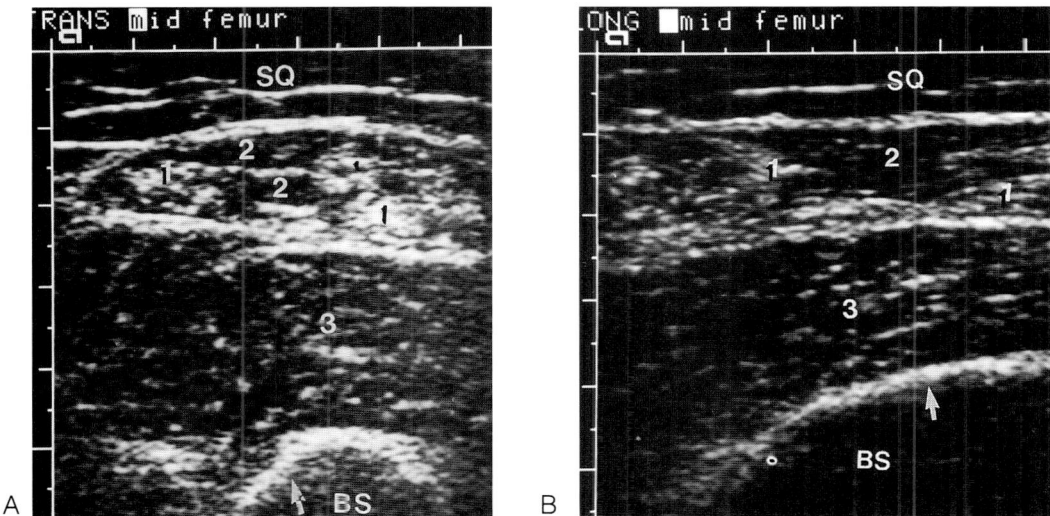

Fig. 2-5 (A) Transverse and **(B)** longitudinal sonograms of the midthigh in an adult with focal neuropathy leading to atrophy of the rectus femoris. 1, hyperechoic areas within the rectus femoris; 2, hypoechoic areas within the rectus femoris; 3, vastus intermedius with normal echogenicity; *arrow*, bone interface; BS, normal bone shadow; SQ, subcutaneous tissue. (From Fischer et al,[9] with permission.)

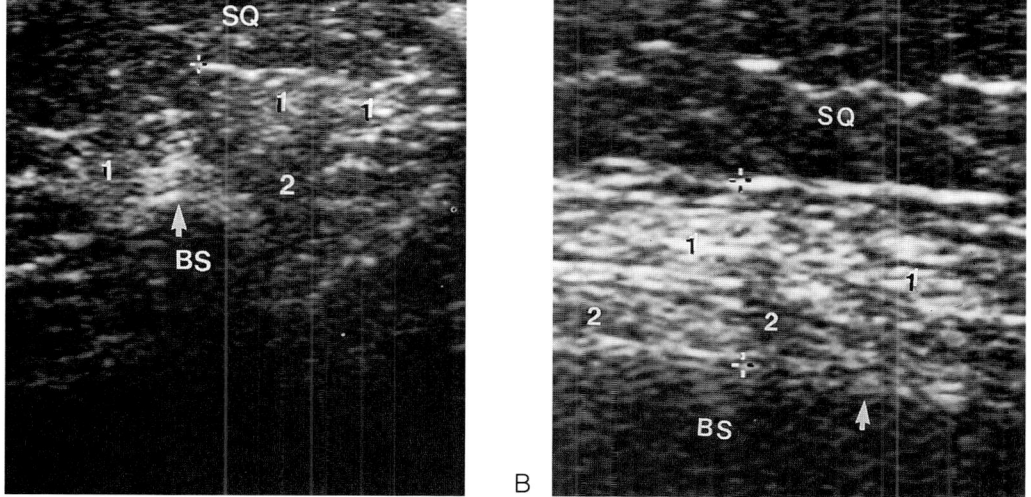

Fig. 2-6 (A) Transverse and **(B)** longitudinal sonograms of the midthigh in a child with spinal muscular atrophy. 1, hyperechoic areas within the quadriceps muscle; 2, areas of relatively normal echogenicity within the quadriceps muscle; *arrow*, smeared bone interface; BS, bone shadow; + *cursor*, muscle–subcutaneous fat interface; SQ, subcutaneous fat. (From Fischer et al,[9] with permission.)

The abnormal ultrasound findings in spinal muscular atrophy are usually found diffusely in all muscle groups. This differentiates it from focal peripheral neuropathies, in which similar sonographic findings of denervation may be found localized to one area of nerve distribution.

DIFFERENTIAL DIAGNOSIS

Several nonneuromuscular disorders can mimic neuromuscular disease in their clinical presentations. Floppy infant syndrome with cerebral palsy, Prader-Willi syndrome, or lax ligaments can mimic symptoms of neuromuscular disease; however, on sonography, the muscles of these patients usually appear normal.[9,10]

CONCLUSIONS

Most progressive myopathies can be visualized by sonography. The intensity and extent of findings depend on the severity of the disease. The majority of metabolic or storage myopathies are not easily identifiable with ultrasound unless significant architectural distortion has taken place.[9] Sonography is most useful in the management of inflammatory myopathies, in which the affected muscles can be followed objectively by sonography during therapy, keeping in mind that the sonographic changes in the muscles may lag behind or precede the clinical changes.

Most neuropathies are also evident on sonography. The abnormal muscles can be identified if the distribution of the affected nerve is known to the examiner. Neuropathies such as spinal muscular atrophy can be identified by the classic moth-eaten pattern of the muscle parenchyma throughout the arms and thighs. Neuromuscular diseases affecting the neuromuscular junction, such as myasthenia gravis, cannot be visualized with current methods of sonography and remain a challenge for ultrasound researchers.

Since most neuromuscular diseases are progressive, an initial normal sonogram may not rule out such a disease. Serial sonograms in patients whose symptoms are progressing are required both for diagnostic purposes as well as for timely planning of muscle biopsy and electromyography. In this age of fiscal restraint, it is best to identify the abnormal muscles sonographically prior to launching an extensive and expensive investigation with other techniques. However, this is feasible only for the diseases for which the sonographic findings are well established and have shown a good correlation with biopsy findings, such as most of the progressive muscular dystrophies, spinal muscular atrophies, and inflammatory myopathies.[11,16,19]

REFERENCES

1. Fischer AQ: Pediatric applications of clinical ultrasound. Neurol Clin 8:759, 1990
2. Heckmatt JZ, Dubowitz V: Diagnosis of spinal muscular atrophy with pulse echo ultrasound imaging. p. 141. In Gamstorp I, Sarnat HB (eds): Progressive Spinal Muscular Atrophies. Raven Press, New York, 1984
3. Hicks JE, Shawker TH, Jones BL et al: Diagnostic ultrasound: its use in the evaluation of muscle. Arch Phys Med Rehabil 65:129, 1984
4. Fincher RME, Jackson MJ, Fischer AQ: Pyomyositis caused by *Citrobacter freundii:* a case and analysis of the diagnostic utility of neuromuscular ultrasound in two additional cases. Am J Med Sci 299:331, 1990
5. Young A, Hughes I, Russel P et al: Measurements of quadriceps muscle wasting by ultrasonography. Rheumatol Rehabil 19:141, 1980
6. Bowen PA II, Wynn JJ, Fischer AQ et al: Nontropical pyomyositis in a renal allograft recipient. Transplantation 47:539, 1989
7. Fischer AQ, Stephens S: Computerized real-time neuromuscular sonography: a new application, techniques, and methods. J Child Neurol 3:69, 1988

8. Fornage BD: Ultrasonography of Muscles and Tendons: Examination Technique and Atlas of Normal Anatomy of the Extremities. Springer-Verlag, New York, 1989
9. Fischer AQ, Carpenter DW. Hartlage PL et al: Muscle imaging in neuromuscular disease using computerized real-time sonography. Muscle Nerve 11:270, 1988
10. Heckmatt JZ, Dubowitz V: Real-time ultrasound imaging of muscles. Muscle Nerve 11:56, 1988
11. Lamminen A, Jaaskelainen J, Juhani R, Suramo I: High frequency ultrasonography of skeletal muscle in children with neuromuscular disease. J Ultrasound Med 7:505, 1988
12. McMenamin JB, Becker LE, Murphy EG: Congenital muscular dystrophy: a clinicopathologic report of 24 cases. J Pediatr 100:692, 1982
13. Fukuyama Y, Osawa M, Suzuki H: Congenital progressive muscular dystrophy of the Fukuyama type: clinical, genetic and pathological considerations. Brain Dev 3:1, 1981
14. Topaloglu H, Gucuyener K, Yalaz K et al: Selective involvement of the quadriceps muscle in congenital muscular dystrophies: an ultrasonic study. Brain Dev 14:84, 1992
15. Fischer AQ, Hartlage PL, Carroll J: Inflammatory myopathies of childhood: diagnosis and follow-up by computerized real-time sonography. Ann Neurol 22:451, 1987
16. Fischer AQ, Longenecker E: Muscle sonography in inflammatory myopathic disease. Ann Neurol 30:502, 1991
17. Fischer AQ, Longenecker E, Trefz J: Muscle sonography and biopsy correlates in inflammatory myopathic disease. Ann Neurol 32:438, 1992
18. Heckmatt JZ, Dubowitz V: Diagnostic advantage of needle biopsy in the diagnosis of selective involvement in muscle disease. J Child Neurol 2:205, 1987
19. Fischer AQ, Longenecker E: Spinal muscular atrophy. In Fleckenstein JL (ed): Imaging Skeletal Muscle in Health and Disease. Springer-Verlag, New York (in press)

3
Soft-Tissue Masses
Bruno D. Fornage

Imaging of soft-tissue masses was long limited to plain radiography or xeroradiography. Although sonography gained acceptance in differentiating between cystic and solid masses, the numerous advantages of magnetic resonance imaging (MRI) in imaging soft tissues has shifted clinicians' interest toward that modality. However, high-frequency sonography provides diagnostic information that is often similar to that obtained with MRI, at a much lower cost and with greater availability. This chapter emphasizes the important role of sonography in the diagnosis of soft-tissue masses.

TECHNICAL CONSIDERATIONS

Instrumentation

Because of their wider field of view and better resolution in the near field, flat linear-array electronic transducers are more appropriate than mechanical or phased-array sector scanners for the evaluation of superficial structures.[1] As the transducer's nominal frequency increases, the depth of the field of view decreases, as usually does its width; thus, scans obtained with most 7.5- or 10-MHz linear-array transducers are only about 3 to 4 cm wide. While 7.5- or 10-MHz probes are needed for the evaluation of very superficial structures such as tendons, distal extremities, and subcutaneous tissues, probes of lower frequency (5 MHz or even 3.5 MHz), which offer a wider field of view and greater penetration, are needed to encompass a large anatomic segment or mass and to visualize a deeply located structure.[2] Also, most real-time scanners have the capability of displaying two and sometimes three adjacent scans on the video monitor (split-screen mode). This permits visualization and measurement of structures or lesions whose size exceeds the width of the probe, although extreme care must be taken when juxtaposing the contiguous scans on the screen.

Standoff pads of anechoic gel-like material are available to aid in visualizing the skin and most superficial structures.[3] Because the pad adds to the beam's pathway, only thin pads should be used with high-frequency transducers.

Color Doppler imaging and spectral analysis are available on most high-end scanners. Evaluation of the degree of vascularity of a mass and analysis of flow patterns are easy to perform. Color Doppler findings may provide invaluable information regarding the pathophysiology of a solid soft-tissue mass. Refinements in the resolution and sensitivity of Doppler equipment may soon provide a close-to-angiography quality of vascular display. This opens the possibility of using the appearance of the vascular tree as a diagnostic criterion, in addition to the mere presence or absence of flow, the overall flow distribution, and the Doppler resistivity and pulsatility indices, to differentiate benign from malignant tumors.

Although the concept of an ultrasound scanner that automatically scans a volume, stores the received echoes, and displays reconstructed three-dimensional views, including coronal scans that would otherwise be impossible to obtain, is appealing,[4] technical advances, particularly in decreasing reconstruction times, are needed before this feature offers practical value. Currently, it is difficult to predict the future of this technique in the evaluation of soft-tissue masses.

Equipment that will permit elasticity mapping of the tissues examined along with conventional B-mode sonography is being developed.

Technique of Examination

It is good practice to begin the examination by taking a careful history and performing a rapid physical examination. Longitudinal and transverse scans are then obtained. Any bulging deformity of the skin surface can be demonstrated on sonograms when a standoff pad is used. The standoff pad also allows optimal contact between a flat linear-array transducer and the skin in areas with uneven surfaces. When a standoff pad is used, it is possible for the operator to slide the fingers of one hand between the pad and the skin while maintaining the transducer over the region of interest so that a mass can be palpated under "sonoscopy," thus establishing a definite correlation between the findings of physical and sonographic examinations. This palpation under sonoscopy also allows the operator to appreciate the elasticity of the mass.[2]

The movements of the mass during dynamic maneuvers such as flexion or extension of the adjacent muscles may also yield valuable information about the relationship between the mass and adjacent structures and therefore about the structure from which it is derived. For example, a nerve sheath tumor in an extremity can be displaced transversely but not along the axis of the nerve and will remain immobile during flexion and extension of the surrounding muscles.

When doubt exists about the presence of an abnormality in a given area, examination of the contralateral area may provide a helpful reference for normal ultrasound anatomy.

A combination of longitudinal and transverse scans is required for three-dimensional localization of lesions and calculation of their volume. The latter usually is done by applying the formula for the volume of a prolate ellipsoid, which is approximated by dividing the product of the three largest diameters of the mass by two.[2]

Ultrasound-Guided Interventional Procedures

Real-time sonography is ideal for guiding needle biopsy of soft-tissue masses. Fluid collections can be aspirated with fine (20- or 22-gauge) needles, although the presence of thick material such as pus may require the use of a larger needle. Tissue diagnosis of solid tumors, particularly the diagnosis of soft-tissue sarcomas, is best achieved with the use of large-core needles and automatic biopsy devices, such as the Biopty gun (Bard, Covington, GA) or the Argovac system (Argon, Athens, TX). To obtain adequate cores, 18-gauge or larger needles should be used. Some devices offer an adjustable (2–4 cm) needle throw; the longer the throw, the better the core.

Sonographically guided localization of nonpalpable masses can be done using any of the techniques routinely used for localization of nonpalpable breast lesions, such as preoperative skin marking, injection of methylene blue or carbon particles, insertion of a localizing needle or hookwire, or intraoperative scanning with either a sterile transducer or a nonsterile transducer that has been gowned in a sterile protective sheath.[2]

PSEUDOTUMORS

Many soft-tissue lumps result from traumatic, inflammatory, or infectious processes or cystic changes and are not true neoplasms.

Masses of Traumatic Origin

Trauma to the superficial soft tissues can result in hematomas, muscular ruptures, muscular hernias, myositis ossificans, and rhabdomyolysis, all of which can present clinically as palpable soft-tissue masses. These conditions are discussed in chapter 1. Retained foreign bodies can be surrounded by an inflammatory reaction, also resulting in a palpable soft-tissue mass. Diagnosis of foreign bodies with sonography is discussed in chapter 8.

Nontraumatic Myositis Ossificans

Nontraumatic heterotopic ossification is associated with a number of neurologic disorders, including spinal cord injuries, closed head injuries, encephalitis, multiple sclerosis, chronic infections, burns, and poliomyelitis, but it can also develop apparently de novo in a healthy individual. The most common locations for this to occur are the shoulder, arm, elbow, buttocks, and thighs.[5] Clinically, there is local swelling, warmth, redness, and limitation of the range of motion of the affected segment. The differential diagnosis includes a hematoma, deep venous thrombosis, cellulitis or abscess, and extraskeletal osteosarcoma.[6] Early in the process, sonography demonstrates a nonspecific hypoechoic area,[7] while color Doppler study confirms the lesion's hypervascularity. Within 3 or 4 weeks, deposition of echogenic calcific material at the periphery of the lesion can be seen on sonograms, first without and then with shadowing (Fig. 3-1).[7-9] The typical rim ossification pattern, confirmed by plain radiography or computed tomography (CT), differentiates the condition from an osteogenic soft-tissue sarcoma, whose calcifications are coarser and located more centrally. By showing the absence of connection between the calcified mass and the adjacent bone, sonography can rule out a parosteal osteosarcoma. Although in many cases the combination of plain films and sonography is sufficient for diagnosis and follow-up, CT may be needed in selected cases because of its superiority in demonstrating ossified material.

Inflammatory Masses

Palpable masses can result from infection or inflammation of muscles, tendons or their sheaths, synovial bursae, and subcutaneous tissues. Evaluation of the response of inflammatory conditions to conservative treatment can be done cost effectively with sonography.

SOFT-TISSUE ABSCESSES

Sonography is sensitive in the detection of soft-tissue fluid collections, including abscesses, which usually appear as anechoic or complex, uniloculated or multiloculated masses with thick, irregular walls.[10,11] When present, gas collection is easily identified through the associated "ring-down" artifact. Color Doppler study demonstrates increased vascularity around and in the wall of the

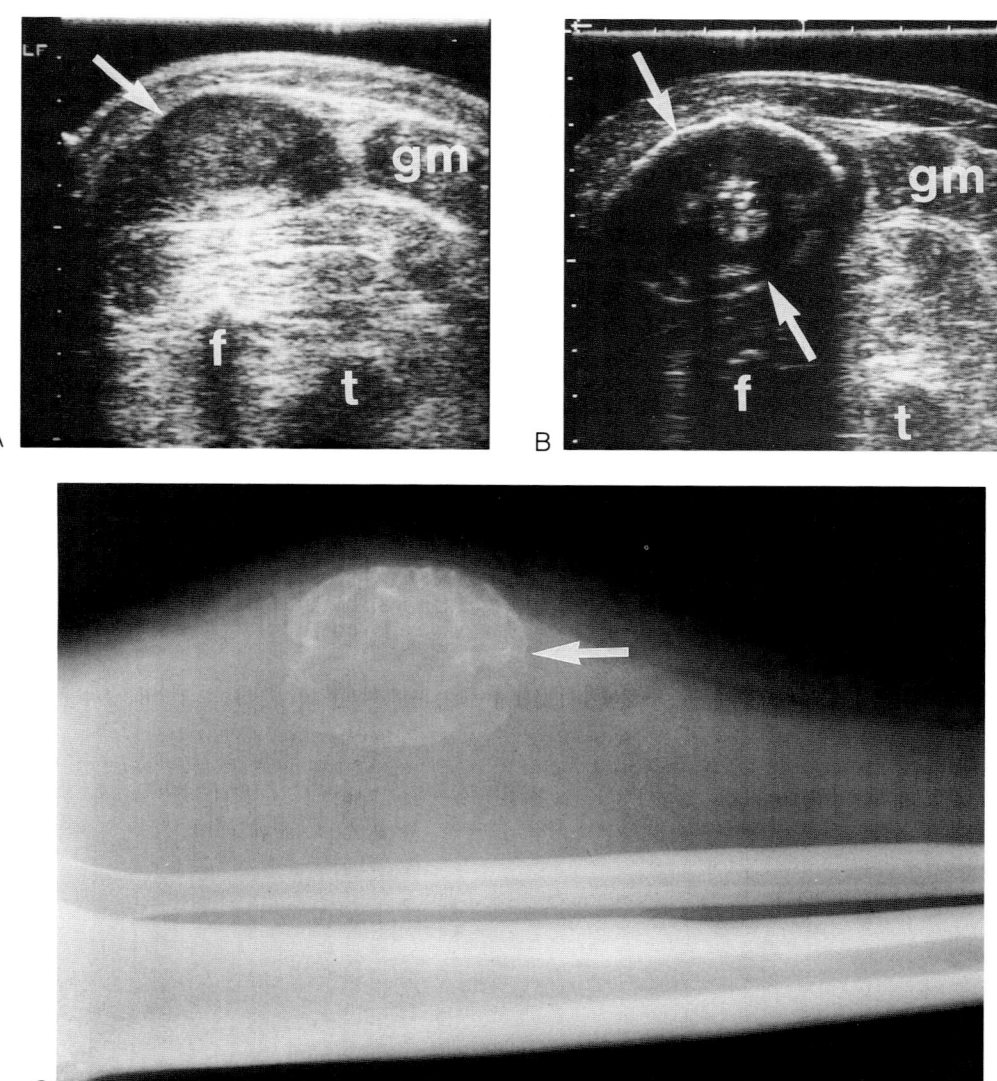

Fig. 3-1 Nontraumatic myositis ossificans of the calf. **(A)** Transverse sonogram obtained 2 weeks after the onset of clinical symptoms shows an ovoid, hypoechoic mass in the gastrocnemius lateralis muscle (*arrow*). f, fibula; gm, gastrocnemius medialis muscle; t, tibia. **(B)** Transverse sonogram obtained 7 weeks later shows a well-defined peripheral reflective ring (*arrows*). Note the marked acoustic shadow. f, fibula; gm, gastrocnemius medialis muscle; t, tibia. **(C)** Lateral radiograph of the leg shows the typical rim calcification (*arrow*). (From Fornage et al,[7] with permission.)

mass (Plate 3-1). Diagnostic aspiration and drainage can be performed under sonographic guidance.[12] Special attention should be given to searching for contact between the soft-tissue collection and the cortex of the underlying bone(s); such contact would strongly suggest osteomyelitis.[13,14]

Pyomyositis, an intramuscular suppuration followed by abscess formation that is most commonly caused by *Staphylococcus aureus*, is common in tropical countries. It may be difficult to diagnose in nontropical countries because of its rarity there and its occasionally deceptive clinical presentation. The most frequent location is the proximal lower extremity and the buttock. Pyomyositis appears as a hypoechoic collection that may contain echogenic debris and occasionally gas bubbles, which cause shadowing, or as a complex mass.[15,16] It sometimes develops quickly, and a repeat sonogram will suggest the diagnosis.[16] On occasion, pyomyositis mimics a malignant tumor.[17]

CELLULITIS

Cellulitis presents clinically as an indurated, erythematous, tender swelling of superficial soft tissues. Sonography shows a thickening and diffuse hyperechogenicity of the subcutaneous fat with blurring of the interface between the echogenic fat and the dermis. Sonography is mostly used to confirm the absence of an abscess. Color Doppler study demonstrates diffuse hypervascularity throughout the inflamed area (see Ch. 7).[18] Fat necrosis and panniculitis also result in abnormal echogenicity of the subcutaneous fat, usually restricted to a more focal area (Fig. 3-2).

Palpable nodules of erythema nodosum in the subcutaneous fat of the lower extremities often appear as elongated masses with a nonspecific hypoechogenicity.

PARASITIC INFECTION

The sonographic appearance of parasitic infestation of soft tissues has rarely been reported. In countries where hydatid disease is endemic, hydatid cysts may be seen in the soft tissues of the extremities (Fig. 3-3); they have the same wide spectrum of sonographic appearances as do visceral lesions, ranging from the classic multivesicular pattern to the misleading predominantly solid pat-

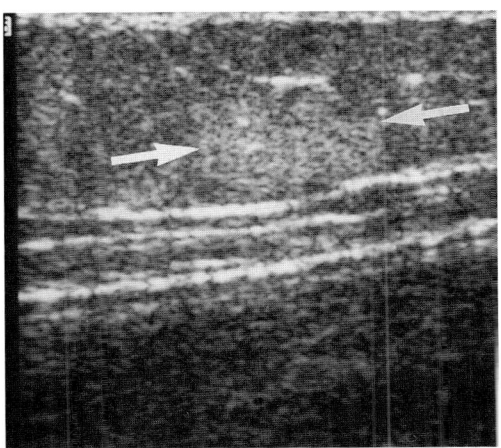

Fig. 3-2 Fat necrosis. Longitudinal sonogram of the calf shows a focal area of increased echogenicity in the subcutaneous fat (*arrows*).

tern, which can mimic a soft-tissue tumor.[19,20]

TENDINITIS

Tendinitis of superficial tendons, such as the patellar and Achilles tendons, can result in a palpable mass. The inflammation can involve the tendon diffusely (e.g., athletic

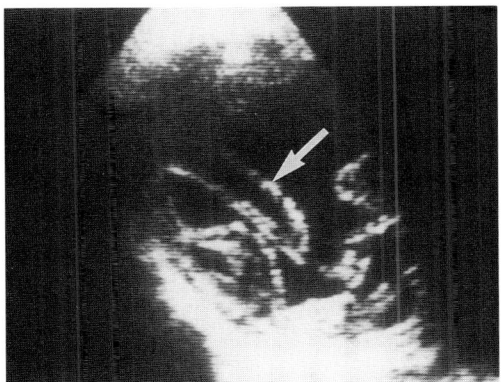

Fig. 3-3 Soft-tissue mass in the thigh of a Tunisian child. Sonogram shows an 8-cm fluid collection with a floating, detached membrane (*arrow*) typical of a hydatid cyst. (Courtesy Dr. H. A. Gharbi.)

overuse) or focally (e.g., focal tendinitis of the lower attachment of the patellar tendon following transposition of the anterior tibial tuberosity). Sonography readily confirms that the palpable mass is the swollen, inflamed, hypoechoic tendon (Fig. 3-4).[21-23] Doppler studies can demonstrate increased echogenicity within and at the periphery of the tendon.[2] Intratendinous calcifications are common in chronic tendinitis.

TENOSYNOVITIS

Tendons that possess a synovial sheath include the biceps brachii tendon and most tendons of the wrist and hand and the ankle. Sonography is the method of choice with which to confirm the diagnosis of tenosynovitis by verifying that the palpable soft-tissue mass is indeed a fluid collection in the tendon sheath (Fig. 3-5).[23-26] Rheumatoid tenosynovitis is a common cause of soft-tissue masses in the hand and foot in patients with rheumatoid arthritis. Sonographically, there is marked thickening of the synovial sheath (the pannus) with a relatively small amount of fluid (Fig. 3-6).[27]

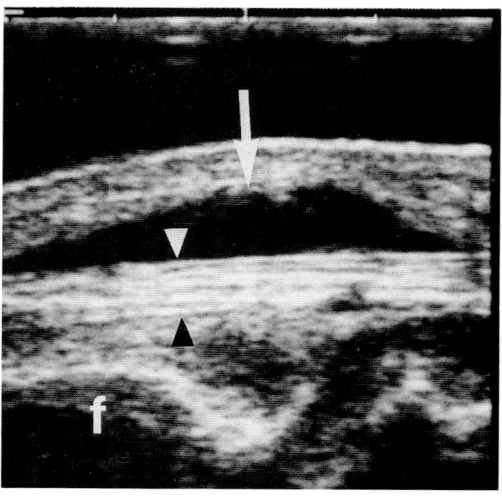

Fig. 3-5 Tenosynovitis of the peroneus longus tendon. Longitudinal sonogram shows that the palpable mass corresponds to a fluid collection in the tendon sheath (*arrow*). The tendon (*arrowheads*) has a normal echogenic appearance. f, fibula.

BURSITIS

Inflammation of a synovial bursa also can give rise to a palpable mass. The diagnosis of bursitis is facilitated by the fact that synovial bursae are located at specific sites: the presence of a fluid collection in one of these sites is virtually diagnostic of bursitis.[2,28] The shape of the fluid-filled bursa depends on the original shape of the bursa and the possibilities for volume expansion in the anatomic region. For example, an inflamed prepatellar bursa will maintain its flat shape, whereas an enlarged iliopsoas bursa assumes a sausage shape and may even expand further in the pelvis.[2,29] In chronic bursitis, the echogenicity of the bursa often becomes mixed, with echogenic debris and occasionally calcifications.

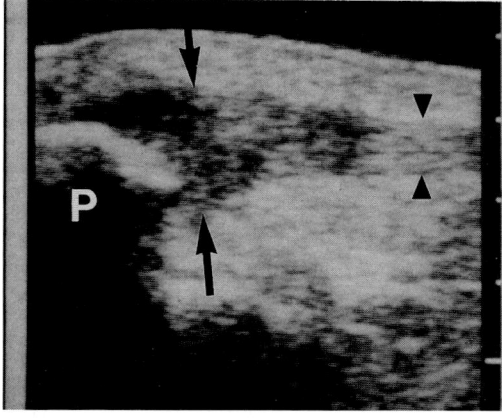

Fig. 3-4 Patellar tendinitis. Longitudinal sonogram shows a markedly swollen, hypoechoic patellar tendon (*arrows*), which accounts for the tender, palpable soft-tissue mass at the apex of the patella. Arrowheads delineate the normal lower half of the patellar tendon. P, patella.

Cysts

Sonography is essential in diagnosing cystic lesions that present as soft-tissue masses.

SOFT-TISSUE MASSES 27

SYNOVIAL CYSTS

Synovial cysts are frequently found in association with pathologic conditions that cause an increase in the intra-articular pressure through overproduction of synovial fluid, capsular sclerosis, or synovial hypertrophy; among these conditions, rheumatoid arthritis is the most common. The best known synovial cyst is the popliteal cyst (Baker's cyst).[30] The typical sonographic appearance is a fluid-filled collection wrapping around the origin of the gastrocnemius medialis muscle (Fig. 3-7) (see Ch. 13). In patients with rheumatoid arthritis, synovial cysts may be completely filled with pannus, thus mimicking solid masses.[2] Osteochondromatosis can also develop in a popliteal cyst, giving rise to hyperechoic loose bodies that cast acoustic shadows when calcified.[31]

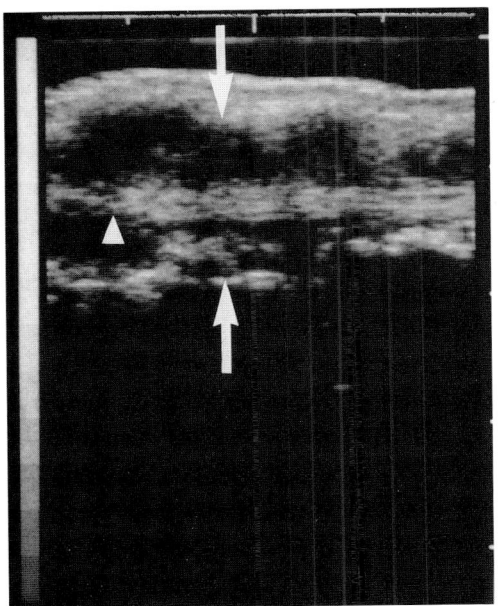

Fig. 3-6 Rheumatoid tenosynovitis involving the flexor tendons of a finger. Longitudinal sonogram of a palpable soft-tissue mass in the palm shows an ill-defined hypoechoic mass (*arrows*) corresponding to the pannus surrounding the flexor tendons of the finger. Note the irregular margins of the eroded tendons (*arrowhead*). (From Fornage et al,[23] with permission.)

Large cysts dissecting into the calf and ruptured cysts produce a swollen, painful limb that mimics thrombophlebitis. When the rupture is recent, sonography can demonstrate the leak as a subcutaneous fluid collection extending distally to the lower calf.[32,33] However, when sonographic examination is deferred, diagnosis may be more problem-

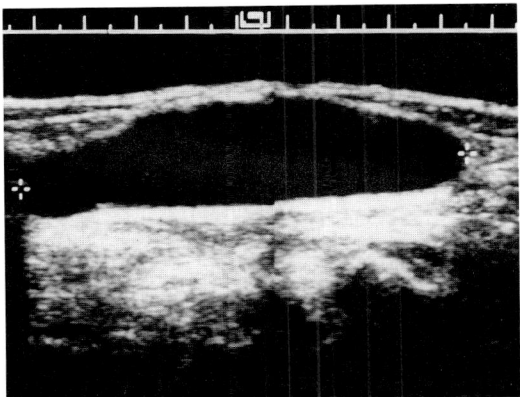
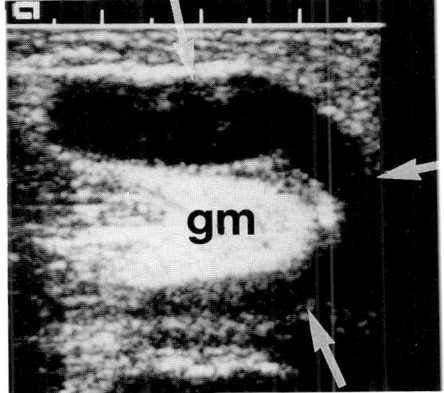

Fig. 3-7 Baker's cyst. **(A)** Longitudinal sonogram of the popliteal fossa shows an elongated fluid collection. **(B)** Transverse sonogram shows the cyst (*arrows*) wrapping medially in a horseshoe shape around the echogenic tendon of the gastrocnemius medialis muscle (gm).

atic because the leaking fluid will have resorbed, leaving only an ill-defined, hypoechoic residual area.[33]

GANGLION CYSTS

Ganglion cysts usually arise from the wrist joint, with which they communicate; they also are found adjacent to tendon sheaths in the hand and foot (Fig. 3-8).[34-36] They are generally sonolucent, but internal low-level echoes can be found in chronic or inflammatory cysts.[37]

MENISCAL CYSTS

Sonography is not routinely used for evaluation of the knee menisci. However, sonography can play a significant role in the diagnosis of meniscal cysts. Meniscal cysts arise from the menisci and are seen in the periarticular soft tissues of the knee; their pathophysiology remains controversial (see Ch. 13). They are more often found in men and usually arise from the lateral meniscus. Clinically, a meniscal cyst presents as a usually tender, firm or fluctuant swelling at the level of the joint space, close to the collateral ligament. Typically, sonograms demonstrate a

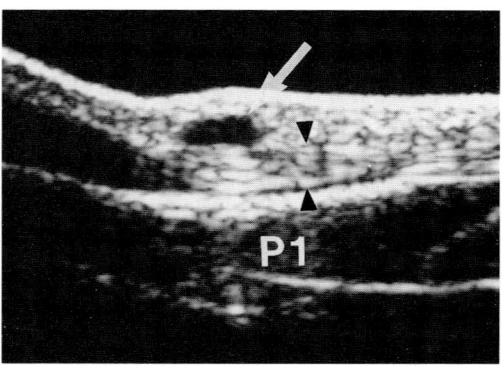

Fig. 3-8 Ganglion cyst at the volar aspect of the first phalanx of a finger. Longitudinal sonogram shows the small cyst (*arrow*) anterior to the flexor tendons of the finger (*arrowheads*). P1, bony phalanx.

loculated fluid-filled collection or a complex mass that connects to a meniscus.[38] Sonographic examination in varus position of a lateral meniscal cyst or valgus position of a medial meniscal cyst may show a change in shape of the cyst, which enters the joint space. However, MRI is superior in demonstrating the frequently associated meniscal lesions.[2]

Joint Effusions

Joint effusions can present as soft-tissue masses. They appear as anechoic intra-articular fluid collections distending the joint cavity. Septic effusions may contain debris, but there is no clear-cut relation between the echogenicity of the fluid and the presence or absence of infection. Recent hemarthrosis may produce an echogenic, sedimented layer of erythrocytes; this may shift with the patient's position.[2] Sonography is ideal for guiding aspiration of small effusions, for example, in the hips of children.

Synovial Proliferations

Proliferative diseases of the synovium, including osteochondromatosis, pigmented villonodular synovitis, rheumatoid arthritis, and changes observed in patients with hemophilic arthritis, can present clinically as soft-tissue masses. These possibilities should be included in the differential diagnosis whenever a soft-tissue mass is related to a joint space. There is often an associated joint effusion, which helps delineate the thickened, hypoechoic synovium on the sonograms (Fig. 3-9).[39-41] In the absence of effusion, however, the thickened synovium may appear as a hypoechoic mass mimicking a soft-tissue tumor. Color Doppler study can visualize some degree of increased vascularity in the abnormal synovium. Synovial chondromas are moderately echogenic, whereas osteochondromas are easily identified by the shadows that they cast.[31] Sonography is useful for preoperative mapping of the lesions.

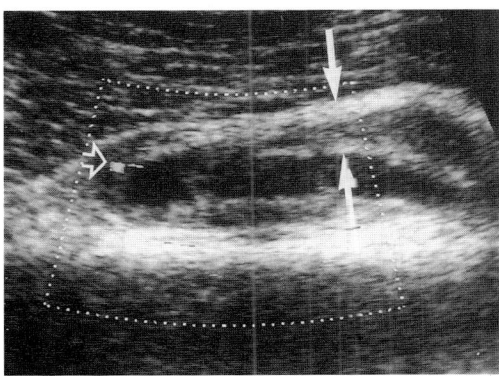

Fig. 3-9 Pigmented villonodular synovitis of the knee. Longitudinal sonogram above the patella shows fluid in the suprapatellar pouch and a marked thickening of the synovium (*arrows*). A color Doppler signal is visible (*open arrow*).

Miscellaneous

Giant cell tumors of tendon sheaths (or xanthomas) represent a circumscribed form of tenosynovitis related to pigmented villonodular synovitis. They involve preferentially the flexor surfaces of the fingers in middle-aged women. Sonographically, these tumors appear as hypoechoic, sometimes lobulated masses.[37]

In palmar fibromatosis (Dupuytren's contracture), sonography shows an ill-defined, hypoechoic mass in the subcutaneous tissues of the palm, associated with skin retraction and usually beginning over the course of the flexor tendons of the fourth or fifth finger.[2] Plantar fibromatosis in the foot also appears as an ill-defined, elongated, hypoechoic mass in the subcutaneous tissues, superficial to the echogenic plantar fascia.[42]

Rheumatoid nodules in soft tissues appear as elongated hypoechoic masses.[27] In patients with hypercholesterolemia, sonography shows the intratendinous xanthomas as hypoechoic masses. Sonography is an ideal modality with which to monitor the effect of therapy on the Achilles tendon's thickness and echotexture.[43,44] In patients with gout, intratendinous tophi appear markedly echogenic with acoustic shadowing.[45]

A rare complication of hemophilia is the development of pseudotumors in soft tissues. These are caused by the encapsulation of hematomas that fail to resolve and frequently affect the psoas and large muscles of the lower extremity. Sonography shows a complex mass and is the method of choice to follow the progression or regression of these hematomas.[46]

Finally, aneurysms, especially aneurysms of the popliteal artery, are a well-known cause of palpable masses, the diagnosis of which relies heavily on sonography (see Ch. 13).[47,48]

SOFT-TISSUE TUMORS

Questions regarding soft-tissue tumors that sonography should answer include the following: Is there a tumor? What kind of tumor is it? Where precisely is it located? How large is it? What adjacent structures are involved? Because of the overlap between the sonographic appearances of benign and malignant tumors, differentiation between the two is rarely achieved with sonography alone.

Soft-tissue tumors must be evaluated for their number, location, shape (e.g., the ratio of the length to the anteroposterior diameter), size, margin regularity, echogenicity and echotexture homogeneity, presence of diagnostic artifacts (e.g., shadowing, sound-through transmission), vascularity on color Doppler mapping, contractility during contraction and relaxation of the muscle(s) involved, and compressibility (elasticity) when pressure is applied with the transducer.

Benign Tumors

Among the benign tumors of the superficial soft tissues, lipomas and hemangiomas are the most common.

LIPOMAS

Lipomas are common benign tumors that usually develop in the subcutaneous fat but can develop in any location where fat is normally present. Most patients with lipomas are in the fifth or sixth decade of life; children are rarely affected. Approximately 5 percent of all patients with lipomas have multiple tumors.[49] Macroscopically, lipomas can be encapsulated or poorly defined with an infiltrative pattern.

Superficial lipomas are usually diagnosed on the basis of clinical history and the palpation of a well-delineated, mobile, and soft superficial mass. When palpation of a superficial soft-tissue mass is inconclusive, sonography can be used to further characterize the mass. It should be kept in mind that nonencapsulated lipomas in subcutaneous fat may not be identified on CT scans.[50] On sonograms, lipomas are elongated, with their greatest diameter parallel to the skin. In a study of subcutaneous lipomas, two-thirds showed a homogeneous echotexture and 60 percent were well defined, with the remainder showing ill-defined margins blending into the surrounding tissues. Twenty-nine percent of the lipomas were hypoechoic, 22 percent were isoechoic, 29 percent were hyperechoic, and 20 percent showed a mixed pattern.[51] The same variability in echogenicity exists for intramuscular lipomas (Fig. 3-10). The presence of internal linear echoes oriented parallel to the skin has been noted in cervical lipomas.[52] Calcifications resulting from fat necrosis can develop inside a lipoma. An elongated isoechoic or hyperechoic mass in the subcutaneous tissues should suggest a lipoma. Low-kilovoltage radiographs or CT scans can be used to confirm the radiolucent fatty tumor.

ANGIOMAS

Hemangiomas are common benign tumors of soft tissues. When they are palpable, they are soft. Intramuscular angiomas may be cir-

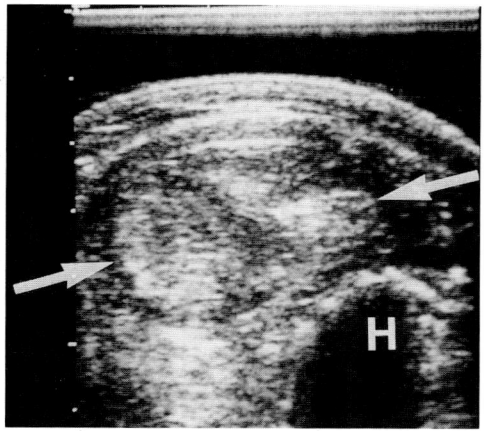
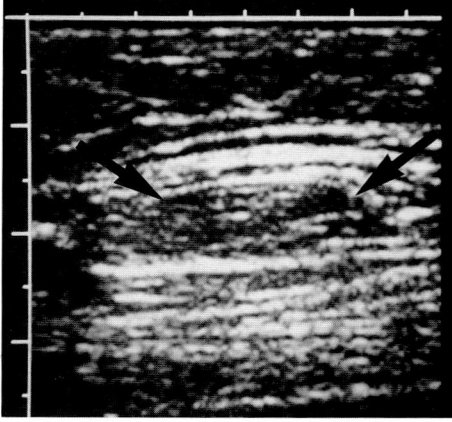

Fig. 3-10 Intramuscular lipomas. **(A)** Transverse sonogram of the posterior arm shows a well-defined, echogenic lipoma (*arrows*) in the triceps muscle. H, humerus. **(B)** Longitudinal sonogram of the distal anterior thigh in another patient shows a hypoechoic lipoma (*arrows*) in the lower quadriceps muscle.

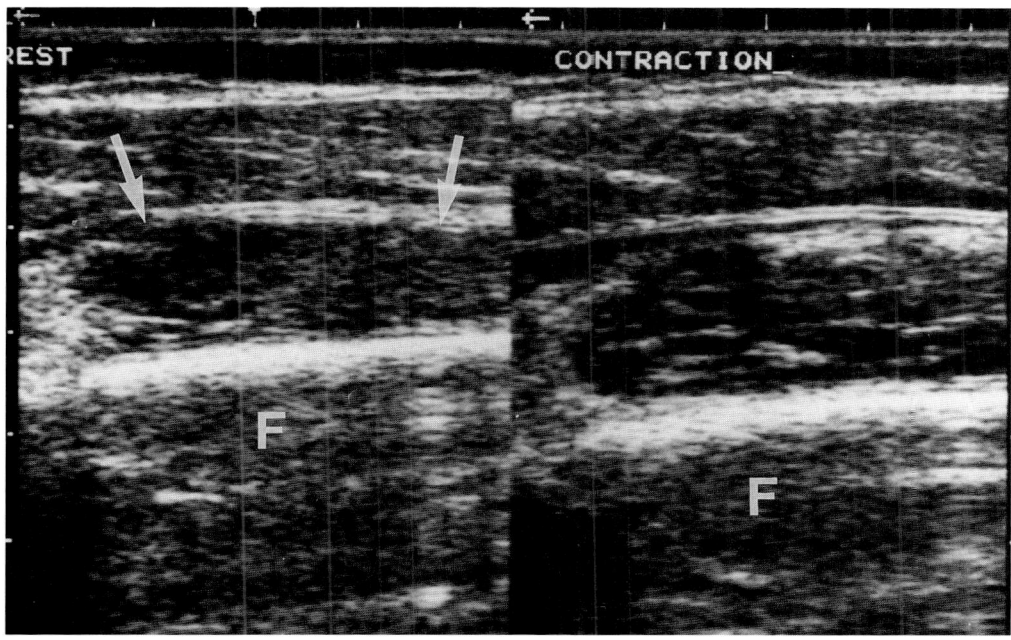

Fig. 3-11 Angioma. Longitudinal sonograms of the anterior thigh at rest (left) and after contraction of the quadriceps muscle (right) show an elongated hypoechoic mass in the vastus intermedius muscle (*arrows*). Note the fluid-filled structures within the mass representing venous lakes, which are more apparent during contraction. F, femur.

cumscribed or infiltrate into the adjacent muscle. Sonographically, angiomas range from markedly hypoechoic to hyperechoic and from homogeneous to multiloculated.[53] They may change in shape during contraction and increase in size after exercising (Fig. 3-11). An important clue to the diagnosis is the demonstration of phleboliths, which appear typically as echogenic foci with acoustic shadowing.[2] MRI is superior to sonography in demonstrating poorly defined, infiltrating hemangiomas. The vascularity of angiomas as seen on color Doppler scans also varies greatly (Plate 3-2), and not uncommonly, no Doppler signals are detectable.

Angiolipomas represent a rare variety characterized by increased echogenicity, as a result of the fat content, and an infiltrating pattern.

NERVE SHEATH TUMORS

Benign nerve sheath tumors of the peripheral nerves include schwannomas (or neurilemomas) and neurofibromas (see Ch. 6). On sonograms, both types of tumors are hypoechoic.[54-58] Often, high-frequency sonography can visualize the junction between the hypoechoic tumor and the normal echogenic nerve, thus confirming the diagnosis of nerve tumor (Fig. 3-12).[2,54] Schwannomas tend to be round or oval, well circumscribed, and eccentric in relation to the nerve axis, with internal cystic cavities and good sound-through transmission; they sometimes mimic a cyst.[54-57] Neurofibromas, on the other hand, are often elongated along the nerve axis and lobulated. In reality, there is considerable overlap between the sonographic appearances of schwannomas and

32 MUSCULOSKELETAL ULTRASOUND

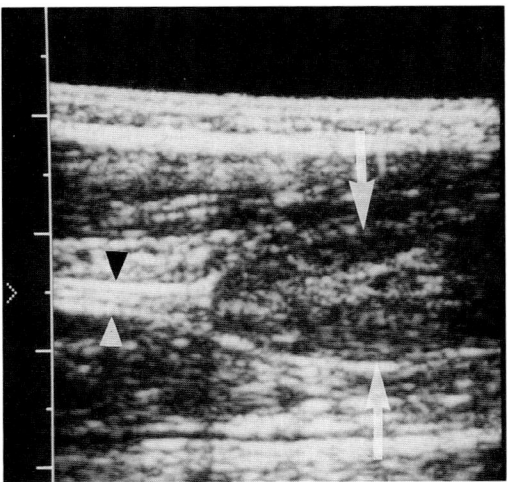

Fig. 3-12 Benign schwannoma of the median nerve in the forearm. Longitudinal scan shows a well-defined hypoechoic mass (*arrows*). Note the flared shape of the proximal nerve. *Arrowheads* point to the normal nerve.

neurofibromas. Color Doppler studies demonstrate increased internal vascularity.

High-frequency sonography is highly sensitive in the detection of nerve sheath tumors of the extremities. Even though a given nerve is not distinctly visualized along its entire course, continuous scanning along its expected course allows detection of a hypoechoic nerve sheath tumor of even minute size.

Occasionally, the inflamed nerve in neurilemmitis results in a palpable mass, as has been reported in a case of leprosy.[59]

Sonographically guided needle biopsy of nerve sheath tumors has occasionally been attempted.[2,58] The insertion of the needle into the tumor may trigger a sharp, excruciating pain, which forces interruption of the procedure but indirectly confirms the neural origin of the tumor.

INTRAMUSCULAR MYXOMAS

Intramuscular myxoma is a rare mesenchymal tumor that arises in skeletal muscle. The most common sites of these tumors are the large muscles of the thigh, followed by those of the buttock, lower leg, shoulder, and upper arm[60]; rare cases of intramuscular myxomas of the head and neck have also been described.[61] The usual clinical presentation is a relatively large, slowly growing, painless mass not associated with a history of trauma. On gross pathologic examination, these tumors show multiple small cysts filled with a myxoid matrix.[62]

On sonograms, intramuscular myxoma appears as a well-demarcated, markedly hypoechoic, intramuscular mass with distal sound enhancement; the tumor contains multiple fluid-filled clefts or cystic areas of various sizes, sometimes only 1 to 2 mm in diameter (Fig. 3-13).[63] Color Doppler examination is negative. Although the aspiration of a myxoid substance narrows down the differential diagnosis, only a large-core needle biopsy of the most solid-appearing component can differentiate preoperatively between an intramuscular myxoma and a well-differentiated liposarcoma.

DESMOID TUMORS

Desmoid tumors are fibromatous lesions arising from muscular aponeuroses, usually in young adults. Association with previous trauma and estrogen stimulation has been reported. In Gardner's syndrome, desmoid tumors are found in association with intestinal polyposis. Muscles of the shoulder, chest and abdominal walls, thigh, popliteal fossa, and calf are most often involved. Desmoid tumors are poorly defined and infiltrate into the muscles. Local recurrences following excision are not rare. On sonograms, desmoid tumors may appear as irregular, ill-defined hypoechoic masses; in many cases, however,

SOFT-TISSUE MASSES

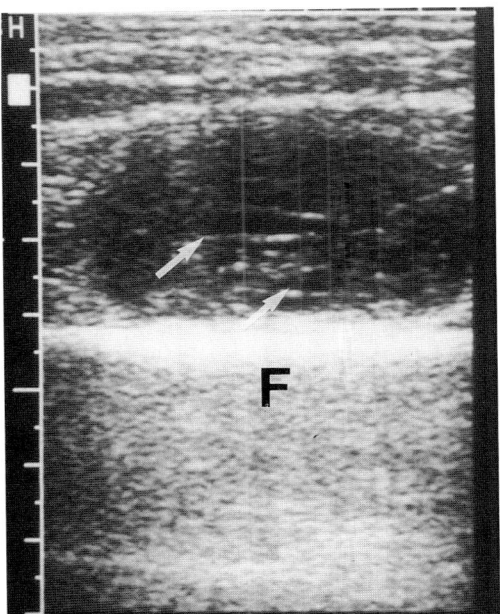

Fig. 3-13 Intramuscular myxoma. Longitudinal sonogram of the thigh shows a markedly hypoechoic mass with fluid-filled clefts (*arrows*). Note the marked distal sound enhancement. F, femur.

marked shadowing from the surface of the lesion, a result of the large amount of dense collagen tissue in the tumor, totally obscures the mass (Fig. 3-14).[2] MRI is the modality of choice in such cases.

MISCELLANEOUS

Lymphangiomas are often poorly defined on sonograms, and MRI should be used instead of sonography, especially for follow-up after excision.

Granular cell tumors, formerly known improperly as granular cell myoblastomas, are rare benign tumors that are thought to derive from the Schwann cells. They can be found anywhere in the soft tissues—including the breast (Fig. 3-15)—although the most frequent site is the tongue. The tumors are firm, small (less than 2 cm in diameter), and poorly defined on gross pathologic examination.[64] Sonographically, they are hypoechoic, roughly round, with irregular margins; some are associated with marked acoustic shadowing, which makes those located in the subcutaneous tissues of the breast seem even more suspicious for malignancy.[65] Granular cell tumors may recur locally if incompletely excised.

Glomus tumors are usually too small to be palpable They are found mostly in the hand. On sonograms, they appear round and markedly hypoechoic. Sonography has proved to be very helpful in the diagnosis and preoperative localization of these lesions.[56]

Malignant Tumors

SOFT-TISSUE SARCOMAS

Soft-tissue sarcomas represent the majority of primary malignant tumors that develop in soft tissues. However, they are relatively rare in adults and represent less than 1% of newly diagnosed cancers in the United States.[67] Soft-tissue sarcomas can develop in any anatomic area, but most develop in the extremities, usually the lower ones. The two most frequent subtypes of sarcoma in adults are malignant fibrous histiocytoma and liposarcoma. Other subtypes include leiomyosarcoma, fibrosarcoma, rhabdomyosarcoma (the most frequent sarcoma in children) synovial sarcoma, angiosarcoma, hemangiopericytoma, epithelioid sarcoma, and malignant peripheral nerve sheath tumors. Soft-tissue sarcomas most commonly metastasize hematogenously to the lungs; regional lymph nodes are rarely involved.

Sonographic Appearances

Most soft-tissue sarcomas have already reached a significant size (several centimeters) at the time of diagnosis. Small satel-

34 MUSCULOSKELETAL ULTRASOUND

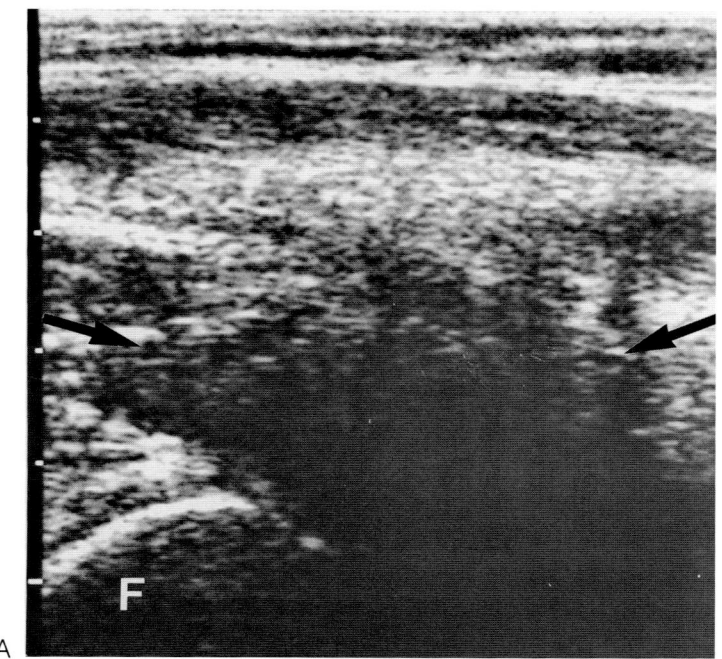

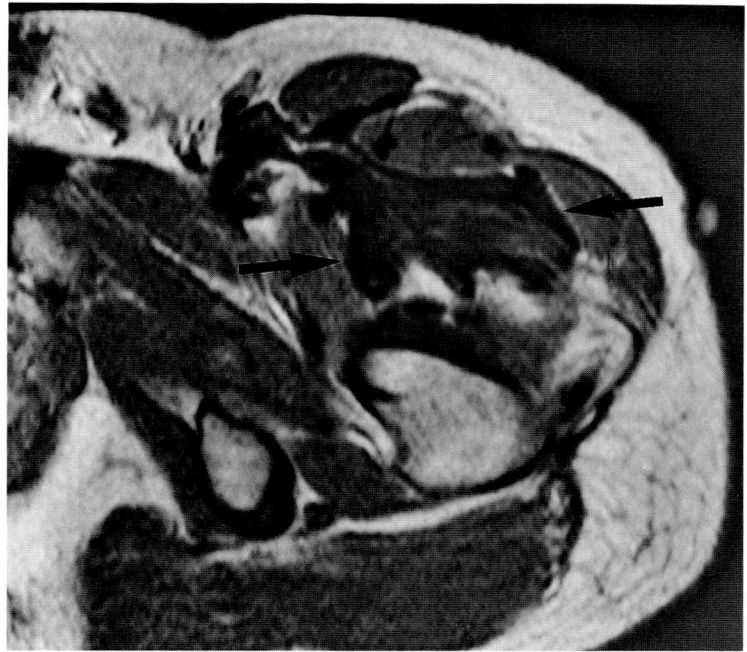

Fig. 3-14 Desmoid tumor of the left proximal thigh and groin in a patient with Gardner's syndrome. **(A)** Transverse sonogram of the anterior aspect of the proximal left thigh shows an ill-defined, irregular, hypoechoic area (*arrows*) that casts an intense shadow. F, part of the femoral head. **(B)** MR scan accurately delineates the extent of the low-signal-intensity fibromatous lesion (*arrows*).

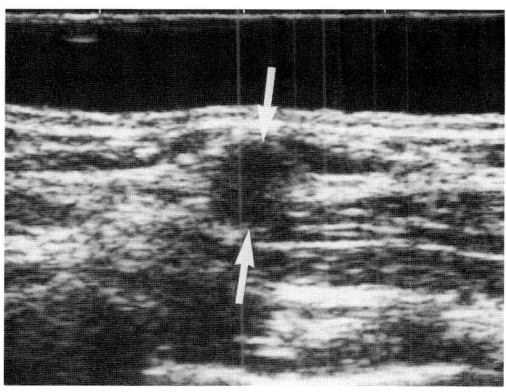

Fig. 3-15 Granular cell tumor of the subcutaneous tissues of the breast. Sonogram shows a superficial, poorly defined, hypoechoic mass (*arrows*). The tumor is round and measures 0.7 cm in diameter.

lite tumors may be present. Except for well-differentiated and myxoid liposarcomas, which are usually echogenic (Plate 3-3), soft-tissue sarcomas appear as hypoechoic, often relatively well-circumscribed, lobulated masses (Fig. 3-16). Areas of necrosis may be present, especially in rhabdomyosarcomas. Calcifications of sufficient size can be visualized on sonograms as bright dots with acoustic shadowing. Up to 30 percent of synovial sarcomas contain calcifications, but calcifications can also be found in epithelioid sarcomas, neurofibrosarcomas, or soft-tissue osteosarcomas.

Color Doppler sonography can demonstrate increased vascularity (Plate 3-3). The more chaotic the distribution of flow within the tumor, the more suspicious for malignancy (Plate 3-4). Spectral analysis usually reveals a low-resistance flow pattern, as in most malignancies.

Differentiation between benign and malignant tumors is rarely achieved with imaging, and tissue diagnosis is usually required prior to starting treatment. Because of its real-time capability, sonography is an ideal imaging technique for guiding needle biopsy of soft-tissue tumors. Although a fine-needle aspiration biopsy is adequate for confirming a local recurrence or metastasis of a known malignancy, only a large-core needle biopsy can establish the histologic subtype and assess the grade of a sarcoma.[2]

Evaluation of Response to Treatment

Sonography can also be used to monitor and quantify the response of a malignant tumor to preoperative therapy such as irradiation and intra-arterial chemotherapy; the tumor volume can be accurately measured on sonograms and its variation quantified (Fig. 3-17).[2] Changes in echotexture due to necrosis may also parallel tumor regression.

Local Recurrences

Soft-tissue sarcomas are associated with a high incidence of local recurrence after surgery. Clinical detection of these recurrent tumors can be difficult, particularly in deep tissues or under the surgical scar. Sonography is extremely sensitive in the detection of early recurrences after surgical excision. Recurrences of soft-tissue sarcomas usually ap-

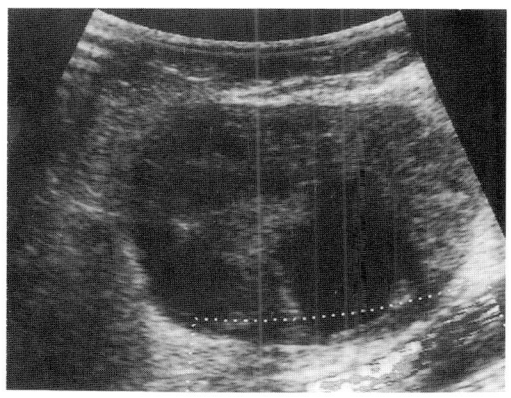

Fig. 3-16 Malignant fibrous histiocytoma of the thigh. Sonogram shows a large, relatively well-defined, ovoid, hypoechoic mass with areas of necrosis.

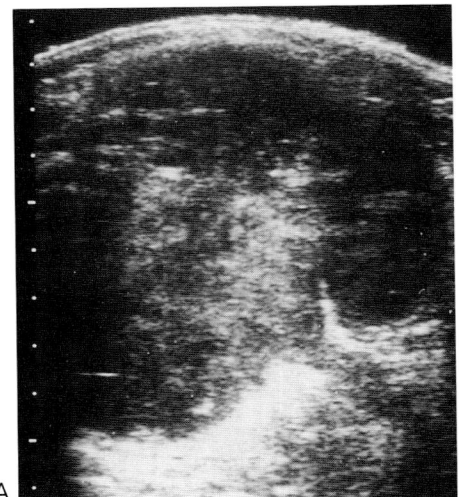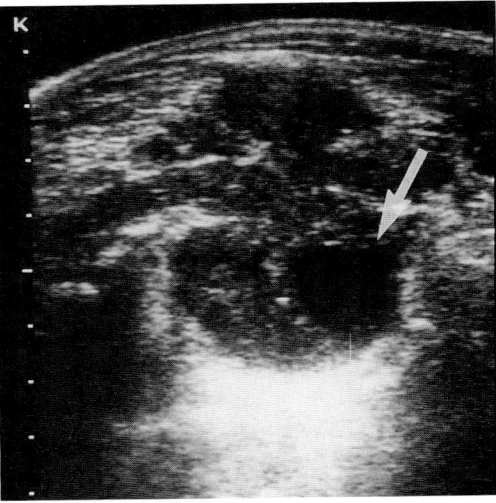

Fig. 3-17 Evaluation of response to treatment of an undifferentiated sarcoma of the gluteus maximus muscle. **(A)** Initial sonogram shows the bulky nonhomogeneous tumor. **(B)** Sonogram obtained after two courses of chemotherapy shows significant shrinkage of the tumor, which contains a fluid-filled necrotic area (*arrow*).

pear as small, round-to-oval, hypoechoic masses (Fig. 3-18). The diagnostic accuracies of MRI and sonography in the detection of locally recurrent tumors after surgery are not significantly different statistically; the sensitivity and specificity are 83 percent and 93 percent, respectively, for MRI and 100 percent and 79 percent, respectively, for sonography.[68] Considering that sonography is less costly and provides easy guidance for needle biopsy, sonography should be used for routine follow-up. To avoid confusing recurrences with postoperative changes, including hematomas, an initial baseline study with both MRI and sonography is recommended 4 to 6 weeks after surgery. When follow-up sonography is inconclusive, MRI should be performed for further evaluation.

A significant advantage of sonography over MRI is the use of ultrasound-guided needle biopsy to evaluate nonpalpable masses suspicious for recurrence. Fine-needle aspiration biopsy can readily document early local recurrences as small as a few millimeters (Fig. 3-18C).[69] After a nonpalpable recurrence has been detected and diagnosed via fine-needle aspiration, preoperative or intraoperative ultrasound-guided localization may be necessary to guide the surgeon's resection.

OTHER MALIGNANT TUMORS

Metastases of carcinomas to soft tissues are rare. They usually derive from primary tumors of the lung and the gastrointestinal tract. Cases of soft-tissue metastases heralding an occult primary have been reported.[70] Sonographically, soft-tissue metastases appear as round-to-oval, hypoechoic masses; margins are often irregular (Fig. 3-19).

BONE-RELATED MASSES

On occasion, a mass thought to be derived from the soft tissues is in fact due to an osseous pathology. In adults, any deformity or discontinuity of the surface of the bony cor-

SOFT-TISSUE MASSES **37**

tex that is directly accessible to the ultrasound beam and sufficiently large can be visualized (Fig. 3-20).[71] Bone lesions that have markedly thinned the cortex (e.g., bone cysts) also can be detected with sonography (Fig. 3-21).[72,73] Osteochondromas are readily identified, and their hypoechoic cartilaginous cap can be seen.[74] In addition to the soft-tissue component of malignant tumors, sonography can visualize the hypoechoic intraosseous tumor if the overlying cortex is destroyed (Fig. 3-22). In such cases, needle biopsy of the tumor is easily performed under sonographic guidance.[75,76]

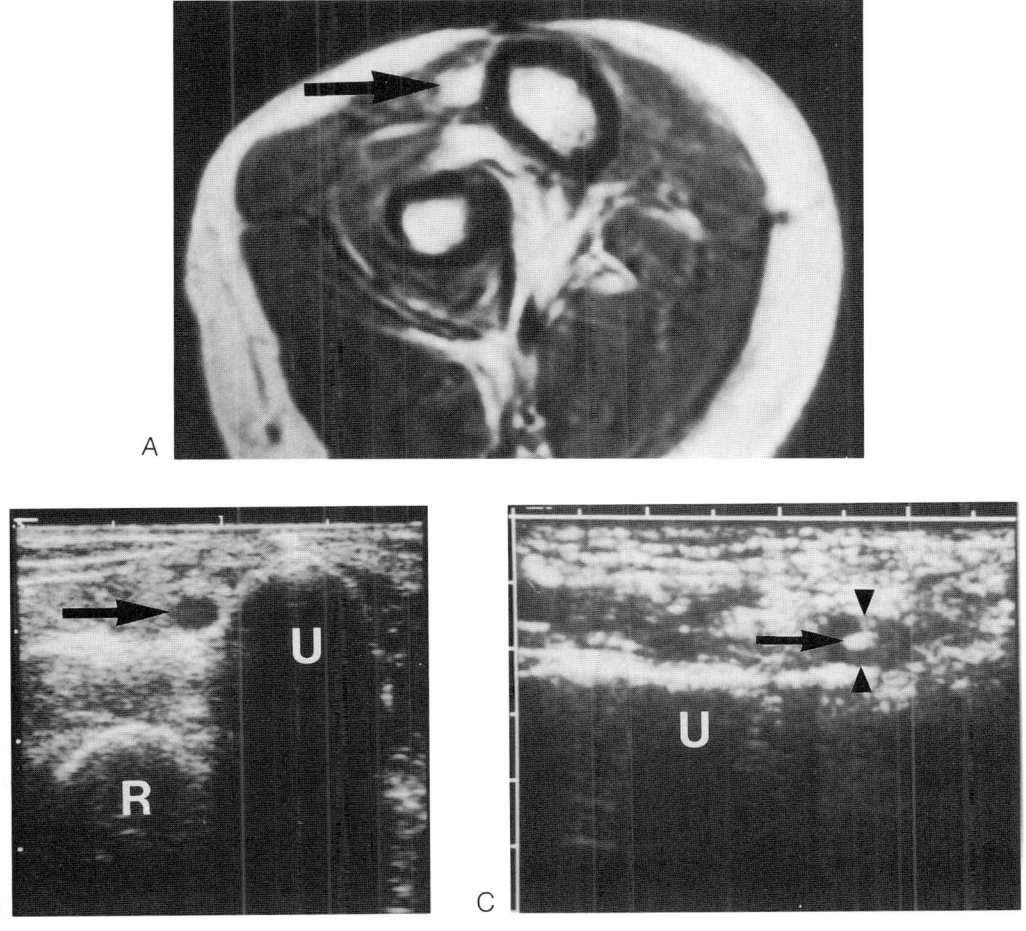

Fig. 3-18 Nonpalpable recurrence of synovial sarcoma at the elbow. **(A)** Contrast-enhanced MR scan shows a hyperintense focus (*arrow*) abutting the ulna, suspicious for recurrence. **(B)** Transverse sonogram shows a well-defined, 0.5-cm, round hypoechoic mass (*arrow*) adjacent to the ulna, correlating with the MRI findings. R, radius; U, ulna. **(C)** Longitudinal sonogram obtained during ultrasound-guided fine-needle aspiration biopsy shows the brightly echogenic tip (*arrow*) of the oblique needle in the center of the mass (*arrowheads*). U, ulna.

38 MUSCULOSKELETAL ULTRASOUND

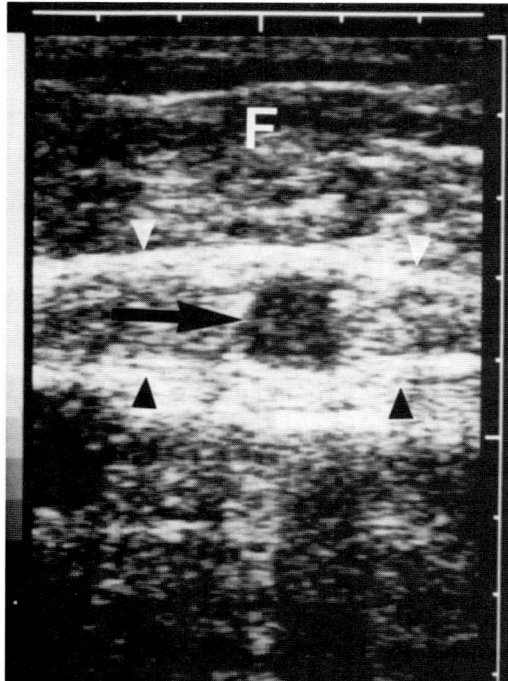

Fig. 3-19 Metastasis from colon carcinoma in the rectus abdominis muscle. Transverse sonogram of the abdominal wall shows a 1.2-cm, roughly round, hypoechoic mass (*arrow*) in the rectus abdominis muscle (*arrowheads*). F, subcutaneous fat.

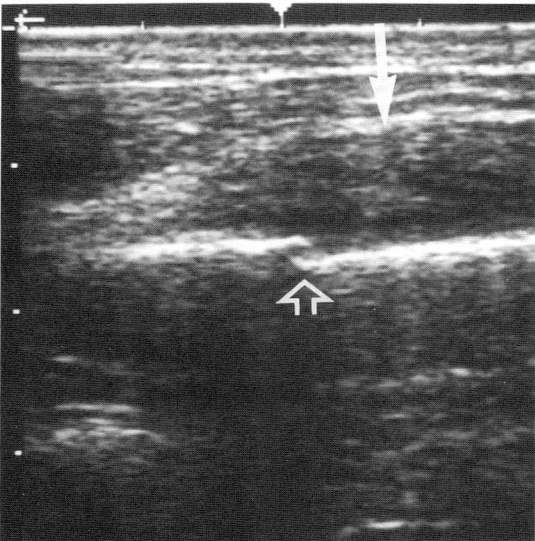
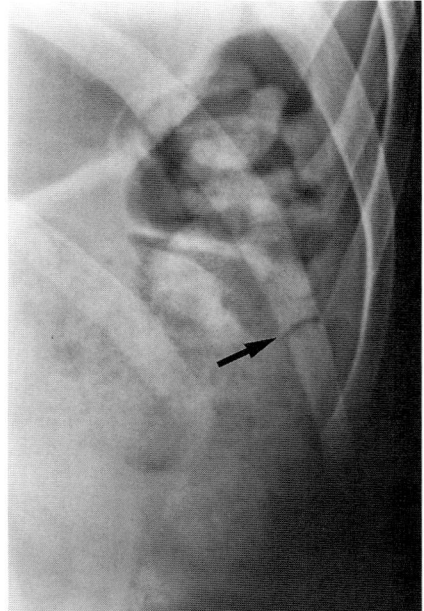

A   B

Fig. 3-20 Tender, soft-tissue mass in the posterior lateral chest wall in a 21-year-old patient with relapsed acute lymphocytic leukemia and severe cough. Rib fracture was detected by sonography. **(A)** Sonogram shows an elongated hypoechoic mass (*arrow*)—corresponding to edema and hemorrhage in the chest wall musculature—and a discontinuity of the cortex of the underlying rib with a steplike displacement typical of rib fracture (*open arrow*). **(B)** Subsequent radiograph with metallic marker at site of pain shows fracture line (*arrow*).

SOFT-TISSUE MASSES **39**

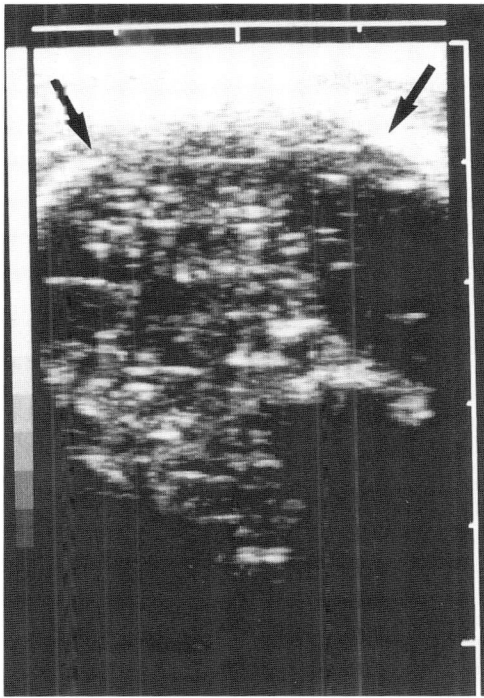

Fig. 3-21 Aneurysmal bone cyst of the cervical spine. Sonogram shows the blood-filled multiloculated cavity through the thinned echogenic cortex (*arrows*).

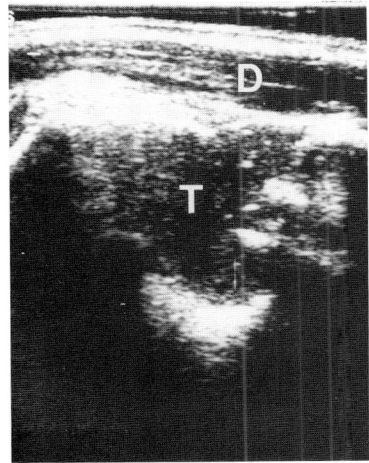
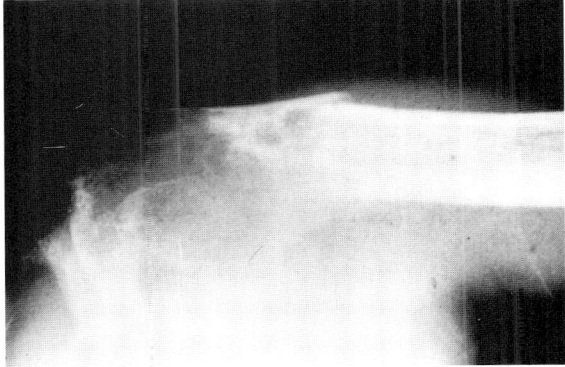

Fig. 3-22 Osteosarcoma of the upper extremity of the humerus with an impacted fracture **(A)** Coronal sonogram shows the very thin, irregular cortex and the hypoechoic tumor. D, deltoid; T tumor. **(B)** Radiograph confirms the osteolytic osteosarcoma and the fracture.

CONCLUSION

In the United States, musculoskeletal sonography has remained in the shadow of MRI. However, sonography's role should be reappraised, especially at a time when the health care system is being reviewed in an attempt to contain costs. For the evaluation of soft tissues, sonography has proved to be very accurate in confirming the presence or absence of a morphologic abnormality, with a very high negative predictive value; in determining the cystic or solid nature of a mass; in guiding percutaneous needle biopsy or drainage; in localizing nonpalpable lesions preoperatively or intraoperatively; in monitoring lesions that are treated conservatively; and in detecting early recurrences of soft-tissue sarcomas.

Limitations of musculoskeletal sonography include the reduced width of view of the scans (which prevents routine staging of malignant tumors), difficulties in reproducing scans on serial examinations, inability of clinicians to read sonograms, and inability of sonography to "see through" bones and to visualize certain structures in areas of complex anatomy. However, provided the examination is done by a well-trained operator using state-of-the-art equipment, the cost-effectiveness of sonography justifies its use as a first-line examination technique in many situations involving soft tissues, with MRI and CT being used as problem-solving tools.

REFERENCES

1. Fornage BD: Ultrasonography of Muscles and Tendons. Examination Technique and Atlas of Normal Anatomy of the Extremities. Springer-Verlag, New York, 1988
2. Fornage BD: Ultrasound of the Extremities. Vigot, Paris, 1991
3. Fornage BD, Touche DH, Rifkin MD: Small parts real-time sonography: a new "water-path." J Ultrasound Med 3:355, 1984
4. Fornage BD: Musculoskeletal evaluation. p. 1. In Mittelstaedt CA (ed): General Ultrasound. Churchill Livingstone, New York, 1992
5. Ogilvie-Harris DJ, Hons CB, Fornasier VL: Pseudomalignant myositis ossificans: heterotopic new-bone formation without a history of trauma. J Bone Joint Surg 62A:1274, 1980
6. Goldman AB: Myositis ossificans circumscripta: a benign lesion with a malignant differential diagnosis. AJR 126:32, 1976
7. Fornage BD, Eftekhari F: Sonographic diagnosis of myositis ossificans. J Ultrasound Med 8:463, 1989
8. Kramer FL, Kurtz AB, Rubin C, Goldberg BB: Ultrasound appearance of myositis ossificans. Skeletal Radiol 4:19, 1979
9. Peck RJ, Metreweli C: Early myositis ossificans: a new echographic sign. Clin Radiol 39:586, 1988
10. Sandler MA, Alpern MB, Madrazo BL, Gitschlag KF: Inflammatory lesions of the groin: ultrasonic evaluation. Radiology 151:747, 1984
11. Yeh HC, Rabinowitz JG: Ultrasonography and computed tomography of inflammatory abdominal wall lesions. Radiology 144:859, 1982
12. VanSonnenberg E, Wittich GR, Casola G et al: Sonography of thigh abscess: detection, diagnosis, and drainage. AJR 149:769, 1987
13. Abiri MM, Kirpekar M, Ablow RC: Osteomyelitis: detection with US. Radiology 172:509, 1989
14. Cartoni C, Capua A, Damico C, Potente GU: Aspergillus osteomyelitis of the rib: sonographic diagnosis. J Clin Ultrasound 20:217, 1992
15. Yousefzadeh DK, Schumann EM, Mulligan GM et al: The role of imaging modalities in diagnosis and management of pyomyositis. Skeletal Radiol 8:285, 1982
16. Quillin SP, McAlister WH: Rapidly progressive pyomyositis. Diagnosis by repeat sonography. J Ultrasound Med 10:181, 1991
17. Ahrens P, Gross-Fengels W, Bovelet K: Zur Differentialdiagnose maligner Weichteiltumoren: Pyomyositis [The differential diagnosis of malignant soft tissue tumors: pyomyositis]. Aktuelle Radiol 1:40, 1991
18. Fornage BD: Sonography of the skin and subcutaneous tissues. Radiol Med (Torino) 85(suppl 1):149, 1993

19. Martin J, Marco V, Zidan A, Marco C: Hydatid disease of the soft tissues of the lower limb: findings in three cases. Skeletal Radiol 22:511, 1993
20. Lamine A, Fikry T, Zryouil B: L'hydatidose primitive des muscles périphériques. A propos de 7 cas [Primary hydatidosis of the peripheral muscles. 7 case reports]. Acta Orthop Belg 59:184, 1993
21. Fornage BD, Rifkin MD, Touche DH, Segal PM: Sonography of the patellar tendon: preliminary observations. AJR 143:179, 1984
22. Fornage BD: Achilles tendon: US examination. Radiology 159:759, 1986
23. Fornage BD, Rifkin MD: Ultrasound examination of tendons. Radiol Clin North Am 26:87, 1988
24. Gooding GAW: Tenosynovitis of the wrist. A sonographic demonstration. J Ultrasound Med 7:225, 1988
25. Jeffrey RB Jr, Laing FC, Schechter WP et al: Acute suppurative tenosynovitis of the hand: diagnosis with US. Radiology 162:741, 1987
26. Stephenson CA, Seibert JJ, McAndrew MP et al: Sonographic diagnosis of tenosynovitis of the posterior tibial tendon. J Clin Ultrasound 18:114, 1990
27. Fornage BD: Soft tissue changes in the hand in rheumatoid arthritis: evaluation with US. Radiology 173:735, 1989
28. Myllymäki T, Tikkakoski T, Typpö T et al: Carpet-layer's knee. An ultrasonographic study. Acta Radiol 34:496, 1993
29. Janus C, Hermann G: Enlargement of the iliopsoas bursa: unusual cause of cystic mass on pelvic sonogram. J Clin Ultrasound 10:133, 1982
30. McDonald DG, Leopold GR: Ultrasound B-scanning in the differentiation of Baker's cyst and thrombophlebitis. Br J Radiol 45:729, 1972
31. Moss GD, Dishuk W: Ultrasound diagnosis of osteochondromatosis of the popliteal fossa. J Clin Ultrasound 12:232, 1984
32. Moore CP, Sarti DA, Louie JS: Ultrasonographic demonstration of popliteal cysts in rheumatoid arthritis. A noninvasive technique. Arthritis Rheum 18:577, 1975
33. Gompels BM, Darlington LG: Evaluation of popliteal cysts and painful calves with ultrasonography: comparison with arthrography. Ann Rheum Dis 41:355, 1982
34. Fornage BD, Schernberg FL, Rifkin MD: Ultrasound examination of the hand. Radiology 55:785, 1985
35. De Flaviis L, Nessi R, Del Bo P et al: High-resolution ultrasonography of wrist ganglia. J Clin Ultrasound 15:17, 1987
36. Bianchi S, Abdelwahab IF, Zwass A, Giacomello P: Ultrasonographic evaluation of wrist ganglia. Skeletal Radiol 23:201, 1994
37. Fornage BD, Rifkin MD: Ultrasound examination of the hand and foot. Radiol Clin North Am 26:109, 1988
38. Peetrons P, Allaer D, Jeanmart L: Cysts of the semilunar cartilages of the knee: a new approach by ultrasound imaging. A study of six cases and review of the literature. J Ultrasound Med 9:333, 1990
39. Kaufman RA, Towbin RB, Babcock DS, Crawford AH: Arthrosonography in the diagnosis of pigmented villonodular synovitis. AJR 139:396, 1982
40. Cooperberg PL, Tsang I, Truelove L, Knickerbocker WJ: Gray scale ultrasound in the evaluation of rheumatoid arthritis of the knee. Radiology 126:759, 1978
41. Wylc PJ, Dawson KP, Chisholm RJ: Ultrasound in the assessment of synovial thickening in the hemophilic knee. Aust NZ J Med 14:678, 1984
42. Reed M, Gooding GAW, Kerley SM et al: Sonography of plantar fibromatosis. J Clin Ultrasound 19:578, 1991
43. Steinmetz A, Schmitt W, Schuler F et al: Ultrasonography of Achilles tendons in primary hypercholesterolemia. Comparison with computed tomography. Atherosclerosis 74:231, 1988
44. Bude RO, Adler RS, Bassett DR: Diagnosis of Achilles tendon xanthoma in patients with heterozygous familial hypercholesterolemia: MR vs sonography. AJR 162:913, 1994
45. Tiliakos N, Morales AR, Wilson CH Jr: Use of ultrasound in identifying tophaceous versus rheumatoid nodules [letter]. Arthritis Rheum 25:478, 1982
46. Hermann G, Gilbert MS, Abdelwahab IF: Hemophilia: evaluation of musculoskeletal involvement with CT, sonography, and MR imaging. AJR 158:119, 1992
47. Silver TM, Washburn RL, Stanley JC, Gross WS: Gray scale ultrasound evaluation of popliteal artery aneurysms. AJR 129:1003, 1977
48. Wu KK: True aneurysm of the dorsalis pedis

artery mimicking a soft tissue tumor. J Foot Surg 30:304, 1991
49. Enzinger FM, Weiss SW: Soft Tissue Tumors. 2nd Ed. CV Mosby, St Louis, 1983
50. Hermann G, Yeh HC, Schwartz I: Computed tomography of soft tissue lesions of the extremities, pelvic and shoulder girdles: sonographic and pathological correlations. Clin Radiol 35:193, 1984
51. Fornage B, Tassin G: Sonographic appearances of superficial soft-tissue lipomas. J Clin Ultrasound 19:215, 1991
52. Gritzmann N, Schratter M, Traxler M, Helmer M: Sonography and computed tomography in deep cervical lipomas and lipomatosis of the neck. J Ultrasound Med 7:451, 1988
53. Derchi LE, Balconi G, De Flaviis L et al: Sonographic appearances of hemangiomas of skeletal muscle. J Ultrasound Med 8:263, 1989
54. Fornage BD: Peripheral nerves of the extremities: imaging with US. Radiology 167:179, 1988
55. Hughes DG, Wilson DJ: Ultrasound appearances of peripheral nerve tumors. Br J Radiol 59:1041, 1986
56. Chinn DH, Filly RA, Callen PW: Unusual ultrasonographic appearance of a solid schwannoma. J Clin Ultrasound 10:243, 1982
57. Hoddick WK, Callen PW, Filly RA et al: Ultrasound evaluation of benign sciatic nerve sheath tumor. J Ultrasound Med 3:505, 1984
58. Cantos-Melian B, Arriaza-Loureda R, Aisa-Varela P: Tibialis posterior nerve schwannoma mimicking Achilles tendinitis: ultrasonographic diagnosis. J Clin Ultrasound 18:671, 1990
59. Fornage BD, Nérot C: Sonographic diagnosis of tuberculoid leprosy. J Ultrasound Med 6:105, 1987
60. Miettinen M, Höckerstedt K, Reitamo J, Tötterman S: Intramuscular myxoma: a clinicopathological study of twenty-three cases. Am J Clin Pathol 84:265, 1985
61. Shugar JM, Som PM, Meyers RJ, Schaeffer BT: Intramuscular head and neck myxoma: report of a case and review of the literature. Laryngoscope 97:105, 1987
62. Kindblom LG, Stener B, Angervall L: Intramuscular myxoma. Cancer 34:1737, 1974
63. Fornage BD, Romsdahl MM: Intramuscular myxoma: sonographic appearance and sonographically guided needle biopsy. J Ultrasound Med 13:91, 1994
64. Morrison JG, Gray GF Jr, Dao AH, Adkins RB Jr: Granular cell tumors. Am Surg 53:156, 1987
65. Scatarige JC, Hsiu JG, de la Torre R et al: Acoustic shadowing in benign granular cell tumor (myoblastoma) of the breast. J Ultrasound Med 6:545, 1987
66. Fornage BD: Glomus tumors in the fingers: diagnosis with US. Radiology 167:183, 1988
67. Boring CC, Squires TS, Montgomery S et al: Cancer statistics, 1994. CA Cancer J Clin 44:7, 1994
68. Choi H, Varma DG, Fornage BD et al: Soft-tissue sarcoma: MR imaging vs sonography for detection of local recurrence after surgery. AJR 157:353, 1991
69. Fornage BD, Lorigan J: Sonographic detection and fine-needle aspiration biopsy of nonpalpable recurrent or metastatic melanoma in subcutaneous tissues. J Ultrasound Med 8:421, 1989
70. Folinais D, Cluzel P, Blangy S et al: Les métastases musculaires: aspects échographiques et tomodensitométriques. J Radiol 69:109, 1988
71. Kratochwil A, Zweymüller K: Ultrasonic examination in orthopedic surgery. p. 343. In Kazner E, de Vlieger M, Müller HR, McCready VR (eds): Ultrasonics in Medicine. Excerpta Medica, Amsterdam, 1975
72. Mukuno DH, Lee TG, Watanabe AS, McIff EB: Aneurysmal bone cyst presenting as a pelvic mass on sonographic examination. J Ultrasound Med 5:215, 1986
73. Fornage BD, Richli WR, Chuapetcharasopon C: Calcaneal bone cyst: sonographic findings and ultrasound-guided aspiration biopsy. J Clin Ultrasound 19:360, 1991
74. Longo JM, Rodriguez-Cabello J, Bilbao JI et al: Popliteal vein thrombosis and popliteal artery compression complicating fibular osteochondroma: ultrasound diagnosis. J Clin Ultrasound 18:507, 1990
75. Civardi G, Livraghi T, Colombo P et al: Lytic bone lesions suspected for metastasis: ultrasonically guided fine-needle aspiration biopsy. J Clin Ultrasound 22:307, 1994
76. Gupta RK, Gupta S, Tandon P, Chhabra DK: Ultrasound-guided needle biopsy of lytic lesions of the cervical spine. J Clin Ultrasound 21:194, 1993

4
Synovial Diseases

Rethy K. Chhem
Germain Beauregard

The synovial membrane, a primordial component of diarthrodial joints, is the site of a large variety of disorders. This chapter reviews the sonographic patterns associated with these conditions and explores the role of sonography as an adjunct to conventional radiography in the diagnosis and initial workup of these diseases.

ANATOMY AND PATHOPHYSIOLOGY OF THE SYNOVIAL MEMBRANE

The synovial membrane covers the inner surface of the articular cavity and its recesses, the bursae, and the tendon sheath. Histologically, the synovial membrane is made up of two layers: a superficial component that is essentially cellular and a deep subintimal layer that contains fibroadipose tissue.[1] Macroscopically, the surface of the synovial membrane is smooth, with some villi and fringe-like folds.[2]

The synovial membrane serves three functions. It secretes fluid to lubricate the articular surface, degrades waste material derived from the joint, and participates in the regulation of hydroelectrolytic transport between the blood and the synovial fluid.[2]

Analysis of aspirated synovial fluid should contribute to the diagnosis of a broad spectrum of disorders affecting the synovium. Routine study of synovial fluid includes determination of the color, viscosity, and quantity; chemical analysis; identification of crystal content; and microbiologic evaluation.

Synovial biopsy is performed as a last resort when clinical signs, noninvasive studies, and synovial fluid analysis results fail to confirm a diagnosis. Biopsy may be the only way to identify synovial disorders with an inflammatory or infiltrating deposition component, e.g., pigmented villonodular synovitis (PVNS), amyloidosis, sarcoidosis, multicentric reticulohistiocytosis, or granulomatous diseases.[3]

TECHNICAL CONSIDERATIONS

High-frequency (5.0- or 7.5-MHz) linear-array transducers are needed to study the extremities. The anatomic area to be scanned is selected on the basis of clinical history, physical findings, abnormalities detected by conventional radiographs or nuclear bone scan, or a combination of these elements. A systematic sonographic study includes a search for synovial effusions or proliferations and assessment of the tendons, bursae, articular cartilage, and juxta-articular bony structures. Comparison with the contralateral side is helpful and allows the detection of subtle abnormalities.

Dynamic studies are used to assess the free motion of tendons within their sheaths[4] or to demonstrate the displacement of fluid from the joint space into the connecting bursa or articular recess, such as occurs during flexion and extension of the knee in a patient with Baker's cyst. This technique is also useful for demonstrating effusion in the subacromial-subdeltoid bursa: the lateral recess fills maximally when the medial recess is compressed under the acromion by forced abduction of the arm.[5] Graded compression of an articular recess evacuates its fluid and should allow detection of hypertrophic synovial folds or a mass trapped between the transducer and the underlying bone. However, excessive compression of an articular recess may result in underestimation of the amount of synovial fluid.

JOINT EFFUSION AND SYNOVITIS

Normal synovium cannot be detected with ultrasound. Any pathologic condition involving the synovium produces edema and stimulates hypersecretion by this membrane, with excess fluid filling the joint space or related structures such as the bursae and tendon sheaths. Therefore, fluid accumulation is the earliest sign of synovial disorders, and a swollen joint is one of the most common presenting symptoms. Sonography can play a major role in the early detection of joint effusions, particularly when clinical examination is difficult or inconclusive.

The approach to sonographic evaluation of a joint effusion varies depending on joint location. In the knee, scanning longitudinally and transversely in the suprapatellar region will demonstrate an accumulation in the suprapatellar recess of the joint. A lateral or medial scan parallel to the femoropatellar joint also may be helpful.

A joint effusion in the hip is evaluated by an anterior approach. The transducer should be placed parallel to the longitudinal axis of the femoral neck, as fluid in the joint will distend the anterior articular recess.[6] Some authors recommend measuring the distance between the capsule and the femoral neck to quantify the effusion.[7] It has been demonstrated in cadavers that as little as 1 ml of effusion may be identified by ultrasound.[7] In children, a tense effusion moves the femoral head anteriorly and displaces the fluid back into the joint. This may cause ultrasound to fail to diagnose a hip effusion.[8]

A glenohumeral joint effusion must be distinguished from the subacromial-subdeltoid bursa. The latter covers the anterior surface of the bicipital groove of the humerus and extends laterally. Because fluid accumulates between the anterior surface of the infraspinatus tendon and the posterior aspect of the humeral head, a posterior sonographic approach is recommended for the detection of a glenohumeral effusion.[9] Some authors suggest an axillary approach for quantitative fluid assessment.[10]

To our knowledge, the role of sonography in the detection of fluid in the ankle has not been specifically studied, but the anterior approach has been successfully used to detect distention of anterior articular recesses.[9,11] Distention of the lateral and medial recesses is limited by the ligaments of the ankle[12]; the posterior recess is deeply located and may be hidden by areolar tissue, the flexor hallucis longus tendon, and the Achilles tendon.

On sonograms, a joint effusion typically appears anechoic, like any other fluid-filled collection; however, posterior enhancement is rarely seen. This is probably because of the presence of bony structures that prevent the ultrasound beam from progressing beyond the posterior wall of the articular recess. Thin or thick septa can be seen in the joint cavity, giving the characteristic loculated appearance of a joint effusion. A wide spectrum of internal echoes can be detected in the joint cavity itself. There may be multiple small particles, homogeneously distributed and moving freely in the joint fluid. Larger echogenic structures also may be seen in the joint space and may represent blood clots, fibrinous deposits, or large synovial masses. A definitive diagnosis cannot be made solely on the basis of these sonographic patterns. These structures must be differentiated from diffuse proliferative synovitis. The latter is manifested as hypoechoic structures of various shapes, such as villous or nodular masses[13] or ribbon-like bands[14]; some float in the joint space, while others remain attached to the joint capsule.

After a joint effusion has been detected, the next step in the diagnosis of synovial disease is to determine the absence or presence of associated synovial proliferation to eliminate or confirm the diagnosis of synovitis. Clinical differentiation between effusion and synovial hypertrophy is often difficult. Synovial thickening may be palpable in the knee or elbow[15]; however, physical examination is less conclusive in the shoulder and hip. Clinical assessment is most difficult at the ankle and wrist because multiple tendons overlie the joint space. In difficult cases, sonography can help discriminate between soft-tissue edema, joint effusion, synovial hypertrophy, and fluid accumulation in the tendon sheath. Graded compression of a swollen joint forces the fluid out of the articular recess while proliferative synovium remains trapped between the transducer and the deep-seated tissues. It is easier to appreciate synovial hypertrophy in the presence of a significant effusion.[13]

The etiology of effusion, with or without associated synovial hypertrophy, varies and includes trauma, internal derangement of the joint, infectious (Fig. 4-1) or inflammatory synovitis, crystal-induced arthropathy (Fig. 4-2), and, less frequently, osteoarthritis (Fig. 4-3).[15] Despite its accurate role in detecting fluid in joint spaces, sonography cannot determine the nature of synovial fluid. The final diagnosis of synovial disorders therefore rests upon correlation of sonographic findings with clinical data and plain radiographic findings. Sonography can play an additional

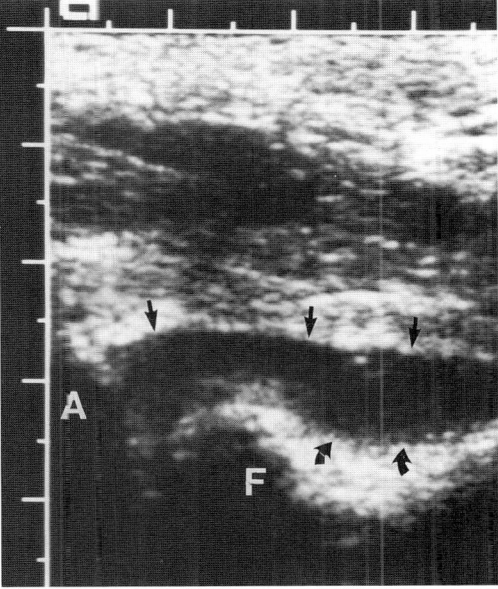

Fig. 4-1 Septic arthritis of the hip. Longitudinal sonogram along the anterior aspect of the hip shows a nonspecific effusion (*arrows*) lifting up the joint capsule from the femoral neck (*curved arrows*). A, acetabulum; F, femur.

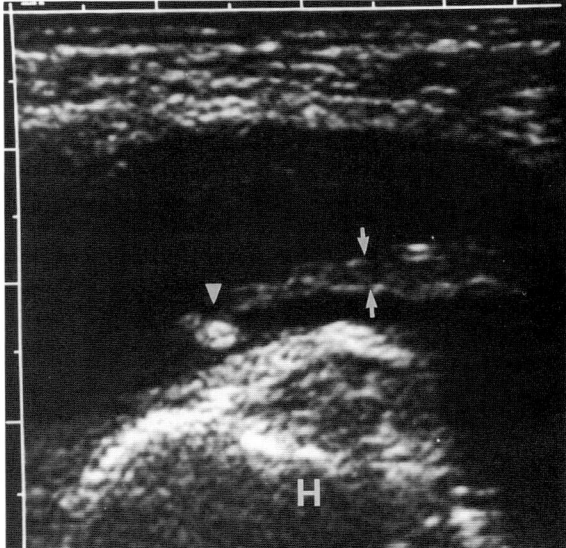

Fig. 4-2 Pyrophosphate arthropathy of the shoulder. Transverse sonogram of the anterior aspect of the shoulder shows a large effusion containing thickened synovial folds (*arrows*) and small calcification (*arrowhead*). Analysis of the synovial fluid revealed calcium pyrophosphate dehydrate crystals. H, humerus.

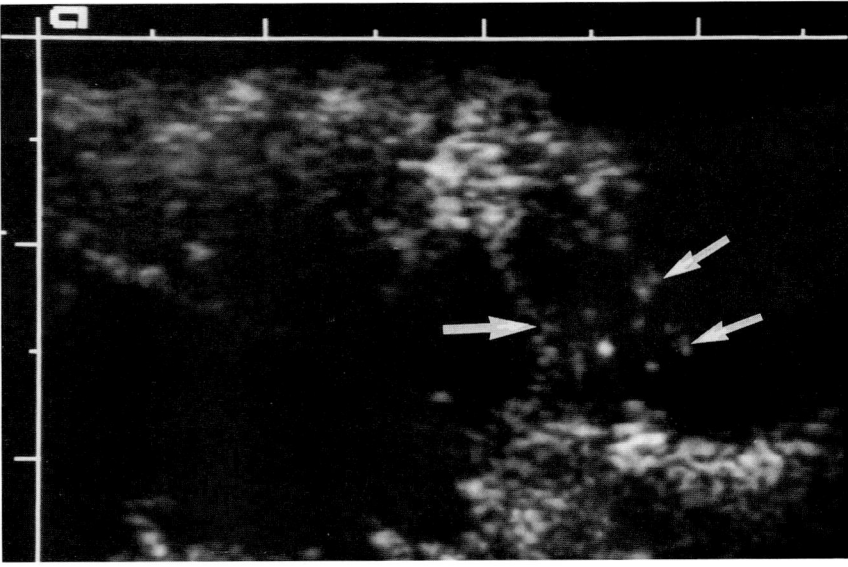

Fig. 4-3 Subnormal synovial membrane. In vitro sonogram of synovial specimen obtained from the hip joint following total joint replacement for osteoarthritis shows multiple, thin, echogenic villi (*arrows*), representing mild hypertrophic synovium.

role in diagnosis by guiding fine-needle aspiration of joint fluid or synovial biopsy.

Because of its ability to discriminate between different components of the soft tissues of the extremities, sonography may contribute to the early diagnosis of synovitis associated with arthropathies, before the onset of radiographically demonstrable osteocartilaginous erosion or joint-space narrowing.

Moreover, because sonography is noninvasive, well tolerated, and fast, it may be repeated during follow-up for inflammatory synovitis to assess disease activity or remission (Figs. 4-4 and 4-5)[17] and to look for complications such as septic arthritis or tendon rupture. Sonography will also demonstrate inflammatory involvement of the related structures of the joint (e.g., bursitis, tenosynovitis, or periarticular soft-tissue nodules).

BURSITIS

Synovial bursae are pouches of variable size and shape, with a synovial membrane forming their inner surface. They facilitate the gliding movement of ligaments, tendons, muscles, and skin. There are three types of bursae: superficial, deep, and adventitious.[18] A normal synovial bursa is detected on sonograms as a hypoechoic cleft in a specific anatomic area.[9] Any pathological condition of the bursa will produce synovial hypersecretion followed by fluid accumulation in the bursal cavity and reactive thickening of the bursal wall.

The sonographic pattern of bursitis is a circumscribed fluid-filled collection; the anatomic location is the key to the diagnosis (Figs. 4-6 and 4-7). Superficial bursitis usually results from trauma. Other causes are infection, gout, and rheumatoid arthritis (Fig. 4-8).[19] Because of its characteristic site, superficial bursitis is easily diagnosed clinically.[19] However, sonography may play an important role when associated soft-tissue edema is present, which may compromise the physical examination. Sonography may also help in the diagnosis of superficial septic bursitis complicating septic arthritis.[19] Untreated superficial inflammatory bursitis in the chronic phase has been reported to simulate a periarticular solid mass on magnetic resonance images.[20] In such cases, sonography can confirm the diagnosis of a fluid collection.

Many deep bursae have been evaluated with sonography, including the subacromial-subdeltoid,[5] iliopsoas,[21] trochanteric,[22] and calcaneal[23] bursae. Because deep bursitis is not palpable, sonography is the examination of choice (Fig. 4-9).

Adventitious bursae develop as a result of preexisting pathologic conditions such as hallux valgus, exostosis, clubfoot, or amputation.[18] The presence of a bursa over the cartilaginous cap of an exostosis may cause overestimation of the thickness of the cap.[24] Sonography may help differentiate bursa formation from malignant transformation.[25]

TENOSYNOVITIS

The synovial membrane of a tendon sheath is histologically identical to that of the diarthrodial joints. The membrane's main functions are to facilitate gliding movement of the tendons and to participate in the nutrition of the tendons.[18] Disorders of a tendon sheath's synovial membrane induce edema of the synovium with increased secretion and fluid accumulation in the tendon sheath. On sonograms, the tendon appears as a hyperechoic central structure with a hypoechoic halo representing effusion or thickening of the synovial membrane (Fig. 4-10). A pannus also may appear hypoechoic, making differentiation from fluid accumulation in the sheath difficult. The complete disappear-

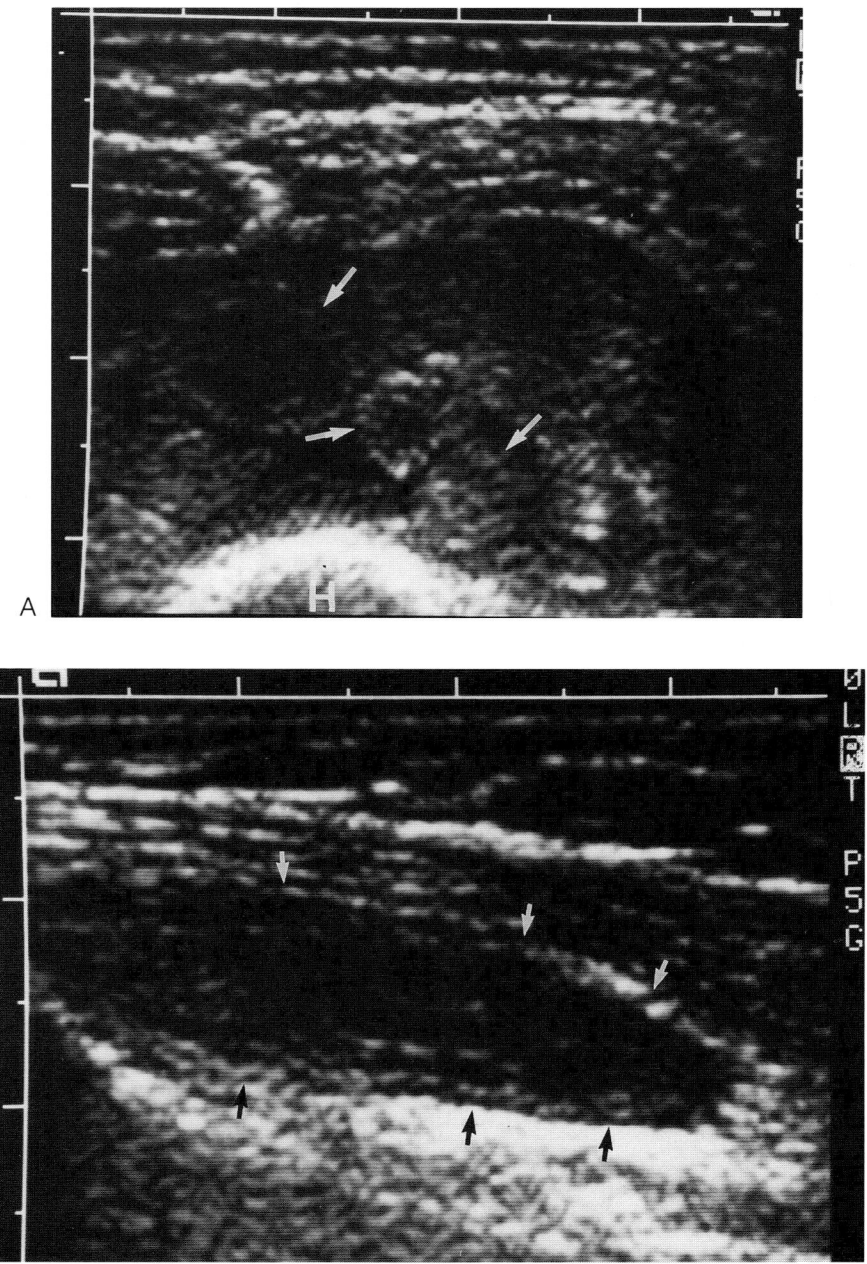

Fig. 4-4 Proliferative synovitis of the shoulder. **(A)** Transverse sonogram shows multiple synovial nodules (*arrows*) filling the joint cavity. A small joint effusion is seen between the coalescent nodules. Clinical history and laboratory tests confirmed the diagnosis of rheumatoid arthritis. H, humerus. **(B)** After adequate therapy, follow-up sonogram shows a decrease in both nodule size and effusion volume (*arrows*).

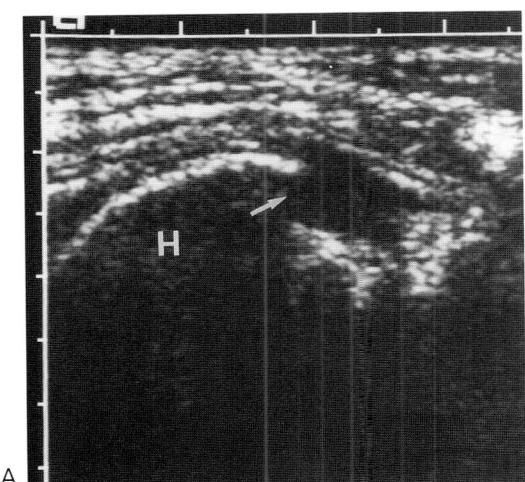 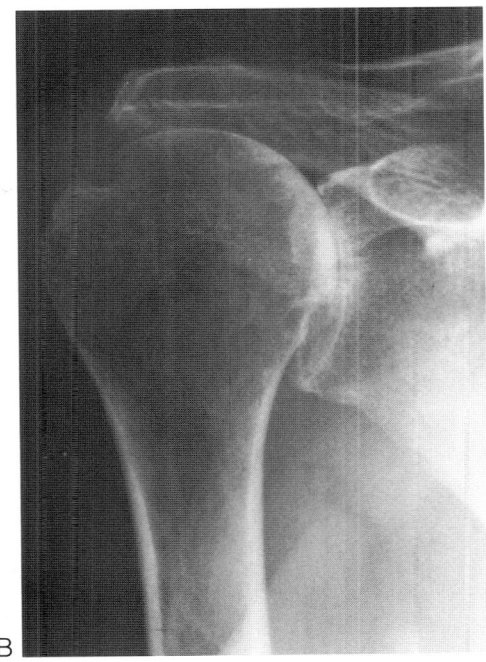

Fig. 4-5 Advanced rheumatoid arthritis of the shoulder. **(A)** Transverse sonogram shows erosion (*arrow*) of humeral head (H) and a hypoechoic mass in the joint space. The latter represents a synovial pannus. The supraspinatus tendon is completely torn. **(B)** Plain film of the shoulder made for correlation.

ance of the hypoechoic halo under strong compression by the transducer usually represents effusion in the tendon sheath. A synovial pannus also can be compressed, but a thin hypoechoic rim remains around the tendon itself. Irregular contour of the tendon is a sign of chronic tenosynovitis.[4] Occupational tenosynovitis or tendinitis is frequently seen in the upper extremities as a result of chronic and repetitive microtrauma to the tendon and its synovial sheath (Fig. 4-11).[26] Inflammatory tenosynovitis is usually associated with rheumatoid arthritis, seronegative spondyloarthropathies, or crystal-induced arthropathies. In rheumatoid arthritis, inflammatory tenosynovitis is seen mostly in the hand and wrist; in the foot, the peroneal tendons are the most often involved site, followed by the tibialis posterior tendon.[27] Edema and peritendinous synovial pannus are also seen in rheumatoid arthritis. The tendon itself must be evaluated for intratendinous rheumatoid nodules and for partial or complete rupture. Sonographic evaluation is useful in the assessment of these complications preceding corrective surgery.[4] In the shoulder, the presence of fluid in the biceps tendon sheath should lead to careful exploration of the rotator cuff for rupture.

Acute septic tenosynovitis is seen mostly in the hand; it must be diagnosed at an early stage for prompt surgical drainage. Sonography readily demonstrates pus in the tendon sheath.[28,29] Infection of the tendon sheath results from an open wound with or without foreign body inclusions. Sonography can help localize such foreign bodies, particularly radiotransparent ones.[23,29] Foreign bodies

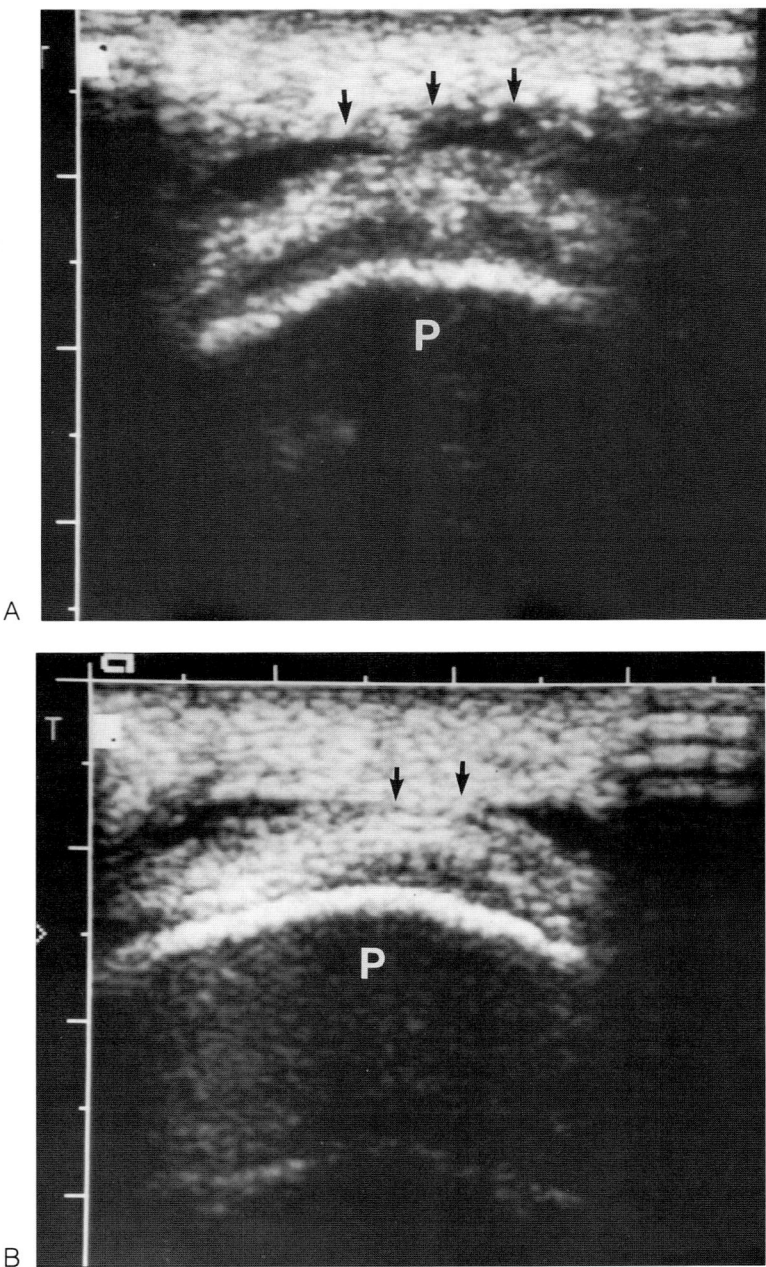

Fig. 4-6 Prepatellar bursitis. **(A)** Longitudinal sonogram of the anterior aspect of the knee shows a small sac (*arrows*) distended by fluid and overlying the patella (P). This represents inflammatory prepatellar bursitis. **(B)** Compression of this area by the transducer has induced collapse of the sac (*arrows*). This represents a technical artifact that could be a source of false-negative findings in superficial bursitis. P, patella.

SYNOVIAL DISEASES

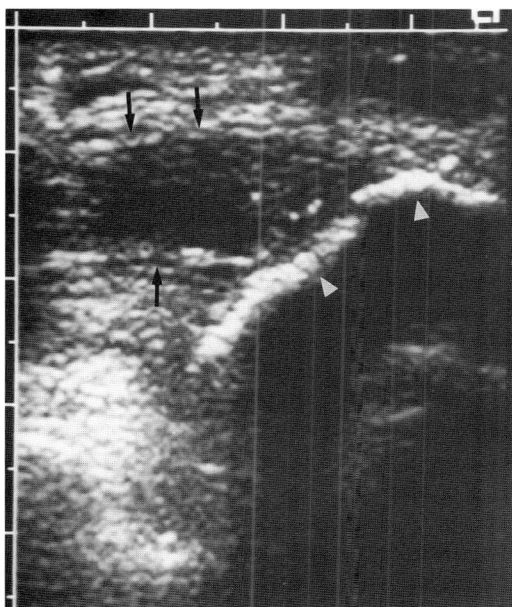

Fig. 4-7 Olecranon bursitis. Longitudinal sonogram of the posterior aspect of the elbow shows fluid collected (*arrows*) between the skin and the olecranon (*arrowheads*). The bursal wall is slightly thickened.

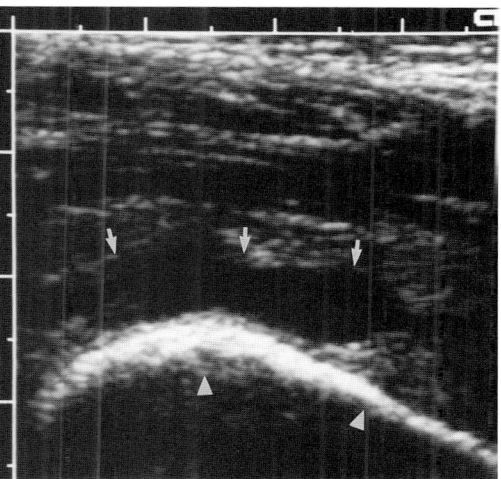

Fig. 4-9 Trochanteric bursitis. Longitudinal sonogram of the lateral aspect of the hip shows a fluid-filled sac (*arrows*) situated between the greater trochanter (*arrowheads*) and the gluteus maximus muscle.

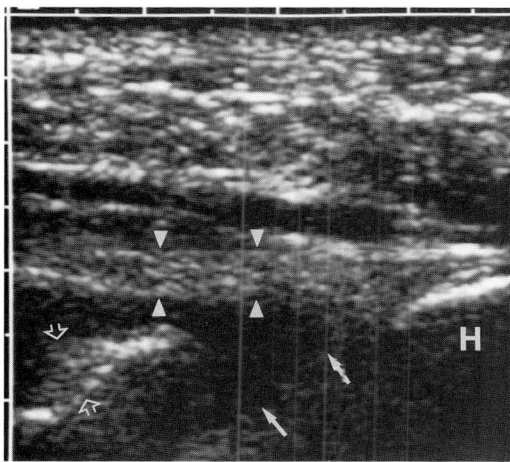

Fig. 4-8 Rheumatoid synovitis of the shoulder with associated inflammation of the subacromial-subdeltoid bursa. Posterior sonogram of the shoulder shows effusion of the glenohumeral joint (*arrows*) in front of the infraspinatus tendon (*arrowheads*). A bursal effusion is located behind the infraspinatus tendon. Note the triangular shape of the posterior labrum (*open arrows*). H, humeral head.

usually appear as hyperechoic foci with acoustic shadowing or comet-tail artifacts,[23] regardless of their radiodensity on plain films. The sonographic appearance of foreign bodies is described in detail in Chapter 8.

INTRA-ARTICULAR SYNOVIAL MASSES

Pigmented Villonodular Synovitis

PVNS is an inflammatory proliferative synovial disorder of unknown origin that can involve the joint, the tendon sheath, or the synovial bursae.[30] Macroscopically, it is made up of brownish-red villous or nodular masses. Microscopic study demonstrates rounded or fingerlike masses made up of a fibrous stroma; the number of masses increases as the disease progresses.[31] Pigmentation of the synovium is a consequence of deposition of hemoglobin degradation products.[31] Two forms of PVNS are recognized: the diffuse type, which involves the synovium of the whole articular surface, and

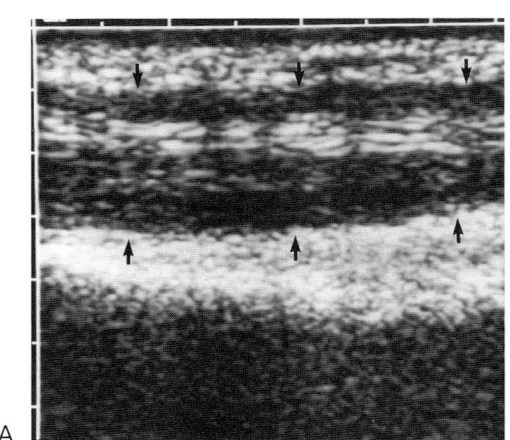
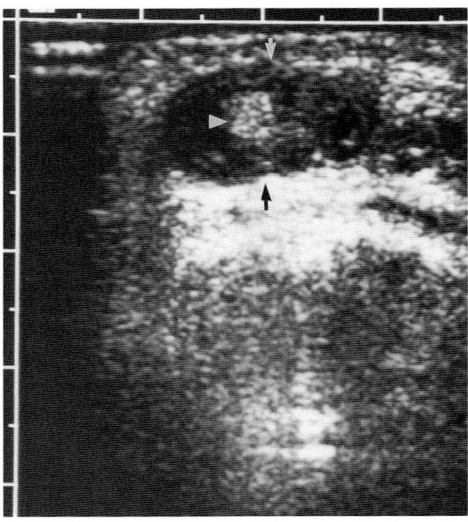

Fig. 4-10 Tenosynovitis of the tibialis anterior tendon. **(A)** Longitudinal and **(B)** transverse sonograms of the anterior aspect of the ankle show synovial thickening of the tibialis anterior tendon sheath (*arrows*) in a patient known to have rheumatoid arthritis. There is no tear of the tendon itself (*arrowhead*). No ankle joint effusion was detected.

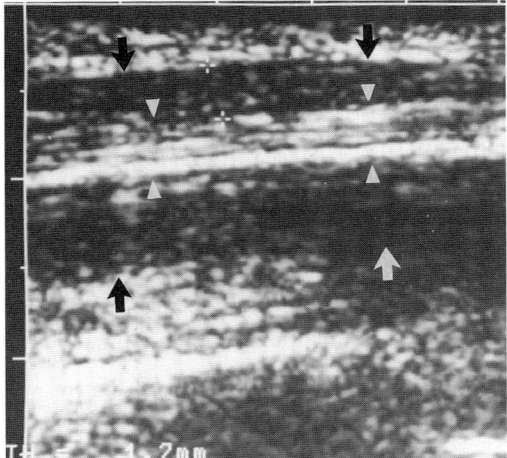

Fig. 4-11 Inflammatory tenosynovitis of the extensor carpi radialis tendon following repetitive microtrauma. Longitudinal sonogram shows thickening of the tendon sheath (*arrows*). Note the smooth contour of the tendon itself (*arrowheads*).

the localized type, which involves only a focal area of the synovial membrane. PVNS is typically monoarticular, accompanied by joint effusion and restriction of motion. The most common sites of involvement are the hip and knee joints; the shoulder and ankle are less commonly affected.[31]

The severity of clinical symptoms depends on the stage of the disease. Clinical examination sometimes reveals painful synovial masses. The synovial fluid is serous but bloodstained. In advanced stages, the synovial hypertrophy produces bony and cartilaginous erosion with articular destruction that is readily detected on conventional radiographs. Plain radiography also may demonstrate intra- or periarticular soft-tissue masses.[32]

In diffuse PVNS, sonography typically demonstrates a large joint effusion and synovial hypertrophy. The effusion has no unique characteristics, but the synovial proliferation appears either as nodules (Fig. 4-12) or thick

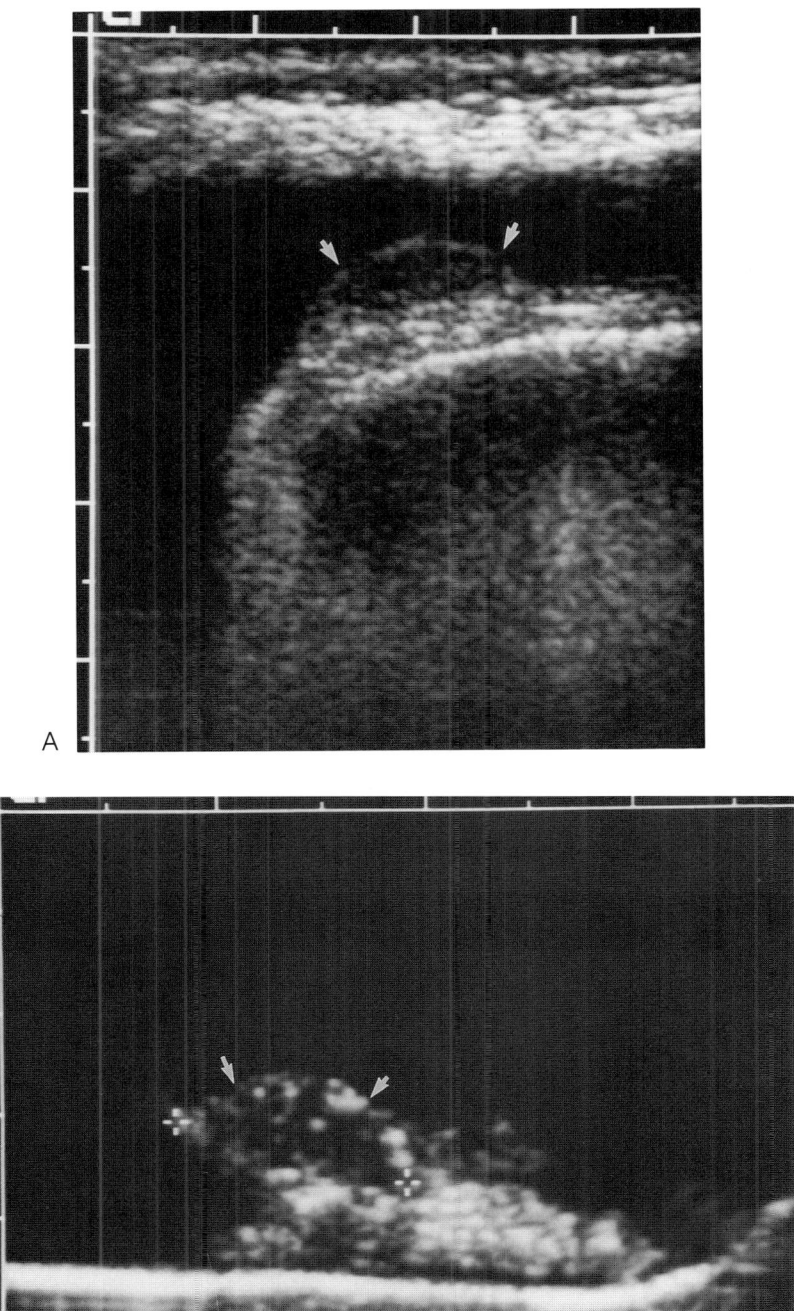

Fig. 4-12 Pigmented villonodular synovitis (PVNS) of the knee. **(A)** Longitudinal sonogram along the anterior aspect of the knee shows a hypoechoic nodule (*arrows*) attached to the posterior wall of the suprapatellar bursa and representing a PVNS nodule. Scanning of the entire knee joint demonstrated similar additional nodules. **(B)** In vitro sonogram of the specimen (*arrows*) shows similar sonographic characteristics.

villous fronds coating the deep surface of the articular capsule or as hypoechoic pedunculated nodular masses floating in fluid. These synovial masses are best evaluated with graded compression of the articular recess with the transducer. Because the hypoechoic masses are solid and resistant to compression, they are trapped under the transducer as the pressure evacuates the fluid from the area being studied. The synovial proliferation seen in PVNS is not a specific sonographic finding; other causes of synovial hypertrophy, such as juvenile or adult rheumatoid arthritis, hemophiliac arthritis, and lipoma arborescens, may be associated with similar findings.[32]

Synovial Osteochondromatosis

Synovial osteochondromatosis is a monoarthropathy with cartilaginous metaplasia of the synovial membrane; its origin is undetermined.[2] The following joints may be involved (in order of decreasing frequency): knee, hip, elbow, wrist, ankle, and shoulder. Clinical signs include limited range of motion or progressively increasing articular pain.[2] The growth of a solid intra-articular mass may be the initial presenting sign. Microscopic examination of the lesion reveals chondromatosis deposits in the synovial membrane.[2] Recurrence after surgical excision is frequent, but malignant transformation is rare.[1] When these masses are calcified, they are readily demonstrated by plain radiography or sonography.[33] Some masses show no calcified component and may suggest effusion or a synovial mass of other origin on lateral plain films. Sonography is useful in these cases because it shows solid hypoechoic masses, therefore ruling out the possibility of joint effusion (Fig. 4-13).

Synovial Lipoma Arborescens

Synovial lipoma arborescens is a rare entity. It represents benign hyperplasia of the synovial membrane secondary to subsynovial infiltration with adipose cells.[34] It is most frequently found in the knee. Synovial lipoma arborescens appears on sonograms as a well-delineated and hypoechoic intra-articular mass.[35] This sonographic appearance is nonspecific. Diagnosis is confirmed by computed tomography scans, which demonstrate the fatty nature of the lesion and the absence of contrast enhancement.[36] Surgical resection is the treatment of choice.

Synovial Hemangioma

Synovial hemangioma is another rare entity. Soft-tissue angiomas are less common than angiomas in bone.[37] The synovial form is a monoarticular lesion developing at the knee in children or adolescents.[2] Clinical symptoms include articular swelling, minimal pain, and restriction in range of motion.[2] Plain films demonstrate an intra-articular soft-tissue mass with or without periarticular extension and sometimes containing phleboliths. Sonography demonstrates a soft, compressible mass made up of tortuous tubular structures. Doppler sonography confirms the diagnosis of hemangioma, with characteristic low-resistance flow within the mass.[37] An angiogram should be obtained only if surgery is planned, to map the lesion and perform preoperative embolization. Synovectomy for this type of lesion is difficult because of widespread extension, and the recurrence rate is high.[2]

SYNOVIAL MASSES IN TENDON SHEATHS OR BURSAE

Most of the above-described masses of the articular synovium are also seen in the tendon sheaths and in bursae. The differential diagnosis of a synovial mass developing within a tendon sheath includes synovial osteochondromatosis, PVNS, lipoma, granulomatous disease, and giant cell tumor.[38] For a bursal mass, the differential diagnosis will

SYNOVIAL DISEASES

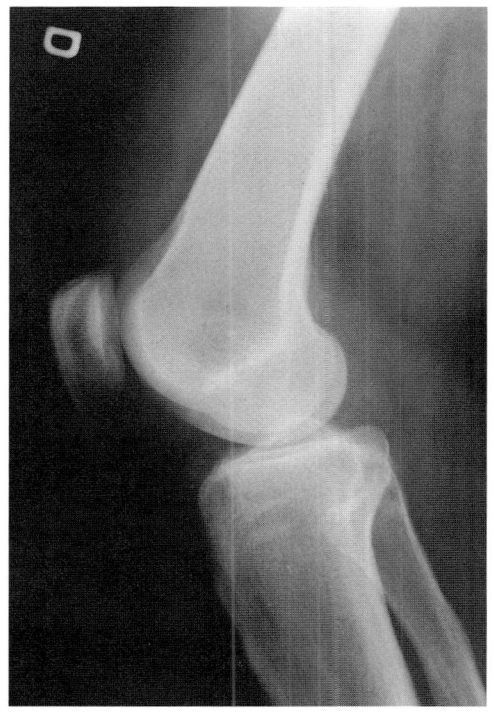

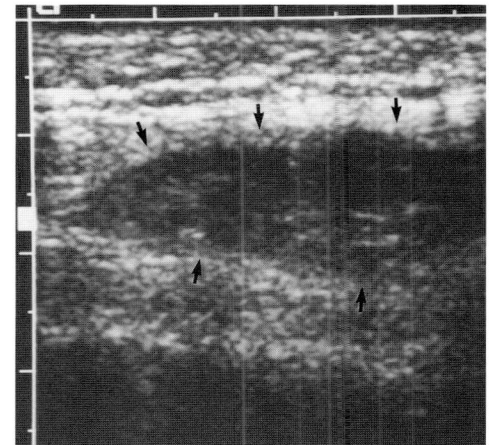

Fig. 4-13 Synovial chondromatosis. **(A)** Lateral radiograph of the knee shows an effusion-like mass in the suprapatellar bursa. **(B)** Longitudinal sonogram through the anterior aspect of the knee shows an incompressible hypoechoic mass (*arrows*) completely filling the suprapatellar bursa. No calcifications are seen. **(C)** In vitro sonogram of the specimen demonstrates an oval hypoechoic mass (*arrows*) representing cartilaginous metaplasia of the synovium. The small bright spots scattered throughout the mass may represent small calcifications or artifacts resulting from interfaces created by the confluence of the cartilaginous lobules.

be limited to PVNS, synovial sarcoma, hemangioma, and chronic bursitis.[23]

SYNOVIAL CYSTS

Sonography plays a fundamental role in differentiating cystic from solid lesions in the initial workup of periarticular masses. Cystic lesions around the synovial joint are easily detected by sonography because of their superficial location and fluid content (Fig. 4-14). Synovial cysts are fibrous or synovium-lined cavities filled with synovial fluid.[59] They are associated with trauma, degenerative joint diseases, and inflammatory arthritis.[16] Baker's cyst is the most commonly seen. Dissection of Baker's cyst down the

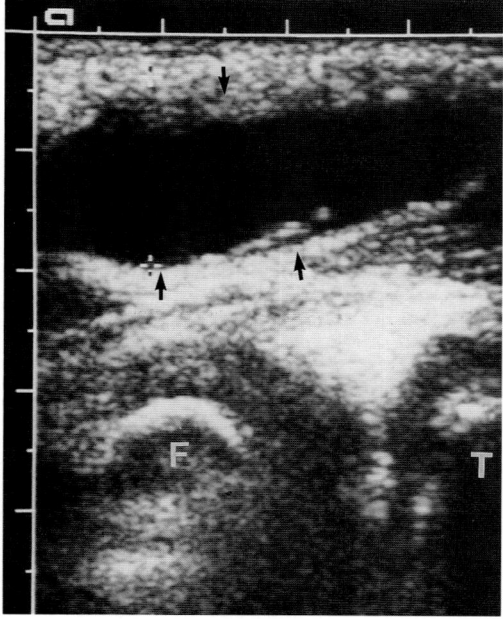

Fig. 4-14 Baker's cyst. Longitudinal sonogram of the popliteal fossa in a 15-year-old boy with juvenile rheumatoid arthritis presenting with a painful mass of the knee shows a cystic mass (*arrows*) developing at the posteromedial aspect of the knee and representing a typical Baker's cyst. F, femoral condyle; T, tibial plateau.

calf can clinically simulate deep vein thrombosis. Sonography helps differentiate these two entities. Synovial cysts can be observed in any joint, including the ankle, wrist, elbow, hip, and shoulder.[16] A meniscal cyst is a collection of mucinous material communicating with the peripheral aspect of the meniscus; sonography typically shows a hypoechoic cyst adjacent to the torn meniscus.[9]

CONCLUSIONS

Sonography is a valuable adjunct to conventional radiography in the evaluation of joint and synovial diseases. In inflammatory arthritis, sonography can help quantify the joint effusion, search for synovial pannus, and assess response to therapy. Related soft-tissue structures such as tendons and bursae also can be effectively studied with sonography. Furthermore, sonography can confirm the cystic nature of a periarticular mass and localize the mass in relation to the joint. However, like other imaging modalities, sonography cannot provide a histologic diagnosis of joint disorders, and correlation of imaging findings with clinical history and laboratory results is often necessary to establish the final diagnosis.

ACKNOWLEDGMENTS

The authors thank Dr. Georges Y. El-Khoury of the University of Iowa for his review of this chapter.

REFERENCES

1. Bogumill GP, Schwamm HA: Orthopaedic Pathology. A Synopsis with Clinical and Radiographic Correlation. WB Saunders, Philadelphia, 1984
2. Bullough PG, Vigorita VJ: Orthopaedic Pathology with Clinical and Radiologic Correlation. JB Lippincott, Philadelphia, 1984
3. Schumacher HR: Synovial biopsy and pathology. p. 648. In Kelley WN, Harris ED, Jr, Ruddy S et al (eds): Textbook of Rheumatology. WB Saunders, Philadelphia, 1985
4. Fornage BD, Rifkin MD: Ultrasound examination of tendons. Radiol Clin North Am 26:87, 1988
5. van Holsbeeck M, Strouse P: Sonography of the shoulder: evaluation of the subacromial-subdeltoid bursa. AJR 160:561, 1993
6. Wingstrand H, Egund N: Ultrasound in hip joint effusion. Acta Orthop Scand 55:469, 1984
7. Marchal GJ, van Holsbeeck MT, Raes M et al: Transient synovitis of the hip in children: role of US. Radiology 162:825, 1987
8. Mathie AG, Benson MK, Wilson DJ: Lessons in the investigation of irritable hip: failure of ultrasound to detect haemarthrosis. J Bone Joint Surg 73B:518, 1991

9. van Holsbeeck MT, Introcaso JH: Musculoskeletal Ultrasound. Mosby-Year Book, St. Louis, 1991
10. Koski JM: Validity of axillary ultrasound scanning in detecting effusion of glenohumeral joint. Scand J Rheumatol 20:49, 1991
11. Chhem RK, Beauregard G, Schmutz GR, Benko AJ: Sonography of the ankle and hindfoot. Can Assoc Radiol J 44:337, 1993
12. Towbin R, Dunbar JS, Towbin J, Clark R: Teardrop sign: plain film recognition of ankle effusion. AJR 134:985, 1980
13. Cooperberg PL, Tsang I, Truelove L, Knickerbocker WJ: Gray scale ultrasound in the evaluation of rheumatoid arthritis of the knee. Radiology 126:759, 1978
14. Lawson JP, Rahn DW: Lyme disease and radiologic findings in Lyme arthritis. AJR 158:1065, 1992
15. Michet CL, Hunde GG: Examination of the joints. p. 369. In Kelley WN, Harris ED, Jr, Ruddy S et al (eds): Textbook of Rheumatology. WB Saunders, Philadelphia, 1985
16. Resnick D: Common disorders of synovium-lined joints: pathogenesis, imaging abnormalities, and complications. AJR 151:1079, 1988
17. Spiegel TM, King W III, Weiner SR, Paulus HE: Measuring disease activity: comparison of joint tenderness, swelling, and ultrasonography in rheumatoid arthritis. Arthritis Rheum 30:1283, 1987
18. Canoso JJ: Bursae, tendons and ligaments. Clin Rheum Dis 7:189, 1981
19. Ho G, Jr, Mikolich DJ: Bacterial infection of the superficial subcutaneous bursae. Clin Rheum Dis 12:437, 1986
20. Zeiss J, Coombs RJ, Booth RL, Jr, Saddemi SR: Chronic bursitis presenting as a mass in the pes anserine bursa: MR diagnosis. J Comput Assist Tomogr 17:137, 1993
21. Toohey AK, Lassalle TL, Martinez S et al: Iliopsoas bursitis: clinical features, radiographic findings, and disease associations. Semin Arthritis Rheum 20:41, 1990
22. Howard CB, Vinzberg A, Nyska M, Zirkin H: Aspiration of acute calcerous trochanteric bursitis using ultrasound guidance. J Clin Ultrasound 21:45, 1993
23. Fornage BD, Rifkin MD: Ultrasound examination of the hand and foot. Radiol Clin North Am 26:109, 1988
24. Malghem J, Van de Berg B, Noël H, Maldague B: Benign osteochondromas and exostotic chondrosarcomas: evaluation of cartilage cap thickness by ultrasound. Skeletal Radiol 21:33, 1992
25. El-Khoury GY, Bassett GS: Symptomatic bursa formation with osteochondromas. AJR 133:895, 1979
26. Thorson E, Szabo RM: Common tendinitis problems in the hand and forearm. Orthop Clin North Am 23:65, 1992
27. Kitaoka HB: Rheumatoid hindfoot. Orthop Clin North Am 20:593, 1989
28. Jeffrey RB, Jr, Laing FC, Schechter WP et al: Acute suppurative tenosynovitis of the hand: diagnosis with US. Radiology 162:741, 1987
29. Groleau S, Chhem RK, Younge D, Basora J: Ultrasonography of foreign-body tenosynovitis. Can Assoc Radiol J 43:454, 1992
30. Jaffe HL, Lichtenstein L, Sutro CJ: Pigmented villonodular synovitis, bursitis and tenosynovitis. Arch Pathol 31:731, 1941
31. Dorwart RH, Genant HK, Johnston WH, Morris JM: Pigmented villonodular synovitis of synovial joints: clinical, pathologic, and radiologic features. AJR 143:877, 1984
32. Kaufman RA, Towbin RB, Babcock DS, Crawford AH: Arthrosonography in the diagnosis of pigmented villonodular synovitis. AJR 139:396, 1982
33. Moss GD, Dishuk W: Ultrasound diagnosis of osteochondromatosis of the popliteal fossa. J Clin Ultrasound 12:232, 1984
34. Schajowitz F: Tumour and Tumourlike Lesions of Bone and Joint. Springer, Berlin, 1981
35. Martinez D, Millner PA, Coral A et al: Case report 745: synovial lipoma arborescens. Skeletal Radiol 21:393, 1992
36. Armstrong SJ, Watt I: Lipoma arborescens of the knee. Br J Radiol 62:178, 1989
37. Greenspan A, McGahan JP, Vogelsang P, Szabo RM: Imaging strategies in the evaluation of soft-tissue hemangiomas of the extremities: correlation of the findings of plain radiography, angiography, CT, MRI, and ultrasonography in 12 histologically proven cases. Skeletal Radiol 21:11, 1992
38. Waggenspack GA, Amparo EG: Case report 466: granulomatous tenosynovitis (left 3rd finger). Skeletal Radiol 17:133, 1988
39. Schwimmer M, Edelstein G, Heiken JP, Gilula LA: Synovial cysts of the knee: CT evaluation. Radiology 154:175, 1985

5
Bone and Articular Cartilage

Ronald S. Adler

The role of gray-scale sonography in evaluating abnormalities of bone and articular cartilage has yet to be entirely defined. Typically, such abnormalities are noted incidentally during the course of real-time examination performed for evaluation of the overlying soft tissues. However, in certain instances sonography has been employed as the principal modality for evaluating a particular bone or joint lesion. The most commonly described application has been assessment of developmental abnormalities prior to complete ossification of the appendicular skeleton, particularly abnormalities involving the centers of ossification.[1-3] The most notable of these is developmental dysplasia of the hip (see Ch. 12).[1] Traumatic lesions involving the cartilaginous epiphyses and growth plates have also been evaluated using gray-scale sonography.[4-6] Less common applications include the evaluation of fractures, osteomyelitis, and abnormalities of hyaline cartilage and fibrocartilage.[7-12] The following discussion will be confined to assessment of the cortical and cartilaginous surfaces in the appendicular skeleton. Evaluation of developmental dysplasia of the hip, vertical talus, and other developmental conditions will not be addressed. A discussion of the use of speed-of-sound and acoustic attenuation as a means of quantifying bone mineral density is largely beyond the scope of this chapter and also will not be considered. The sonographic evaluation of the corticotomy site in patients with the Ilizarov device will be discussed briefly since sonography may play an increasingly important role in the monitoring of orthopaedic procedures, particularly because sonography is not subject to the artifacts evident in computed tomography (CT) and magnetic resonance imaging (MRI) in the presence of orthopaedic hardware.

TECHNIQUE

The methods employed are similar to those used for other musculoskeletal applications. It is essential to use a linear-array transducer at the highest center frequency (typically 7.5 or 10 MHz) allowable by the amount of overlying soft tissue. A thin standoff pad may be of value in placing the area of interest within the focal zone of the transducer. Whenever possible, the transducer face should parallel the surface of

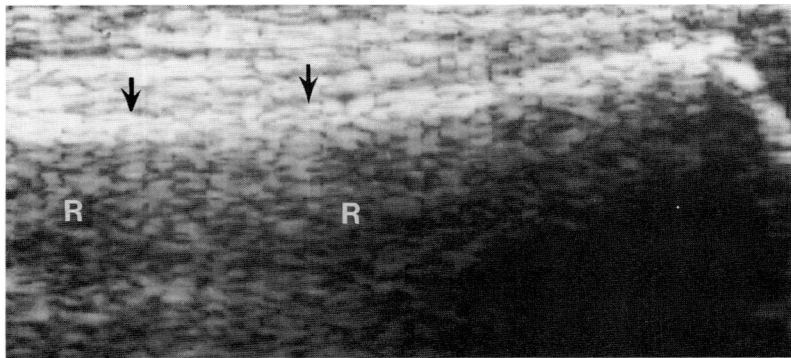

Fig. 5-1 Normal cortical bone. Longitudinally oriented sonogram of the normal distal radial metadiaphysis in a child. The smooth, mildly curved specular reflector (*arrows*) denotes the cortical surface. The periosteum is not readily distinguished from the remaining overlying soft tissue. Note the strong reverberation echoes (R) deep to the portion of the cortical surface that is most parallel to the footprint of the transducer.

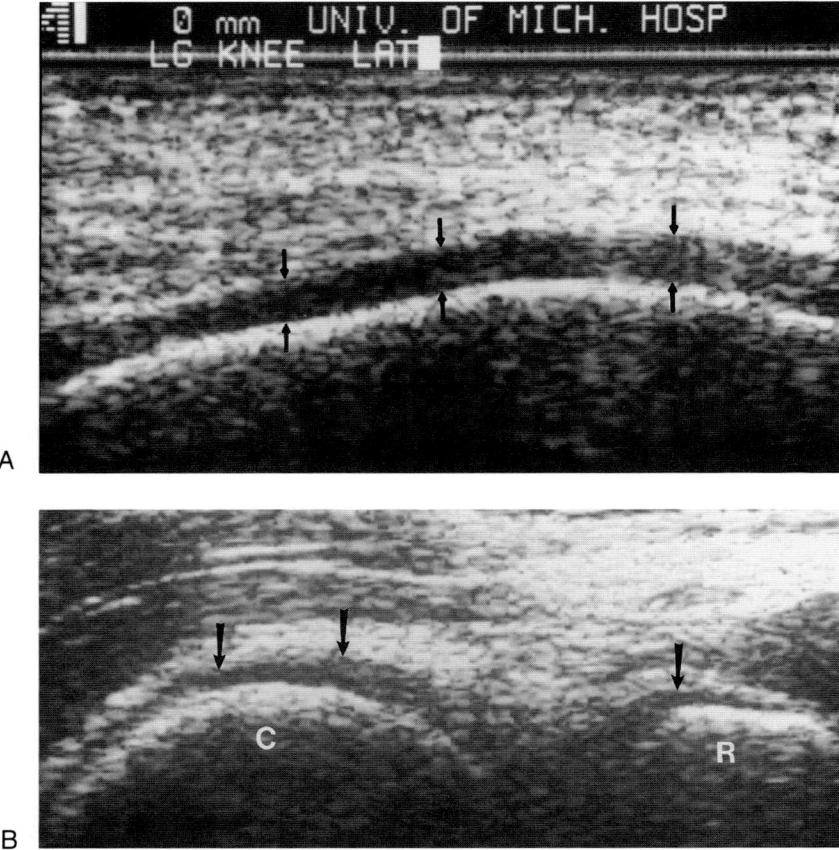

Fig. 5-2 (A, B) Arrows indicate normal articular cartilage, seen as a hypoechoic band paralleling the underlying cortex. In the absence of any superimposed pathologic process, gross assessment of cartilage integrity and thickness is possible. **(A)** Longitudinal sonogram of the lateral femoral condyle with the knee in flexion. **(B)** Longitudinal sonogram of the radiocapitellar joint. C, capitellum; R, radial head.

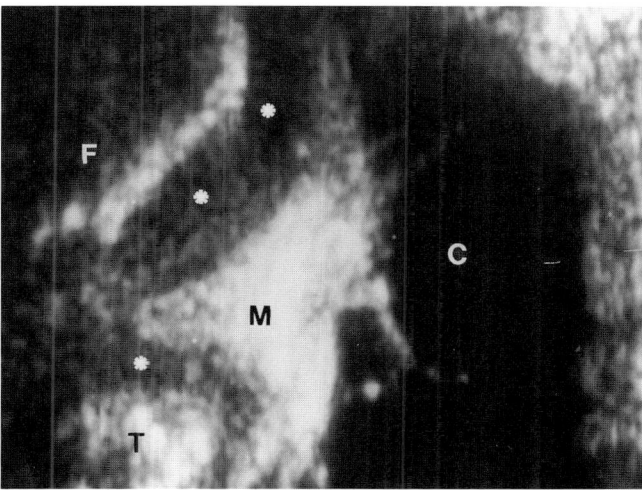

Fig. 5-3 Hyaline cartilage in developing ossification centers of the knee in a child with a Baker's cyst (C). Longitudinal sonogram of the posteromedial aspect of the knee shows the developing epiphyseal cartilage of the femur (F) and tibia (T) as thick hypoechoic bands (*asterisks*). This is in contrast to the adjacent meniscal fibrocartilage (M), which appears as an echogenic triangle situated between the articular cartilage and the cyst.

interest, although on occasion sonograms performed off-angle may provide useful information.

Normal cortical bone appears as a smooth, essentially perfect specular reflector (Fig. 5-1). Echoes appearing deep to the cortical surface result from the strong cortical reverberations scattering off of the overlying soft tissue. The periosteum adheres closely to the bone and normally cannot be recognized as a separate structure. Alterations of the surface characteristics and/or visualization of the periosteum may, therefore, indicate underlying pathology. An exception to this rule occurs at the articular surface, where hyaline cartilage appears on gray-scale sonograms as a 2- to 3-mm hypoechoic band superficial to the subchondral bone (Fig. 5-2).[13] This appearance is similar to that of the hyaline cartilage in developing ossification centers in infants and young children (Fig. 5-3).[1] At other sites, fibrocartilage appears echogenic and so is readily distinguished.[3] The thin layer of overlying synovial fluid is usually not distinguishable from the cartilage.

TRAUMATIC LESIONS

Radiography is the examination of choice when a fracture is suspected clinically. However, inasmuch as soft-tissue injuries may be accompanied by traumatic lesions of bone, fractures may be recognized during real-time sonographic examination and should be recorded. Although the resolutions of radiography and sonography are not comparable, the tomographic nature of sonography may, on occasion, allow detection of a radiographically occult fracture. This has been frequently observed during the assessment of rotator cuff injuries.[8] Abrupt discontinuities, angulation deformities, or both can be seen at the cortical margin in acute or subacute bone fractures (Figs. 5-4 and 5-5).[7,8] Other abnormalities that have similar sonographic features, including inflammatory erosions

62 MUSCULOSKELETAL ULTRASOUND

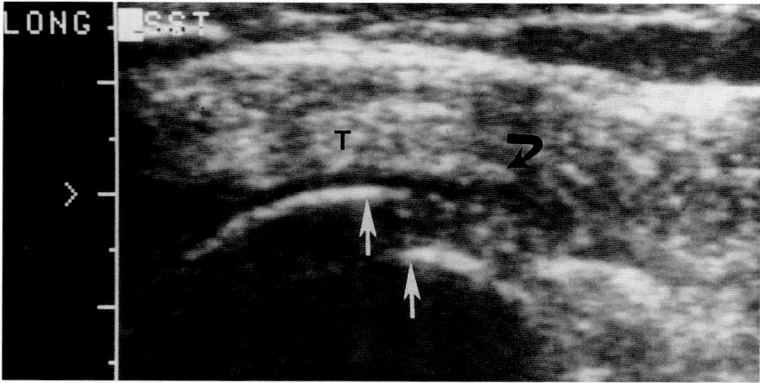

Fig. 5-4 Humeral fracture. Longitudinal sonogram of the region of the greater tuberosity in a patient with a previously unsuspected fracture of the anatomic neck of the humerus. An abrupt discontinuity is present at the fracture site (*arrows*). A continuous supraspinatus tendon (T) is not evident, and the tendon was, in fact, torn (*curved arrow*).

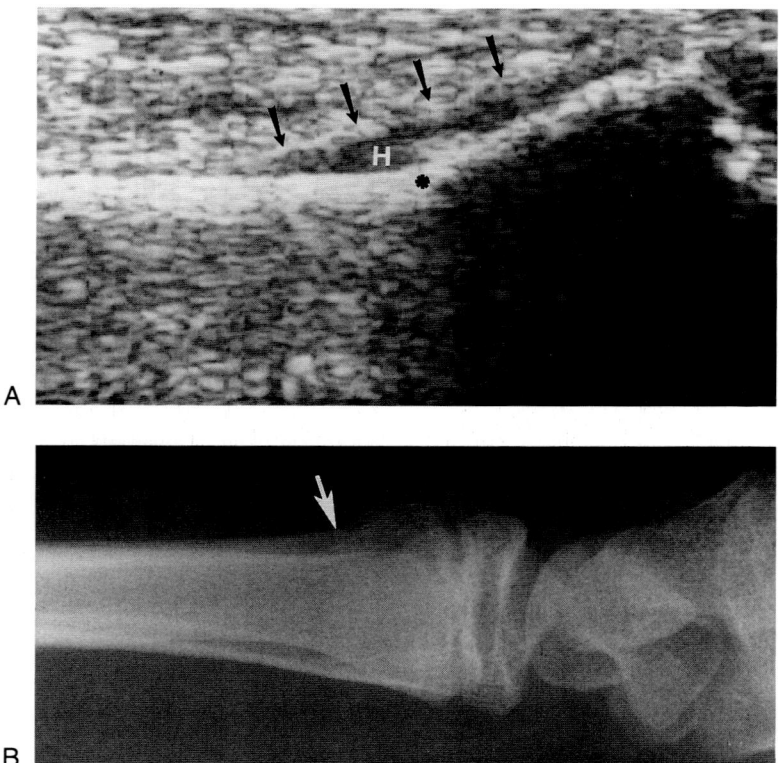

Fig. 5-5 Acute buckle fracture of the distal radial metadiaphysis in a child. **(A)** Longitudinal sonogram. *Asterisk*, apex of the fracture; *arrows*, displaced but intact periosteum; H, subperiosteal hematoma. Figure 5-1 shows the contralateral extremity in the same child for comparison. **(B)** Corresponding radiograph. The fracture site is indicated by an arrow.

and calcific tendinitis, represent potential pitfalls (Figs. 5-6 and 5-7).

Situations in which sonography plays a primary clinical role are in newborns and in young children with incomplete ossification at the site of trauma. Sonographic examination of fractures or dislocations—or both—involving the cartilaginous ossification centers and the corresponding growth plates eliminates the need for arthrography because sonography directly visualizes displaced cartilage fragments, cartilaginous growth centers, or both.[3-6]

The appearance of fractures in long bones is similar to those involving intra-articular structures.[7,8] When a fracture occurs along the cortical surface, the periosteum, if intact, will be displaced by a subperiosteal hematoma (Fig. 5-5A). The appearance is somewhat analogous to that of a joint capsule displaced by an effusion. The presence of an intact periosteum may be important to document since this can have prognostic significance. The lack of an appropriate tangential beam radiograph may preclude plain film detection, whereas a corresponding sonogram, again due to its tomographic nature and ability to evaluate the overlying soft tissues, may identify the previously unsuspected fracture.

INFLAMMATORY AND INFECTIOUS LESIONS

The ultrasound detection of inflammatory or infectious lesions of bone generally occurs during evaluation for soft-tissue lesions such as abscesses, bursitis, or joint effusions. Sonography may, however, provide one of the earliest indications of a more extensive abnormality involving the bone, and in these cases, additional appropriate imaging evaluation should follow. The earliest sonographic sign of osteomyelitis in long bones has been described as a hypoechoic band immediately adjacent to the soft-tissue–bone interface (Fig. 5-8)[9-11]; this can precede the corresponding radiographic abnormalities. Differentiation of this finding from the appearance of other infiltrative lesions (e.g., cellulitis) is generally not possible. The earliest associated abnormalities of the cortex are those that alter the nature of the otherwise smooth specular surface. As a result, the significant amount of reverberation artifact typically present deep to the bone on the corresponding gray-scale image is altered. The loss of phase coherence in the ultrasound beam reflected from a roughened surface, as may occur in patients with cortical erosion, has been shown experimentally to alter the nature of this posterior "dirty" shadow to a "cleaner" one (Fig. 5-8).[14] Some component of isotropic scattering of the acoustic beam off of the surface irregularities may similarly contribute to a cleaner shadow. When these irregularities are sufficiently large, they become apparent on gray-scale images. Furthermore, in the case of a subperiosteal abscess, displacement of the periosteal membrane as well as infiltration of the overlying soft tissues may be present.[10] If the level of cortical destruction is sufficient to produce clear communication with the underlying marrow cavity, a corresponding gap in the cortex is evident sonographically, with associated enhancement of the underlying marrow space (relative to the attenuated appearance behind the intact cortex). Although such discontinuities can be evident in long bones, they are more frequently observed in the joints of patients with erosive disease from an underlying inflammatory arthritis (Fig. 5-6). Again, other abnormalities that mimic erosions by producing an apparent discontinuity in the cortical echo must be kept in mind (e.g., overlying calcification, see Fig. 5-7), and correlation with plain films can be invaluable.

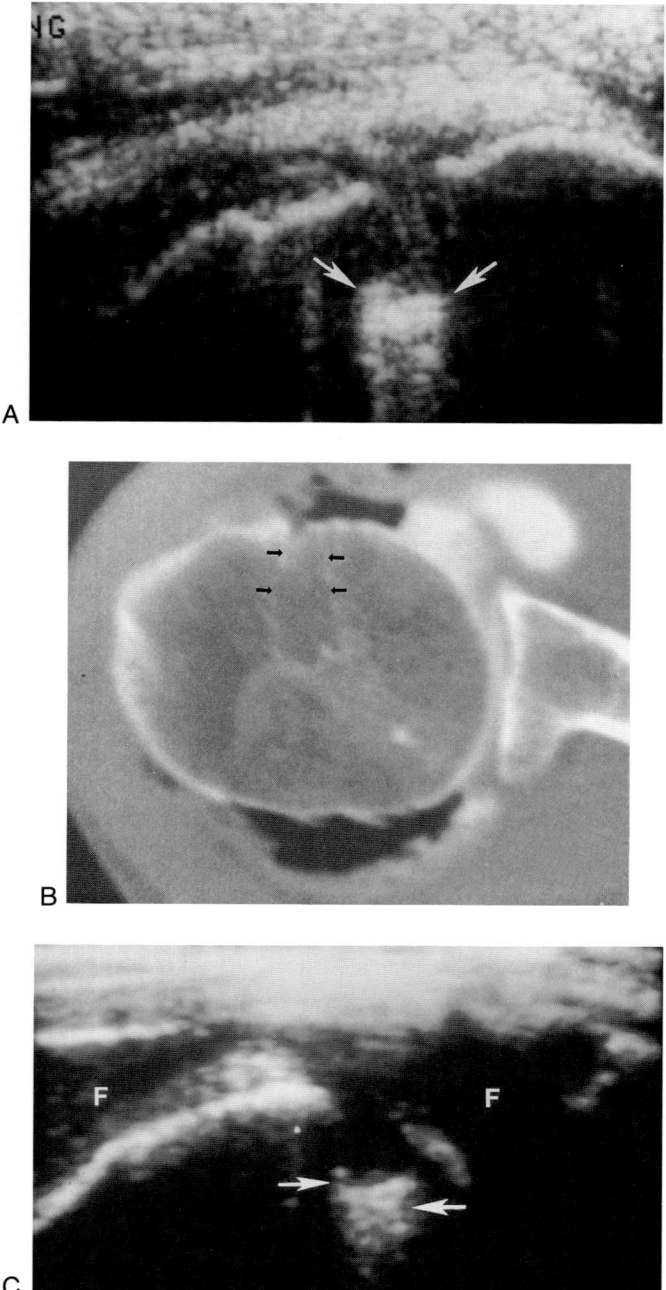

Fig. 5-6 Inflammatory erosions of bones. **(A)** Acute erosion of the humeral head in a patient with a seronegative arthritis. Sonogram shows an abrupt discontinuity of the humeral cortex with acoustic enhancement within the subjacent marrow (*arrows*). **(B)** CT arthrogram shows the same erosion (*small arrows*). **(C)** Sonogram demonstrates humeral head erosion in a patient with advanced rheumatoid arthritis. Note acoustic enhancement distal to the lesion (*arrows*). There is complete absence of the overlying rotator cuff, which is replaced by a large subdeltoid bursal effusion (F).

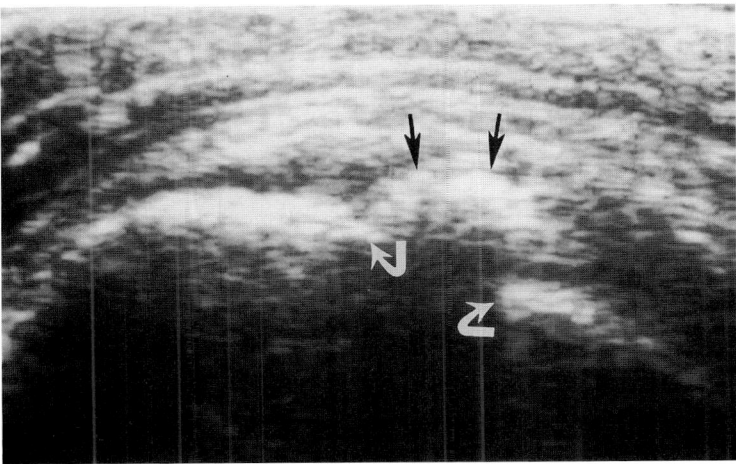

Fig. 5-7 Calcific tendinitis. Transverse sonogram shows an intratendinous calcification of supraspinatus tendon (*arrows*). The associated shadow produces an apparent discontinuity in the underlying bone (*curved arrows*). This can be differentiated from an erosion because of the absence of acoustic enhancement in the marrow. A displaced fracture fragment can also have this appearance, and correlation with clinical history and/or plain films is necessary to exclude that possibility.

NEOPLASTIC INVOLVEMENT OF BONE

An aneurysmal bone cyst may be seen on sonograms when the cortex is sufficiently thinned to allow beam penetration.[15] However, the specificity of fluid-fluid levels has been questioned recently, and sonography has little use in the assessment of tumors. One possible exception is the assessment of the cartilage thickness of an osteochondroma, as increasing thickness correlates with increased likelihood of malignancy. The soft-tissue extension of tumor at sites of cortical disruption also may be evaluated sonographically (Fig. 5-9),[16] but as with other modalities, separation of tumor from the secondary inflammation that it may elicit can be difficult. Percutaneous biopsy of an associated soft-tissue mass may be facilitated by sonographic guidance, although fluoroscopically directed biopsy will be adequate in most cases.

ORTHOPAEDIC APPLICATIONS

The detection of an infective process in patients with indwelling arthroplasty devices can be difficult. Joint aspiration with concomitant arthrographic examination has been the primary means of investigation in these patients. Scintigraphy can play an important role in this population, but the findings often lack specificity. CT and MRI are both limited by the large amount of image distortion created by metallic hardware in the field of view. Sonography, on the other hand, enables differentiation between a primary soft-tissue-related and joint-related process. Joint effusions in patients with infections around prostheses often can be detected sonographically, and aspiration can be performed with ultrasound guidance.[17] The diagnosis of osteomyelitis can be made using the criteria described above, although subject to some lack of specificity. Therefore, aspi-

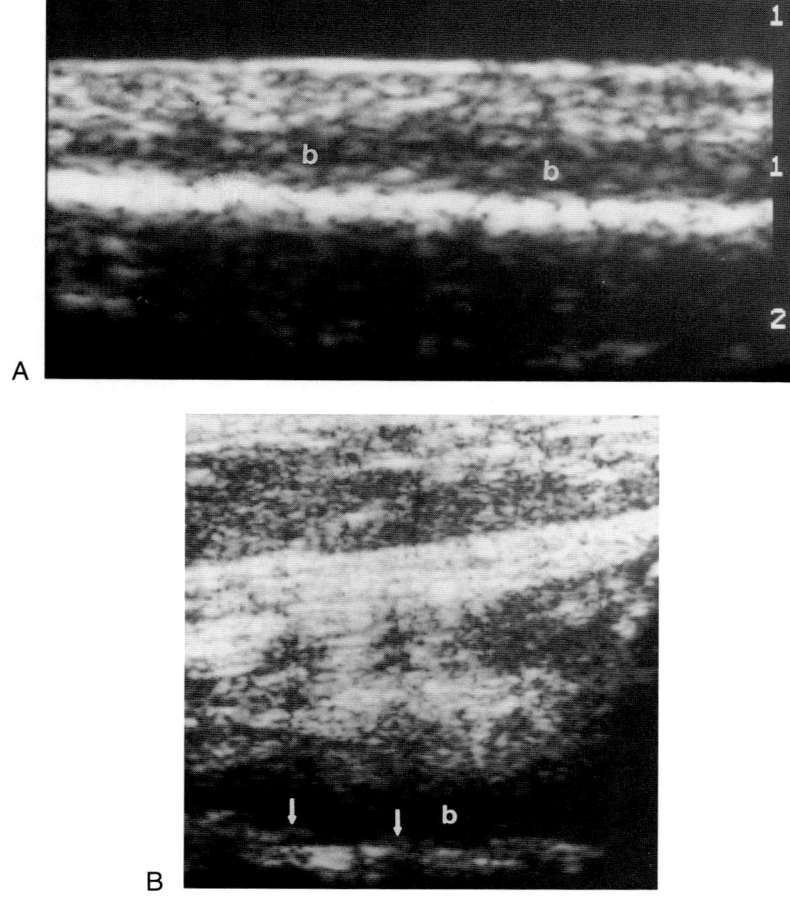

Fig. 5-8 Osteomyelitis. **(A)** Longitudinal sonogram of the tibia in a diabetic patient with osteomyelitis. The sonogram was performed to look for an associated abscess. Mild periosteal resorption was evident on the corresponding radiograph. A hypoechoic band (b), which parallels the cortex, is seen in the overlying soft tissues. The paucity of echoes behind the cortex may relate in part to the hypoechogenicity of the overlying soft tissues and in part to the periosteal erosion evident on the plain films. **(B)** Sonogram shows more extensive osteomyelitis in the distal femur of a patient with a previously documented septic knee. In addition to the hypoechoic band (b), the associated cortical erosion/periostitis is evident (*arrows*).

ration of infected fluid will continue to be the standard for diagnosing an infection around a joint prosthesis.

A more important orthopaedic application may be in patients undergoing limb lengthening by the Ilizarov technique.[18–21] A number of developmental and posttraumatic abnormalities result in limb length discrepancies that have been corrected in our institution using this technique. Following a brief latency period after placement of the device, distraction is performed at a relatively fixed rate over a period of weeks until the desired length is reached and final consolidation occurs. A number of problems can

BONE AND ARTICULAR CARTILAGE **67**

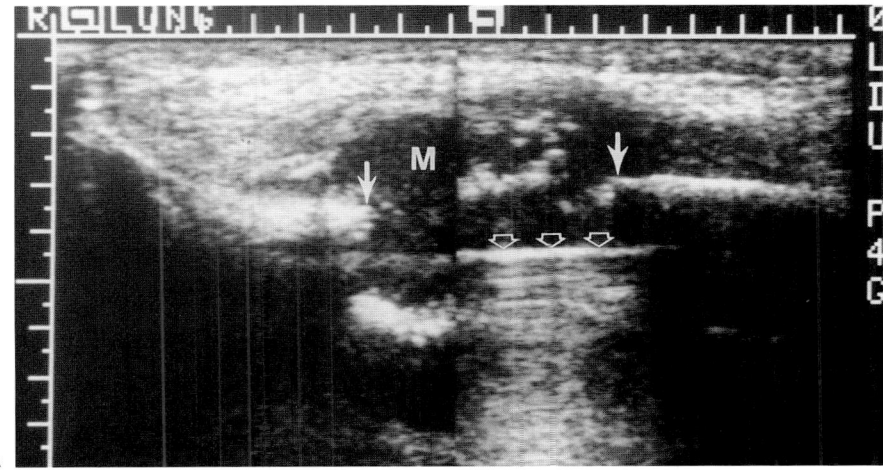

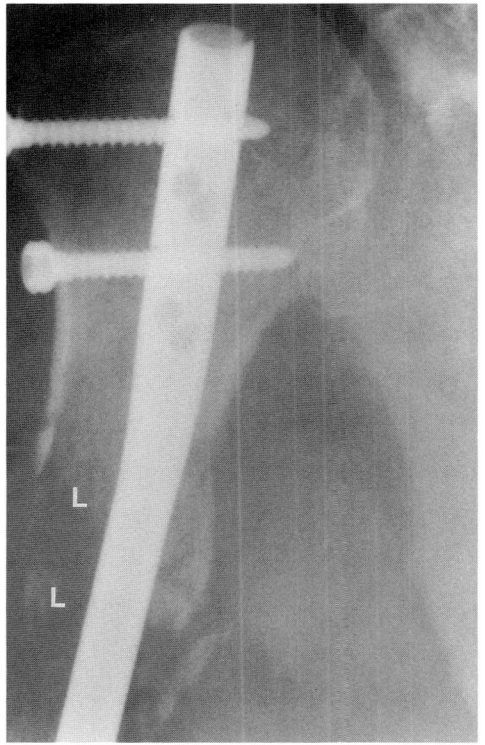

Fig. 5-9 Metastasis from lung carcinoma to the humerus. **(A)** Longitudinal sonogram of the proximal humeral diaphysis shows cortical discontinuity (*arrows*) and a large soft-tissue mass (M). The extent of the mass is easily demonstrated. A central, echogenic linear structure (*open arrows*) that has strong posterior reverberations represents an artifact from an indwelling intramedullary rod traversing the associated pathologic fracture; this was placed earlier for purposes of palliation. **(B)** Corresponding radiograph illustrates the destructive lesion (L).

delay or prevent consolidation or possibly cause premature consolidation. Plain film evaluation can measure the corticotomy gap, with correction for magnification, and so provide gross assessment of the gap. Periosteal and endosteal new bone is eventually apparent radiographically after sufficient mineralization. However, the sensitivity of radiography in detecting new bone formation relative to the size of the gap is limited at best, and the adjustment rate for corticotomy lengthening has been largely empiric.

Further, factors such as intragap cyst formation that impair bony union are not evident on plain films. Sonography is an excellent means of demonstrating such a cyst and has been shown to provide accurate assessments of the gap size in multiple projections (Fig. 5-10).[18,19,21] Sonography has also been shown to detect the development of relatively early periosteal and endosteal new bone formation.[18–20] This appears sonographically as nonshadowing, bright linear echoes that parallel the bone axis. However, such evaluations are at best qualitative, and consequently the rate of adjustment is still largely empiric.

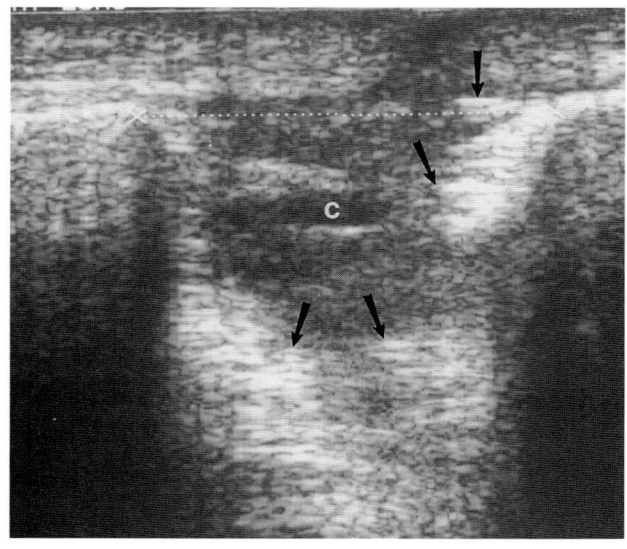

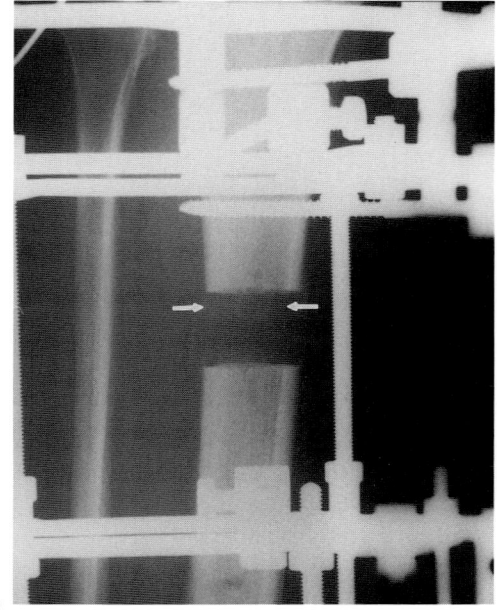

Fig. 5-10 Limb lengthening with the Ilizarov technique. **(A)** Longitudinal sonogram of the corticotomy gap (*dashed line*) in a patient with an Ilizarov device in place. A small cyst (c) is seen in the marrow space of the corticotomy gap. Dense linearly oriented echoes (*arrows*) correspond to the periosteal and endosteal new bone formation. **(B)** New bone formation is also apparent (*arrows*) on the corresponding plain film.

ASSESSMENT OF ARTICULAR CARTILAGE

As the range of motion of a joint increases, the amount of cartilage that can be examined sonographically increases. The use of sonography in the evaluation of early degenerative changes, inflammatory arthritides, and osteochondral fractures has been described.[3,5-7] Such indications have not found widespread clinical acceptance, largely because of limited acoustic access and confounding processes that further limit interpretation. In the absence of superimposed pathologic processes, the articular cartilage is well seen in certain locations (e.g., femoral condyles, radiocapitellar joint, see Fig. 5-2), and areas of focal or diffuse thinning are evident (Fig. 5-11). However, the presence of complex effusions with or without superimposed synovial inflammation can obscure the cartilage surface, limiting assessment (Fig. 5-12). Furthermore, in addition to limited acoustic access, the curvature of the surface and aberration of the ultrasound beam by subcutaneous fat may limit one's ability to detect the cartilage surface via a transcutaneous approach. Similar limitations apply to the assessment of menisci in the knee and the triangular fibrocartilage in the wrist.

In the context of intra-articular scanning, sonography has the potential to detect some of the earliest changes in osteoarthritis. Alterations of the cartilage early in this disease process involve changes in proteoglycan and collagen composition, changes in the level of hydration, and development of surface fibrillations.[22,23] Sonography may be able to identify alterations in the articular cartilage that are due to early osteoarthritis through changes in the speed of sound and attenuation, which depend strongly on the collagen composition, and through acoustic backscatter off of the cartilage surface, which relates to the degree of surface fibrillation.[24-26] De-

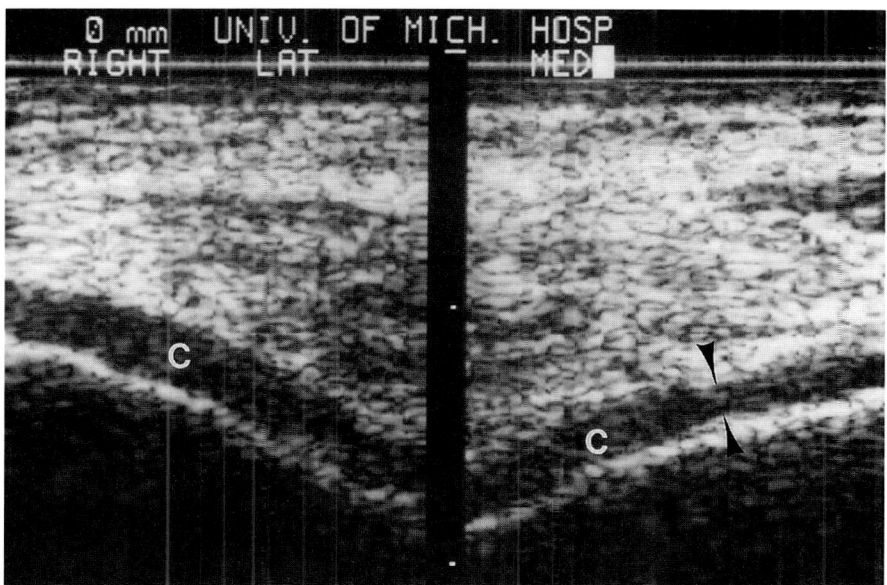

Fig. 5-11 Focal thinning of articular cartilage. Split sonogram of lateral (**Left**) and medial (**Right**) femoral condyles. Hypoechoic bands correspond to the articular cartilage (C) over both these surfaces. An area of focal cartilage thinning (*arrowheads*) is evident at the surface of the medial condyle.

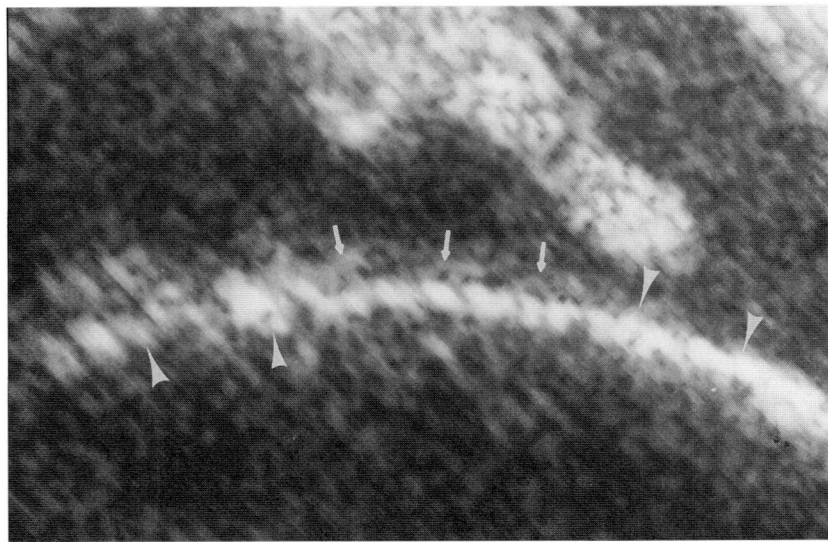

Fig. 5-12 Transverse sonogram of the posterior humeral head in a patient with rheumatoid arthritis and marked synovial inflammation. The cartilage (*small arrows*) is markedly irregular and poorly defined because of the extensive adjacent synovial inflammation in the posterior recess of the joint. *Arrowheads* indicate the subchondral cortex, which is also irregular.

velopment of such techniques requires advances in transducer technology, although characterization of surface fibrillation may be more readily accessible with the current generation of small intraluminal ultrasound systems that can be used in conjunction with arthroscopy.[27] The principle of altered acoustic backscatter with surface roughening can be illustrated on gray-scale images. In a normal specimen, the smooth surface results in a strong specular echo at the point of tangency to the perpendicular ultrasound beam (Fig. 5-13), while off-normal points are less distinct and are of diminished intensity. The presence of surface roughening tends to increase scatter in all directions, making the off-normal points subjectively brighter. This gray-scale appearance actually has a quantitative analogue that may serve to estimate the early changes of osteoarthritis in vivo.[25,28]

CONCLUSIONS

Significant improvements in image quality demonstrated by the newer generation of high-resolution linear-array transducers, as well as the cost-effectiveness of sonography in general, have been important factors in the increasing popularity of this modality for evaluating musculoskeletal abnormalities. Although sonography has definite indications in evaluating soft-tissue lesions, its role in assessing bone and cartilage abnormalities is not entirely clear. As discussed in the previous sections, gray-scale sonography can clearly define a host of bony lesions, but often this occurs incidentally during examination for another indication. Perhaps most significant is sonography's potential utility in assessment of symptomatic joint replacements and other orthopaedic procedures, as well as in the intra-articular evaluation of os-

BONE AND ARTICULAR CARTILAGE **71**

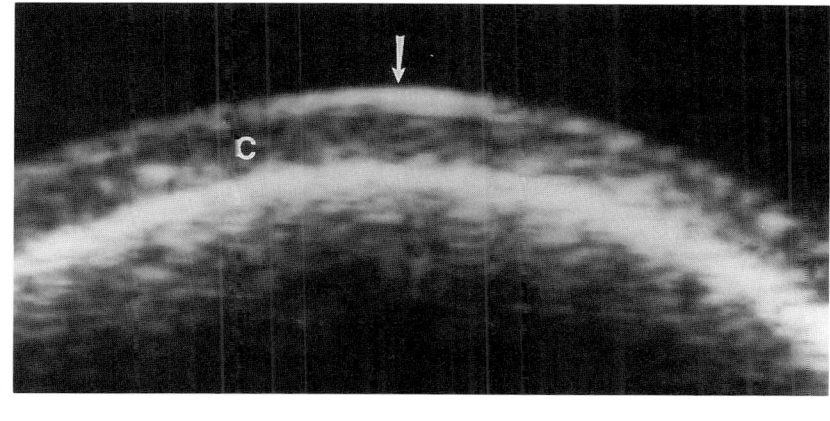

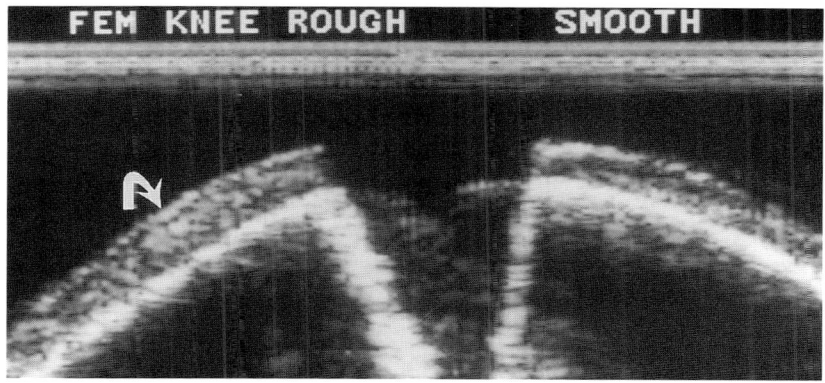

Fig. 5-13 In vitro sonograms of normal cadaveric articular cartilage scanned in a water bath **(A)** Sonogram shows the cartilage (C), which normally contains low-level echoes. The apex of the curved cartilage surface shows a strong specular echo (*arrow*) with gradual fall-off of the echo intensity along the curved surface in either direction. **(B)** Sonograms of two specimens of normal cadaveric articular cartilage. The specimen labeled "rough" (left) has had its surface roughened with 100-grit emery paper; the specimen labeled "smooth" (right) has not been roughened. A fall-off in surface echo intensity is evident along the smooth specimen (right), similar to that noted in Figure A. In comparison, increased surface backscatter (*curved arrow*) at off-normal points in the specimen on the left indicates its roughened nature.

teochondral abnormalities. In the latter case, sonography may indeed play a unique role because of its ability to achieve very-high-resolution characterization of intra-articular abnormalities.

REFERENCES

1. Graf R: Classification of hip joint dysplasia by means of sonography. Arch Orthop Trauma Surg 102:248, 1984
2. Schlesinger AE, Deeney VFX, Cooker PF: Sonography of the nonossified tarsal navicular cartilage in an infant with congenital vertical talus. Pediatr Radiol 20:134, 1989
3. van Holsbeeck M, Introcaso J: Musculoskeletal Ultrasound. Mosby-Year Book, St Louis, 1991
4. Dias JJ, Lamount AC, Jones JM: Ultrasonic diagnosis of neonatal separation of the distal humeral epiphysis. J Bone Joint Surg 70B:825, 1988
5. Broker FHL, Burbach T: Ultrasonic diagnosis of separation of the proximal humeral

epiphysis in the newborn. J Bone Joint Surg 72A:187, 1990
6. Welk LA, Adler RS: Case report 725. Skeletal Radiol 21:198, 1992
7. Graif M, Stahl-Kent V, Ben-Ami T et al: Sonographic detection of occult bone fractures. Pediatr Radiol 18:383, 1988
8. Pattern RM, Mack LA, Wang KY, Lingel J: Non-displaced fractures of the greater tuberosity of the humerus: sonographic detection. Radiology 18:201, 1992
9. Abiri M, Kirpekar M, Ablow RC: Osteomyelitis: detection with US. Radiology 172:509, 1989
10. Howard CB, Einhorun MS: Ultrasound in the detection of subperiosteal abscesses. J Bone Joint Surg 73B:175, 1991
11. Abiri MM, DeAngelis GA, Kirpekar M et al: Ultrasonic detection of osteomyelitis. Pathologic correlation in an animal model. Invest Radiol 27:111, 1992
12. Jonsson K, Buckwalter K, Helvie M et al: Precision of hyaline cartilage thickness measurements. Acta Radiol 33:234, 1992
13. Aisen AM, McCune WJ, MacGuire A et al: Sonographic evaluation of the cartilage of the knee. Radiology 153:781, 1984
14. Rubin JM, Adler RS, Bude RO et al: Clean and dirty shadowing at US: a reappraisal. Radiology 181:231, 1991
15. Mukuno DH, Lee TG, Watanabe AS, McIff EB: Aneurysmal bone cyst presenting as a pelvic mass on sonographic examination. J Ultrasound Med 5:215, 1986
16. Chhem RK, Schmutz GR, Huynh HH, Bao TB: Ultrasonography of a bone metastasis. Can Assoc Radiol J 43:138, 1992
17. Sherman L, Lombardi T, van Holsbeeck M: Sonographic detection of septic hip arthroplasties. Radiology, suppl. 181(P):385, 1991
18. Blane CE, Herzenberg JE, DiPietro MA: Radiographic imaging for Ilizarov limb lengthening in children. Pediatr Radiol 21:117, 1991
19. Young JRW, Kostrubiak IS, Resnik CS, Paley D: Sonographic evaluation of bone production at the distraction site in Ilizarov limb-lengthening procedures. AJR 154:125, 1990
20. Eyres KS, Bell MJ, Kanis JA: Methods of assessing new bone formation during limb lengthening. J Bone Joint Surg 75B:358, 1993
21. Zynamon A, Crabbe JP, Rubin JM, Adler RS: Usefulness of an acoustic edge artifact in assessment of the Ilizarov corticotomy interval. Skeletal Radiol 21:293, 1992
22. Minns RJ, Stevens FS, Hardinge K: Osteoarthritic articular cartilage lesions of the femoral head observed in the scanning electron microscopy. J Pathol 122:63, 1977
23. Bland JH, Cooper SM: Osteoarthritis: a review of the cell biology involved and evidence for reversibility. Management rationally related to known genesis and pathophysiology. Semin Arthritis Rheum 14:106, 1984
24. Agemura DH, O'Brien WD, Olerud JE et al: Ultrasonic propagation properties of articular cartilage at 100 MHz. J Acoust Soc Am 87:1786, 1990
25. Adler RS, Dedrick DK, Laing TJ et al: Quantitative assessment of cartilage surface roughness in osteoarthritis using high frequency ultrasound. Ultrasound Med Biol 18:51, 1992
26. Senzig DA, Forster FK, Olerud JE: Ultrasonic attenuation in articular cartilage. J Acoust Soc Am 92:676, 1992
27. McDonnell CH, Jeffrey RB Jr, Bjorkengren AG, Li KCP: Intraarticular sonography for imaging the knee menisci: evaluation in cadaveric specimens. AJR 159:573, 1992
28. Chiang EH, Adler RS, Meyer CR et al: Quantitative assessment of surface roughness using backscattered ultrasound: the effects of finite surface curvature. Ultrasound Med Biol (in press)

6
Peripheral Nerves
Moshe Graif

In patients with symptoms of nerve lesions, the current evaluation process is based on clinical history, neurologic examination, and electrophysiologic studies. Electrodiagnostic tests play an important role in the investigation of nerve abnormalities; however, because of the unique advantages and disadvantages of these tests, no such test can be considered sufficiently sensitive by itself. For example, nerve conduction measurements reflect the status of the best surviving nerve fibers; if a few fibers are unaffected by the disease or injury, the test results may appear normal. Focal compression of a nerve may produce local slowing of conduction through narrowing of axons and demyelination at the site of compression. However, the finding of normal conduction times does not exclude the presence of compression. Some pathologic conditions such as idiopathic polyneuropthy, diphtheria, metachromatic leukodystrophy, and hypertrophic neuropathy also can slow conduction velocities as a result of segmental demyelination.

The role of conventional radiology in the imaging of peripheral nerves was (and still is) very limited and, for the most part, confined to the demonstration of secondary skeletal changes resulting from a peripheral nerve lesion.[1] The introduction of computed tomography (CT) and, later, magnetic resonance imaging (MRI) enabled direct demonstration of nerve tumors along the presumed course of the nerve. Although direct imaging of the nerve structure itself is possible only at the level of the nerve roots,[1] CT with longitudinal reconstruction can be useful in examining changes in size or attenuation along the course of the sciatic nerve. This application, however, remains relatively limited. Recently, Filler et al presented the first image neurogram, a method for producing nerve images using a commercially available MRI system.[2] The image displays the nerve by itself, much like a blood vessel on a subtraction angiogram.

In recent years, sonography has proved to be a useful tool for studying the morphologic aspects of peripheral nerves, through its ability to demonstrate the presence or absence of a focal nerve lesion.[3,4] This chapter addresses peripheral nerve sonography, including technique, sonographic appearances, pitfalls, and limitations.

TECHNIQUE

Instrumentation

Because of their wider near-field of view, linear-array transducers of 3.5 to 7.5 MHz are preferred to sector transducers. Most examinations of the peripheral nerves can be performed within the 5.0 to 7.5 MHz range. However, in obese patients, when the sciatic nerve is examined at the gluteal level, a 3.0- or 3.5-MHz transducer may be necessary to obtain an adequate depth of ultrasound beam penetration.

Technique of Examination

The nerves are examined along their anatomic course. Extremities should be systematically scanned in the longitudinal and transverse planes. Transverse scans are essential to distinguish a nerve from a flat intramuscular fibrous septum in an adjacent muscle; the latter will appear as a linear echo on transverse scans. A dynamic (real-time) examination during active or passive flexion and extension maneuvers is helpful in confirming the identity of a nerve, which will remain immobile in relation to the surrounding musculotendinous structures. The following guidelines will help ensure the success of sonographic examination.

1. Always compare both limbs.
2. Use the lowest gain setting.
3. Use a combination of longitudinal and transverse scans to avoid confusing nerves with echogenic aponeuroses.
4. Check the mobility of the structure observed to differentiate nerves from tendons.

NORMAL ANATOMY AND SONOGRAPHIC APPEARANCE

The nerve fibers (axons) and their ensheathing Schwann cells are grouped into fascicles of a widely varying number. The spaces between nerve fibers contain endoneurium, which is a loose, delicate connective tissue. Each fascicle is surrounded by thin, concentric layers of connective tissue forming the perineurium. The entire nerve trunk is surrounded by the epineurium, a dense, irregular connective sheath containing collagenous fibers, arranged mainly longitudinally. Fat cells also may be found here (Fig. 6-1).

On longitudinal sonograms, normal peripheral nerves appear as markedly echogenic structures with parallel internal linear echoes (Figs. 6-2 and 6-3). The outer layer (nerve sheath) is often more echogenic. On transverse sonograms, nerves are displayed as oval-to-round structures with internal punctate echoes. During real-time scanning, the nerves appear immobile in relation to the surrounding tissues. The relatively high echogenicity of the nerves provides good contrast with the hypoechoic surrounding muscles (Fig. 6-3). However, in elderly patients and in patients who have undergone surgery in an adjacent area, visualization of the nerves may be more difficult because of the increased echogenicity of the surrounding tissues.[3,4]

The nerve diameter varies according to anatomic location. The sciatic nerve, which is the largest nerve in the human body, may reach a diameter of 9 mm (mean, 7.1 ± 1.0 mm).[4] When both lower extremities are compared, nerve thickness may appear asymmetric. In a study of 43 lower extremities, the mean diameter of the left sciatic nerve was slightly less (by 8%) than that of the right (all patients examined were right-handed).[4]

NERVE SHEATH TUMORS

Peripheral nerve tumors are relatively uncommon. Most tumors of peripheral nerves, excluding those of the sympathetic nervous

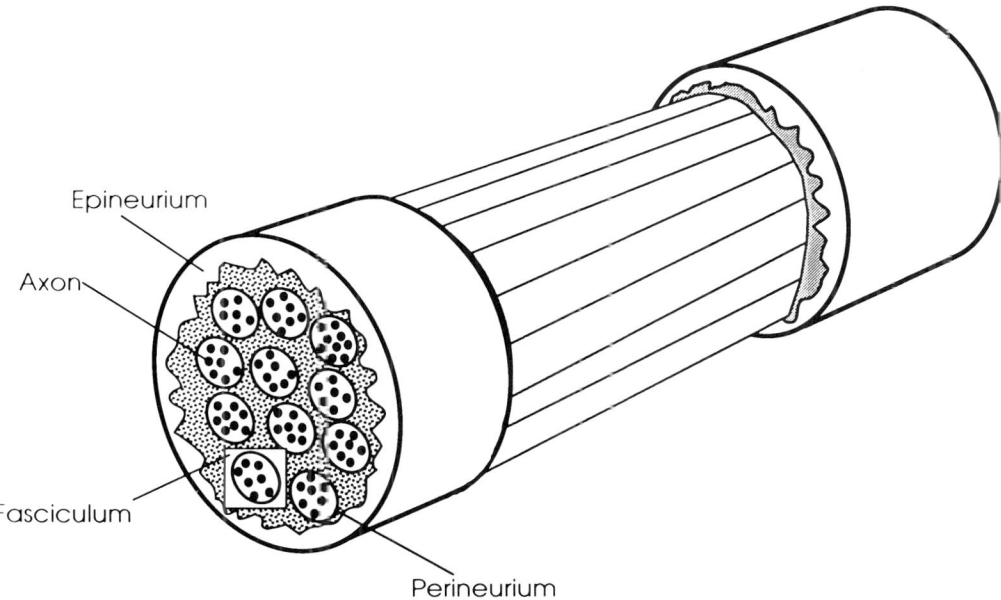

Fig. 6-1 Schematic representation of the cross section of a peripheral nerve.

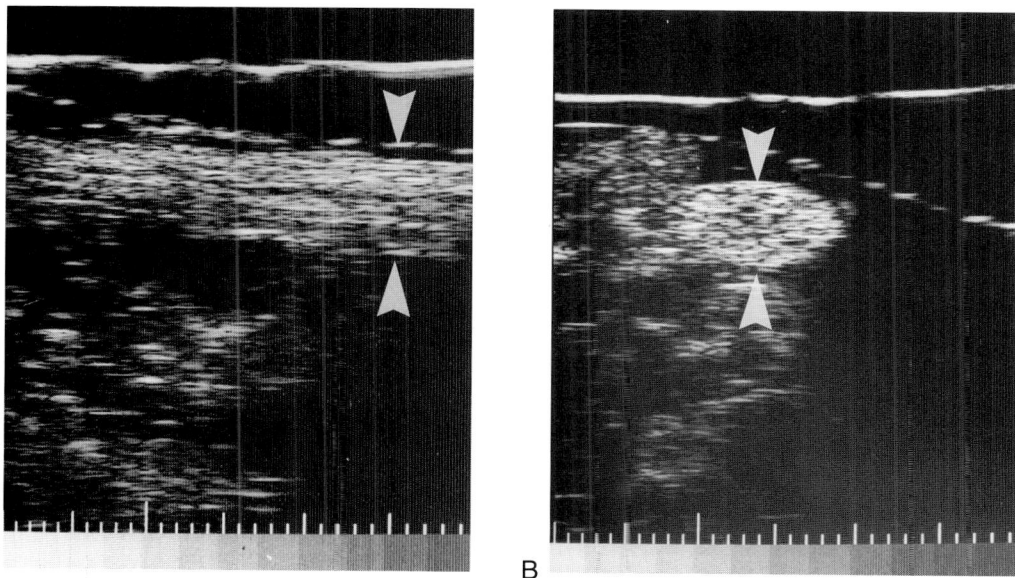

Fig. 6-2 (A) Longitudinal and **(B)** transverse sonograms of an anatomic specimen of a normal sciatic nerve (*arrowheads*) immersed in a water bath. Images were obtained with a 10-MHz transducer. (From Graif et al,[4] with permission.)

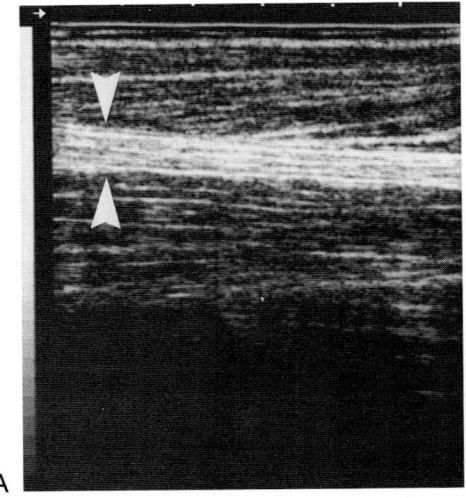 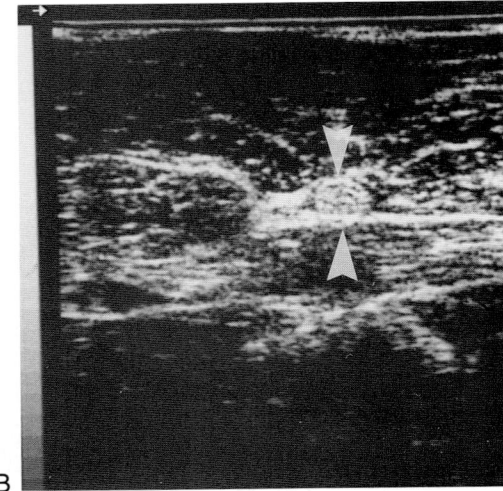

Fig. 6-3 **(A)** Longitudinal and **(B)** transverse sonograms of a normal sciatic nerve (*arrowheads*). (From Graif et al,[4] with permission.)

system, are the result of a usually benign neoplastic proliferation of Schwann cells. There are many synonyms for peripheral nerve tumors, apparently because of confusion about the cell of origin of these tumors. However, most peripheral nerve tumors can be classified as nerve sheath tumors, including schwannomas and neurofibromas.

Tumors of peripheral nerves that have been described sonographically include schwannomas, neurofibromas, and neurofibrosarcomas.[3–10] These tumors are usually in a relatively superficial location and therefore suitable for sonographic assessment. Most have a similar sonographic appearance: a hypoechoic, grossly homogeneous echotexture with well-defined contours (Fig. 6-4). The uniform cellular pattern seen in schwannomas is responsible for their markedly decreased echogenicity, which is occasionally associated with moderate-to-marked distal acoustic enhancement that may result in a pseudocystic appearance.[5,11] It must be emphasized that although very rare, true intraneural cystic lesions may occur (Fig. 6-5).[12] Color Doppler study may reveal the presence of vessels within large neurofibromas (Fig. 6-4B).

Tumors affecting the following nerves have been detected sonographically: median nerve, radial nerve, ulnar nerve, digital nerves, sciatic nerve, tibial nerve, and peroneal nerve. Nerves also can be invaded by adjacent, usually malignant, tumors. In these cases, the mass tends to appear more irregular in contour, with an inhomogeneous echotexture; a "Codman's-like triangle" created by tumor infiltration between the epineurium and the fascicles can be observed. Such changes have been reported in association with desmoid tumors[4] and extranodal non-Hodgkin's lymphoma (Fig. 6-6).

NEUROMAS

Traumatic Neuromas

The formation of a bulbous neuroma is a physiologic attempt at spontaneous axon repair after the severing of a nerve. The severed nerve fibers undergo wallerian degener-

PERIPHERAL NERVES 77

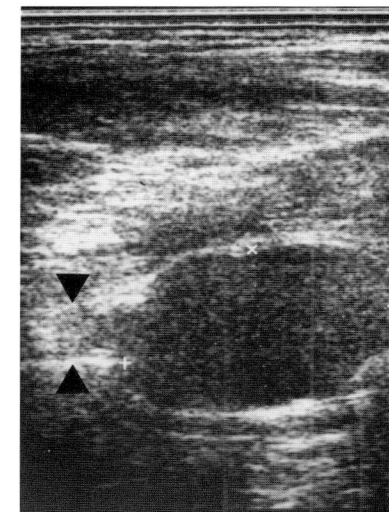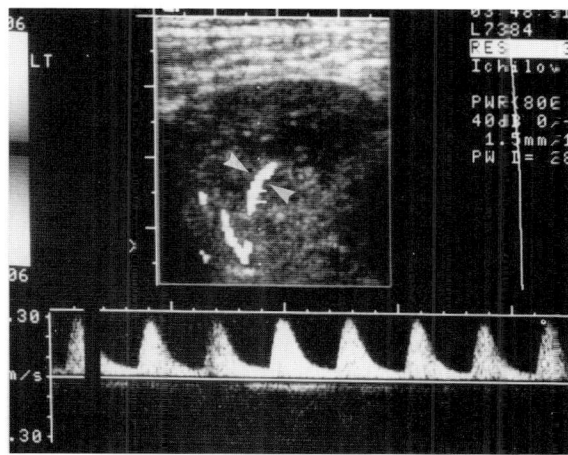

Fig. 6-4 Neurofibroma. **(A)** Longitudinal sonogram obtained with a 5-MHz transducer shows a well-defined, hypoechoic mass (*calipers*) in continuity with the cordlike echogenic nerve (*arrowheads*). **(B)** A color Doppler study shows the presence of internal vasculature (*arrowheads*). (Fig. A from Graif et al,[4] with permission.)

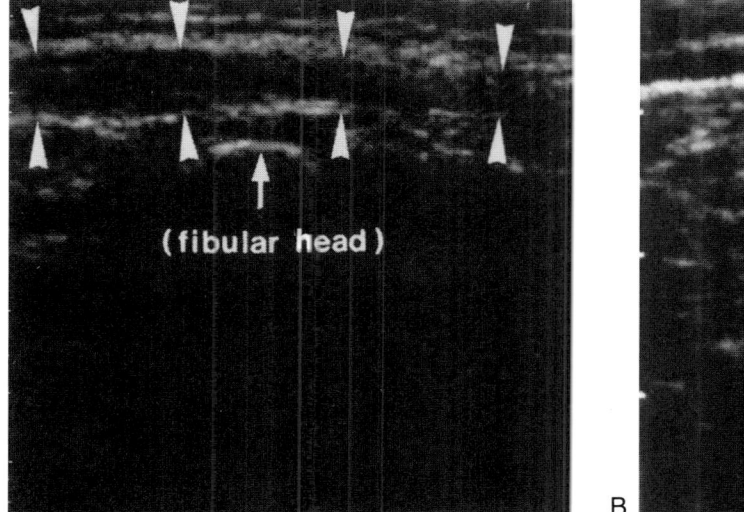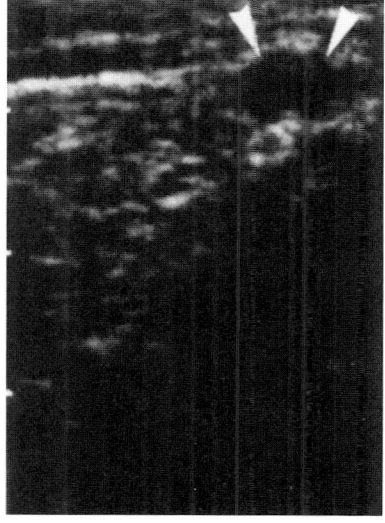

Fig. 6-5 Intraneural ganglion cyst. **(A)** Longitudinal and **(B)** transverse sonograms reveal a spindle-shaped anechoic structure (*arrowheads*) that follows the course of the peroneal nerve at the level of the fibular head (*arrow*). At its largest point, the cyst is 7 × 5 mm. (From Leijten et al,[12] with permission.)

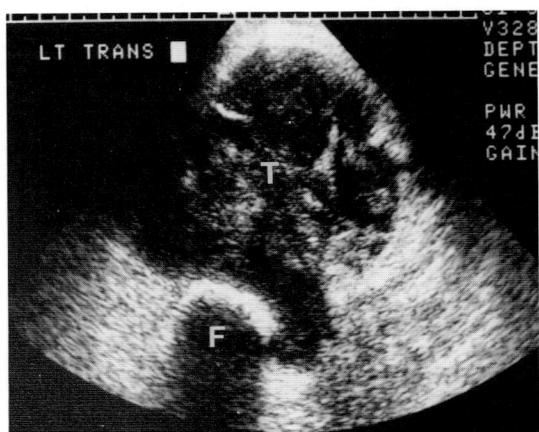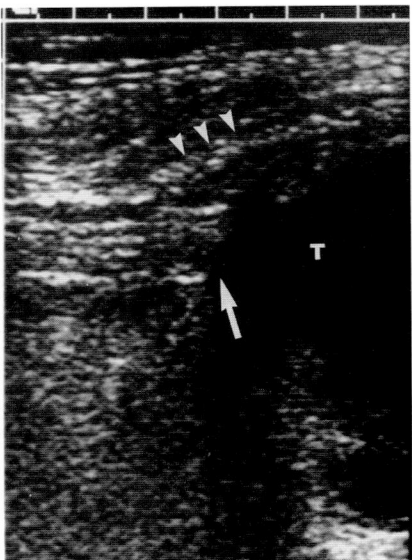

Fig. 6-6 Extranodal non-Hodgkin's lymphoma of the thigh. **(A)** Transverse sonogram of the thigh shows the tumor (T). **(B)** Longitudinal sonogram shows the junction between the nerve and the tumor (T). Note the abrupt discontinuity of the nerve (*arrow*); the nerve sheath is splayed over the mass (*arrowheads*). F, femur.

ation. As the repair proceeds, fine axonal sprouts grow distally from the cut end of the nerve and meet with a tangled mass of fibroblasts and regenerating Schwann cells. If these sprouts fail to enter the endoneurial tubules of the distal part of the cut nerve, they may form a neuroma. Traumatic neuromas contain poorly vascularized connective tissue infiltrated by large numbers of sprouts from the parent axons. The neuroma may be well encapsulated in a fibrous sheath, or it may be firmly attached to the surrounding structures by fibrous adhesions.[13]

Sonographically, a traumatic neuroma appears as a well-defined, ovoid, hypoechoic mass. The mass is usually well demarcated from the normal nerve structure (Fig. 6-7).[3,4,13] Occasionally, increased echogenicity is observed internally; this is likely due to the presence of marked fibrosis.

Morton's Neuromas

Morton's neuroma is a fibrotic pseudotumor arising from the interdigital plantar nerves and located near the heads of the metatarsal bones.[14] Its sonographic appearance is described in chapter 14.

COMPRESSION INJURIES

Long-lasting mechanical pressure with distortion, such as chronic entrapment neuropathy, is probably associated with some degree of localized ischemia at the site of deformation. Ischemia can produce a wide range of nerve fiber lesions and, when severe and prolonged, will result in axonal loss and wallerian degeneration.[15] The most common causes of local compression are fracture-dislocations of the hip joint and complications

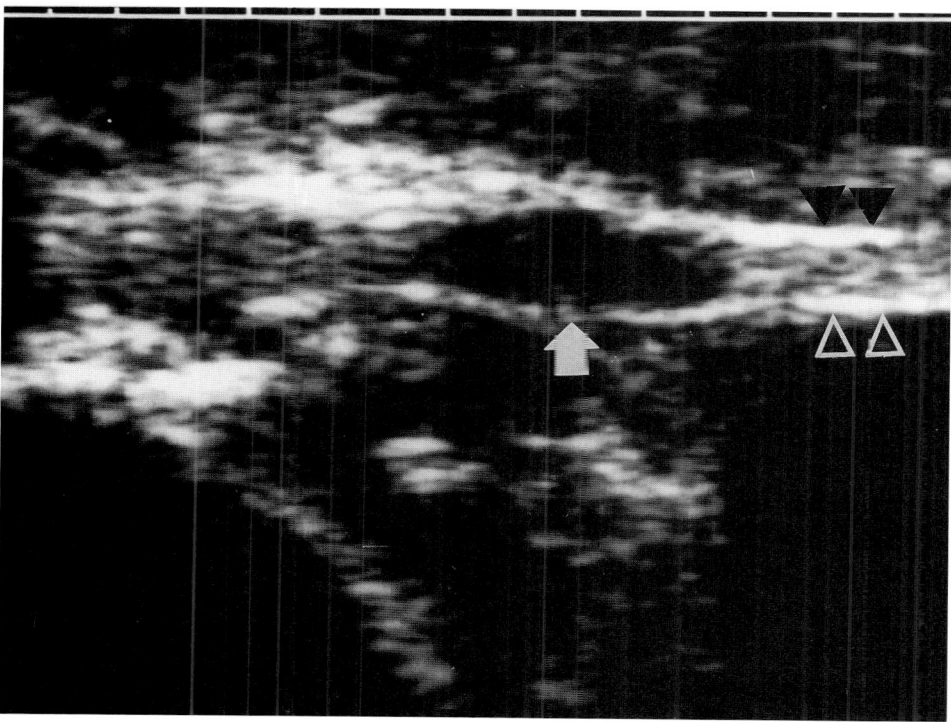

Fig. 6-7 Traumatic neuroma. Longitudinal sonogram of a sciatic nerve shows a hypoechoic oval mass (*arrow*), measuring 2.0 × 0.9 cm and contained within the nerve sheath (*arrowheads*).

resulting from hip replacement. Other causes of long-lasting pressure are anesthetic palsies, improper application of plaster casts, tourniquet paralysis, and malposition during prolonged comatose states. The presence of a space-occupying lesion, such as a hematoma, lipoma, or aneurysm, also may be responsible for compression, as may postoperative formation of excessive scar tissue.

Sonographic assessment of nerve compression includes screening the soft tissues for a mass and evaluating the nerve in question for distortion or displacement. The sonographic features of nerve displacement caused by hematoma, scar tissue, vascular aneurysm (Fig. 6-8), or muscular edema have been described in the literature.[4] A hematoma can appear either as an anechoic fluid-filled collection or as a focal mass with relatively decreased echogenicity, depending on the state of aggregation. Scar tissue displays increased echogenicity, probably because of the presence of fibrotic tissue. An aneurysm appears as a pulsatile fluid-filled mass, with or without internal thrombus. An edematous muscle shows diffuse thickening with or without some decrease in echogenicity; the internal architecture is usually preserved. The presence of swelling is readily recognized when the sonogram is compared with one of the contralateral limb.

INJECTION INJURIES

Injury to a peripheral nerve may occur as a complication of injection with a needle.[16,17] This type of injury is not uncommon in the buttocks. Thin or chronically ill patients

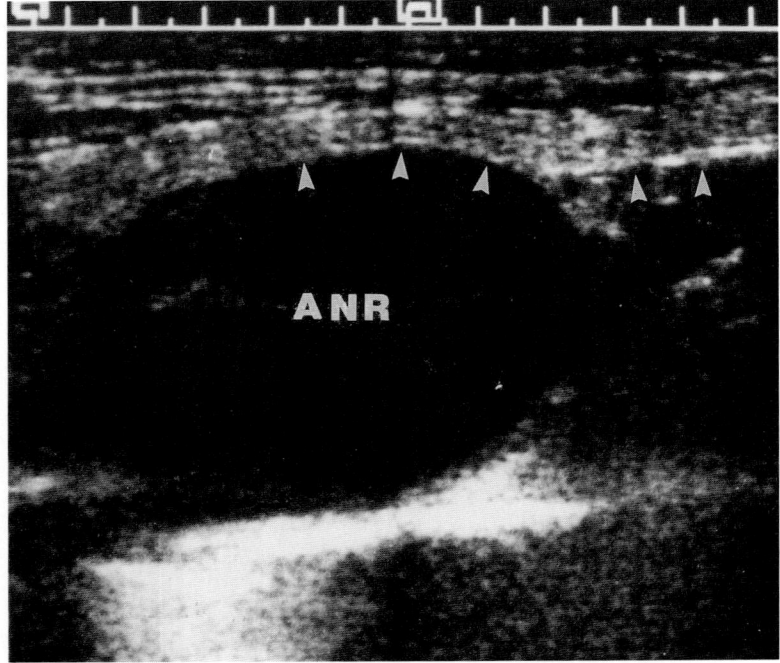

Fig. 6-8 Sciatic nerve displacement by aneurysm. Longitudinal sonogram of the posterior aspect of the lower thigh shows the echogenic nerve (*arrowheads*) and the adjacent femoral artery aneurysm (ANR).

with poor gluteal covering are predisposed to this type of injury. The use of a long needle, forceful placement of a needle, or injection in an area other than the outer upper quadrant of the buttock are risk factors for development of injection injury. The postulated mechanisms of injury include direct needle trauma, secondary constriction by scar tissue, and direct nerve fiber damage caused by the chemical neurotoxicity of the agent injected. Among the common drugs with severe neurotoxic effects are benzylpenicillin, diazepam, chlorpromazine, procaine, and hydrocortisone sodium succinate.[15] Neurologic symptoms can range from minor, transient sensory disturbance to severe sensory disturbance and motor paralysis with poor response.

Sonography can be used to locate the focal fluid collection (Fig. 6-9) or the nodule of scar tissue responsible for local compression. In the case of deeply located nerves such as the sciatic nerve in the buttocks, intraneural degenerative changes are difficult to detect because the thickness of interposed tissues requires the use of low-frequency transducers, thereby limiting resolution.

INFLAMMATORY CHANGES

Sonographic detection of pathologic processes such as neurilemmitis, tuberculoid leprosy, and fibrous degeneration of nerves has been described.[3] Neurilemmitis and degeneration appear sonographically as diffuse or segmental thickening of the nerve with decreased internal echogenicity and disappearance of the parallel fibrillar echotexture (Fig. 6-10). In a case of tuberculoid leprosy, sonography demonstrated a hypoechoic

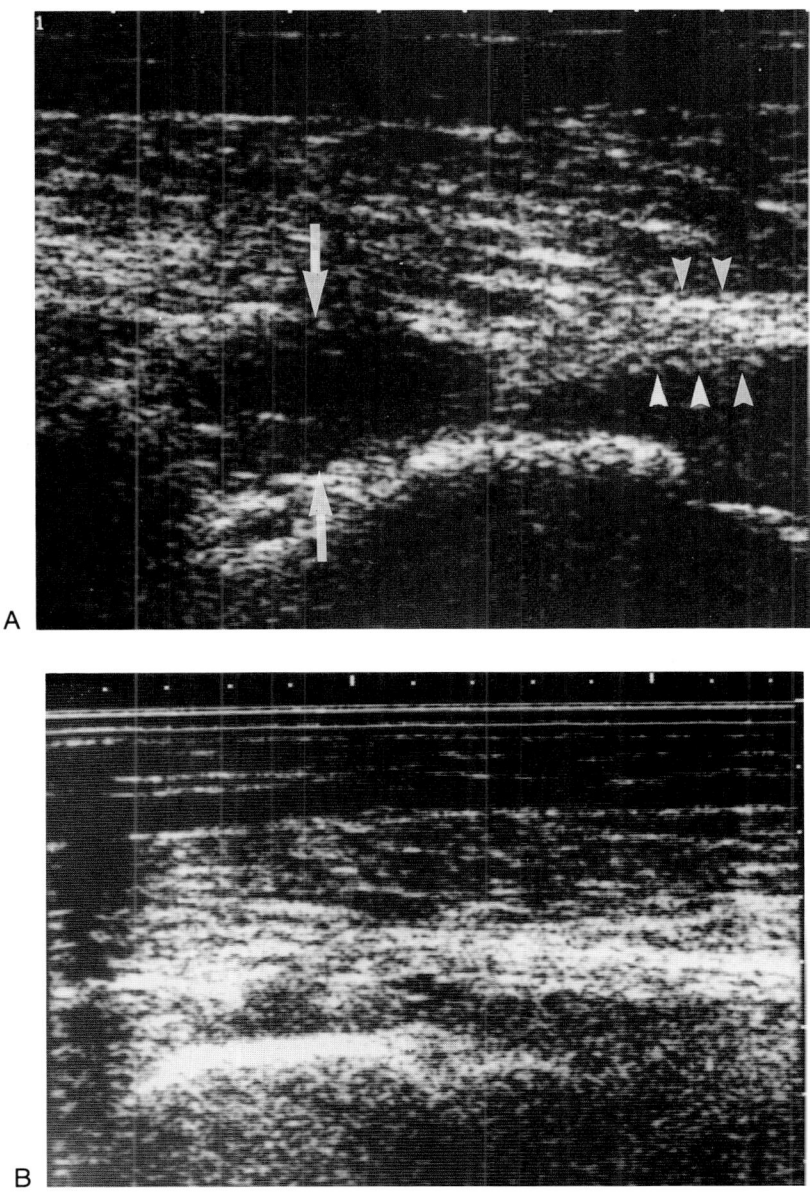

Fig. 6-9 Injection injury. **(A)** Longitudinal sonogram of the buttock 1 day after injection shows a 3.2- × 1.9-cm hypoechoic mass (*arrows*) with no posterior enhancement and the disrupted course of the sciatic nerve (*arrowheads*). **(B)** A follow-up sonogram performed 3 weeks later confirms the disappearance of the mass.

82 MUSCULOSKELETAL ULTRASOUND

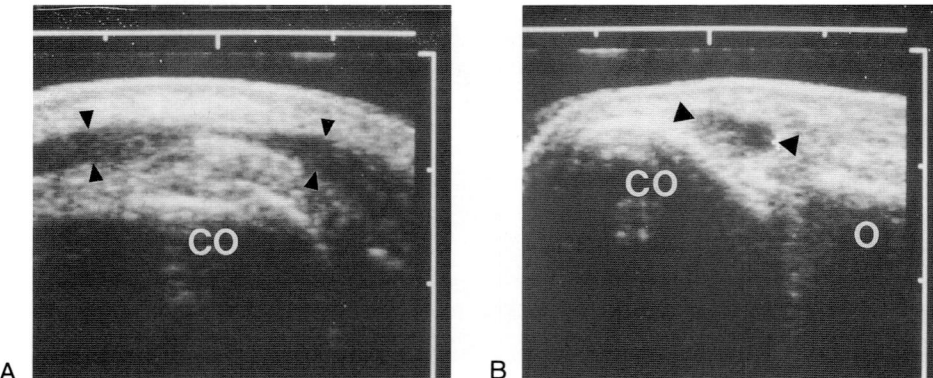

Fig. 6-10 Neurilemmitis of the ulnar nerve at the elbow. **(A)** Longitudinal and **(B)** transverse sonograms of the elbow show the markedly hypoechoic, thickened ulnar nerve (*arrowheads*) posterior to the medial condyle (CO). O, olecranon. (From Fornage,[3] with permission.)

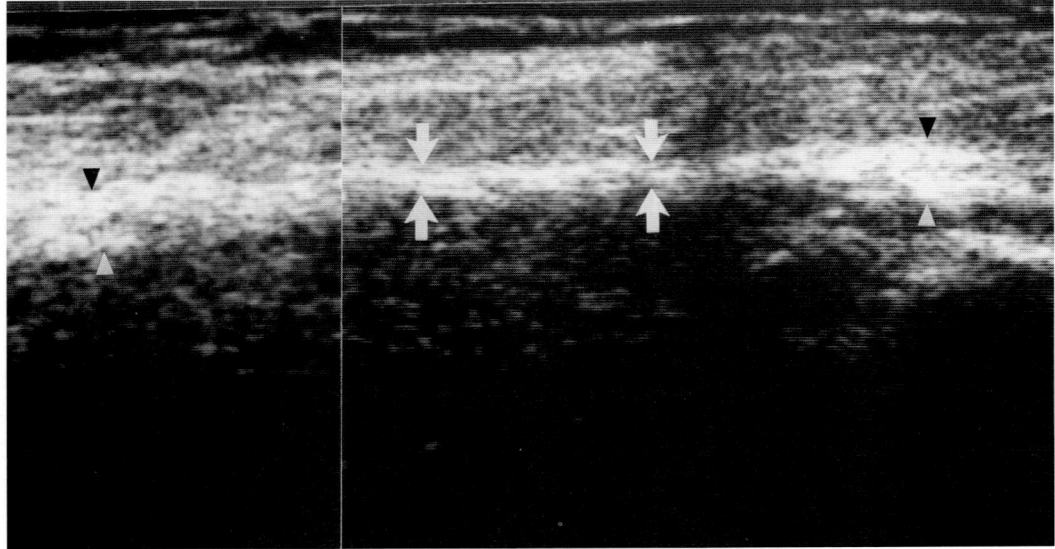

Fig. 6-11 Nerve graft. Longitudinal sonogram (at 3.5 MHz) of a sciatic nerve graft shows the interposed sural nerve (*arrows*), which is narrower than the sciatic nerve (*arrowheads*).

mass that correlated with the caseous pouch surrounding the thickened nerve.

NERVE RECONSTRUCTION SURGERY

Nerve reconstruction surgery can fail because of infection at the suture site or excessive tension at the suture line, resulting in disruption of the line and formation of fibrous tissue at the anastomosis. Clinical and electrophysiologic tests may not be able to provide early detection of these complications. Sonography of a nerve repaired surgically by graft interposition allows early evaluation of the continuity of the interposed segment with the proximal and distal edges of the injured nerve and of the formation of a bulbous fibrous tissue reaction at the suture site. If there is no disruption of the suture line, the change of caliber between the original nerve and the interposed graft may be the only sonographic finding at the site of anastomosis (Fig. 6-11).[4]

REFERENCES

1. Fahr LM, Sauser DD: Imaging of peripheral nerve lesions. Orthop Clin North Am 19:27, 1988
2. Filler AG, Howe FA, Hayes CE et al: Magnetic resonance neurography. Lancet 341:659, 1993
3. Fornage BD: Peripheral nerves of the extremities: imaging with US. Radiology 167:179, 1988
4. Graif M, Seton A, Nerubai I et al: Sciatic nerve: sonographic evaluation and anatomic-pathologic considerations. Radiology 181:405, 1991
5. Chinn DH, Filly RA, Callen PW: Unusual ultrasonographic appearance of a solid schwannoma. J Clin Ultrasound 10:243, 1982
6. Hoddick WK, Callen PW, Filly RA et al: Ultrasound evaluation of benign sciatic nerve sheath tumors. J Ultrasound Med 3:505, 1984
7. Cravioto H: Neoplasms of peripheral nerves. p. 1894. In Wilkins RH, Rehgachary SS (eds): Neurosurgery. Vol. 2. McGraw-Hill, New York, 1985
8. Hughes DG, Wilson DJ: Ultrasound appearances of peripheral nerve tumours. Br J Radiol 59:1041, 1986
9. Obayashi T, Itoh K, Nakano A: Ultrasonic diagnosis of schwannoma. Neurology 37:1817, 1987
10. Cantos-Melian B, Arriaza-Loureda R, Aisa-Varela P: Tibialis posterior nerve schwannoma mimicking Achilles tendinitis: ultrasonographic diagnosis. J Clin Ultrasound 18:671, 1990
11. Goss SA, Johnston RL, Dunn F: Comprehensive compilation of empirical ultrasonic properties of mammalian tissues. J Acoust Soc Am 64:423, 1978
12. Leijten FSS, Arts WF, Puylaert JBCM: Ultrasound diagnosis of an intraneural ganglion cyst of the peroneal nerve. Case report. J Neurosurg 76:538, 1992
13. Tindall SC: Painful neuromas. p. 1884. In Wilkins RH, Rehgachary SS (eds): Neurosurgery. Vol. 2. McGraw-Hill, New York, 1985
14. Redd RA, Peters VJ, Emery SF et al: Morton neuroma: sonographic evaluation. Radiology 171:415, 1989
15. Gentili F, Hudson AR: Peripheral nerve injuries: types, causes, grading. p. 1802. In Wilkins RH, Rehgachary SS (eds): Neurosurgery. Vol. 2. McGraw-Hill, New York, 1985
16. Kline DG: Diagnostic approach to individual nerve injuries. p. 1833. In Wilkins RH, Rehgachary SS (eds): Neurosurgery. Vol. 2. McGraw-Hill, New York, 1985
17. Steward JD, Aquayo AJ: Compression and entrapment neuropathies. p. 1446. In Dyck PJ, Thomas PK, Lambert EH et al (eds): Peripheral Neuropathy. 2nd Ed. Vol. 2. WB Saunders, Philadelphia, 1984

7
Skin and Subcutaneous Tissues

Bruno D. Fornage

Sonography of the skin and subcutaneous tissues is performed more often by dermatologists than by radiologists. Initial research on dermatologic applications of sonography was limited to the measurement of skin thickness using A-mode equipment.[1,2] Interest in sonography of the skin was later revived with the availability of B-mode scanners equipped with 7.5- or 10-MHz transducers[3–6] and more recently with the advent of very-high-frequency (20- and 30-MHz) scanners.[7] Examination of the subcutaneous tissues requires no dedicated equipment and is a simple and very effective—though underused—application of sonography.

TECHNICAL CONSIDERATIONS

Sonography of the skin can be performed with 7.5- or 10-MHz transducers, which are typically the highest frequency transducers that can be connected to conventional real-time ultrasound scanners.[3–6,8,9] Recently, 15-MHz mechanical sector probes have become commercially available for use with general purpose ultrasound scanners. The use of a soft, thin standoff pad (about 1 cm thick) is required to avoid direct contact between the skin and the transducer.[10] However, when a standoff pad is used, it is crucial to check the perpendicularity of the ultrasound beam continuously; the beam must remain perpendicular to the skin at all times to avoid artifacts due to beam scattering.

Dedicated B-mode scanners with a single 20- or 30-MHz mechanical transducer are now commercially available. The probes house a single crystal, which is moved mechanically (Fig. 7-1).[11] In addition to grey-scale imaging, some scanners offer the option of displaying the echoes on a color-coded scale, although this has proved of no benefit in our experience. The unit used at our institution (Dermascan, Cortex Technology, Denmark) has a theoretical axial resolution of 80 microns and a lateral resolution of 200 microns. The field of view is 1.2 cm wide and about 1.5 cm deep. C-mode scanning (providing coronal or horizontal reconstructed scans) and three-dimensional reconstruction of skin lesions are also available, although they are used mostly for research purposes.[12]

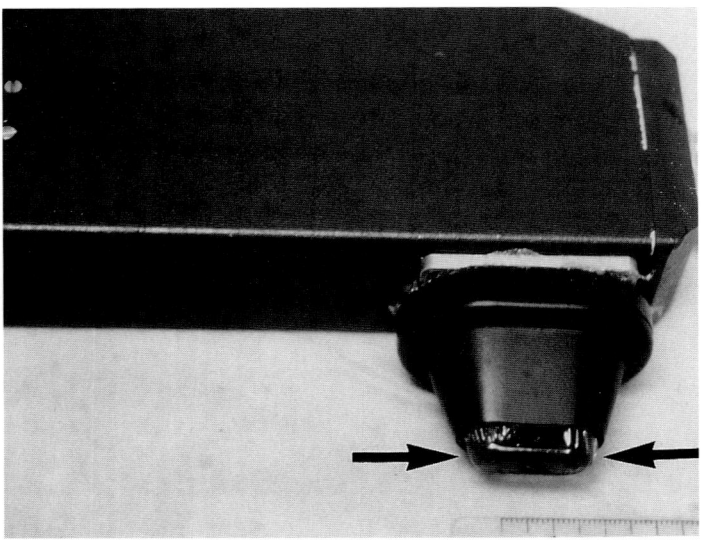

Fig. 7-1 View of a dedicated 20-MHz hand-held probe for examination of the skin (Dermascan, Cortex Technology, Denmark). *Arrows* point to the narrow footprint of the water chamber housing the transducer.

Doppler examination of skin lesions is most sensitive when a high-frequency continuous-wave (e.g., 10 MHz) transducer is used. However, skin lesions that are sufficiently thick to be visualized with 7.5- or 10-MHz linear-array transducers can also be evaluated with color Doppler flow imaging and spectral analysis, which is not possible with the dedicated 20- or 30-MHz units.

NORMAL ANATOMY AND SONOGRAPHIC APPEARANCE

The skin consists of the epidermis and dermis (Fig. 7-2). The epidermis comprises several cell layers that cannot be resolved by sonographic imaging; there are no blood or lymphatic vessels. The dermis represents the bulk of the skin and consists of a thin, superficial papillary dermis and a thicker, more coarse reticular dermis. The dermis contains a rich network of blood vessels, organized in plexuses, as well as hair follicles, sebaceous glands, and sweat glands. The subcutaneous tissues consist primarily of fat with a variable amount of connective tissue in the form of scattered fibrous strands that are oriented grossly parallel to the skin.

At frequencies of up to 10 MHz, the skin appears as a regular, moderately echogenic tissue layer (Fig. 7-3).[4,6] The epidermis and dermis cannot be differentiated, nor can any skin appendage be visualized. The echogenicity of the dermis is probably due to the dense network of collagen fibers. The interface between the echogenic dermis and the hypoechoic subcutaneous fat appears as a slightly irregular line, parallel to the skin surface. The thickness of normal skin varies significantly depending on the site examined. A study using a 10-MHz transducer has shown that the thickness of the skin ranges from 1.4 ± 0.3 mm at the dorsum of the hand to 4.8 ± 0.6 mm at the sole of the heel, with a mean value of 2.4 ± 1.0 mm.[6] The subcutaneous fat is hypoechoic and contains thin linear echoes representing strands of connective tissue. Large subcutaneous veins appear as

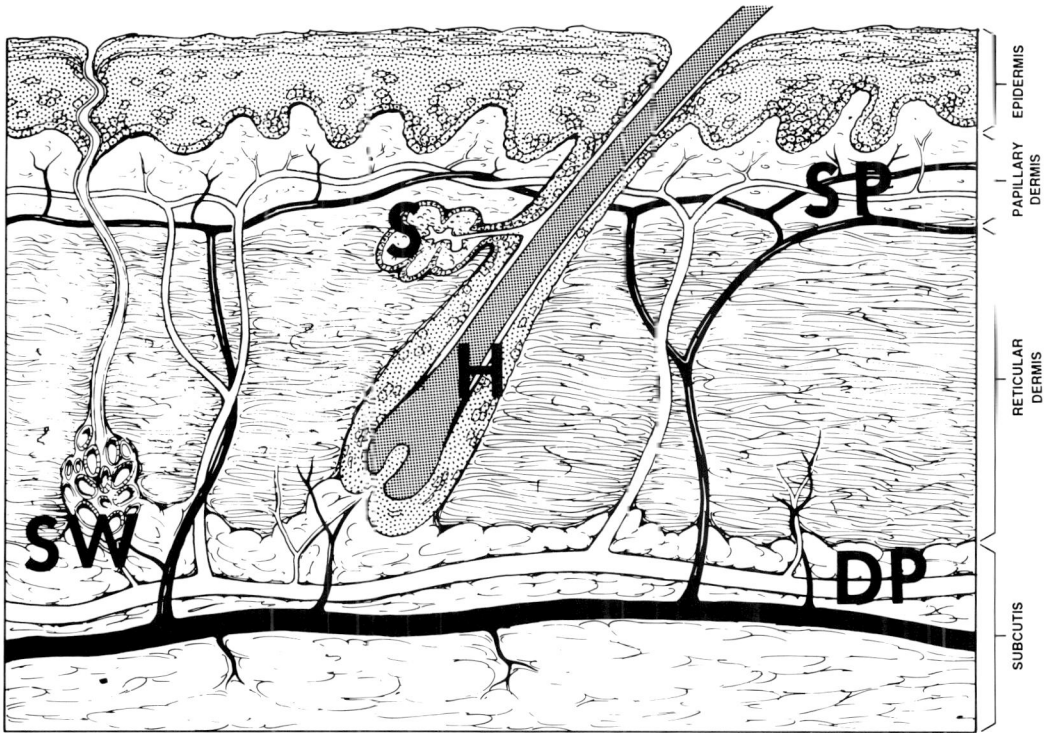

Fig. 7-2 Diagram showing the anatomy of the normal skin. DP, deep vascular plexus; H, hair follicle; S, sebaceous gland; SP, superficial vascular plexus; SW, sweat gland. (From Fornage et al,[11] with permission.)

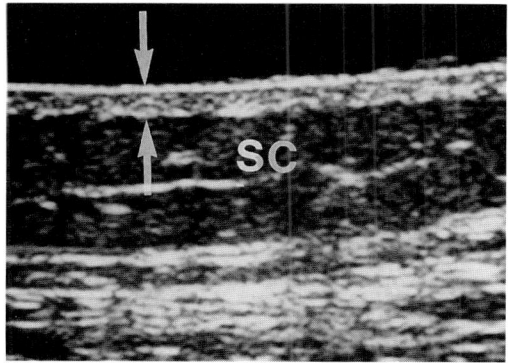

Fig. 7-3 Normal skin examined at 7 MHz. The skin appears as a layer of moderately echogenic tissue (*arrows*) sharply demarcated from the underlying hypoechoic subcutaneous fat (SC). Skin appendages are not visualized.

anechoic tubular structures that collapse easily under minimal pressure with the transducer.

On 20- or 30-MHz scans, the echogenic dermis appears markedly echogenic and sharply demarcated from the hypoechoic subcutaneous fat. The epidermis is still too thin to be resolved, except at the sole of the foot and at the hypothenar area, where it can be seen as a hypoechoic layer up to 1 mm thick between two bright, parallel, undulating lines, the more superficial of these representing the surface of the skin and the deeper representing the dermoepidermal junction (Fig. 7-4).[11] Hair follicles appear as parallel, hypoechoic bands traversing the skin obliquely when they are scanned longitudinally and as hypoechoic round structures when they are

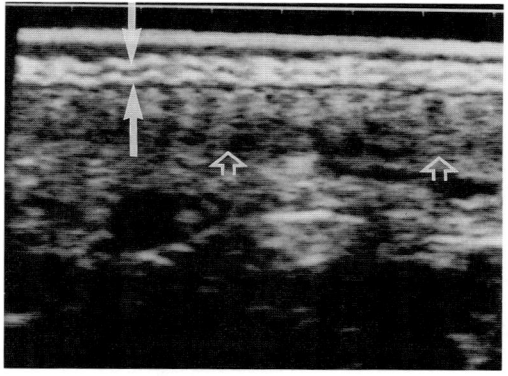

Fig. 7-4 Normal skin at the hypothenar area examined at 30 MHz. The epidermis (*arrows*) appears as a thin, superficial, hypoechoic layer between two echogenic, undulating, parallel lines. The deeper line is the dermoepidermal junction. *Open arrows* indicate the deep margin of the skin. The distance between ticks on the scale at top represents 1 mm. (Scan obtained with UX-02 scanner from Rion Co., Tokyo, Japan.)

area of the nail and the subungual space, which is a site of predilection for glomus tumors. Because of the limited penetration of 20- and 30-MHz equipment, only the most superficial portion of the subcutaneous tissues can be visualized.

PATHOLOGIC CONDITIONS OF THE SKIN

Because the vast majority of lesions that arise from the skin are hypoechoic, sonography cannot contribute significantly to their differential diagnosis. However, high-resolution sonography makes visible, for the first time, the third dimension (thickness) of lesions, which allows greater accuracy in assessing their extension into the dermis or subcutaneous tissues and greater reliability in follow-up measurements, as compared with palpation.

Benign Lesions

scanned transversely (Fig. 7-5). A thin, hypoechoic, superficial band can be seen in elderly subjects and corresponds to the so-called solar elastotic band. The nails appear as a double echogenic line with a thickness of about 0.6 mm for the third fingernail (Fig. 7-6).[11] Longitudinal scans show the matrix

CYSTS

Cysts that develop in the skin (epidermal inclusion or epidermoid cysts and pilar or sebaceous cysts) exhibit well-defined, usually

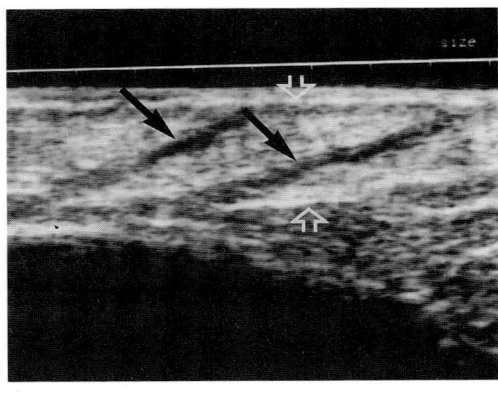
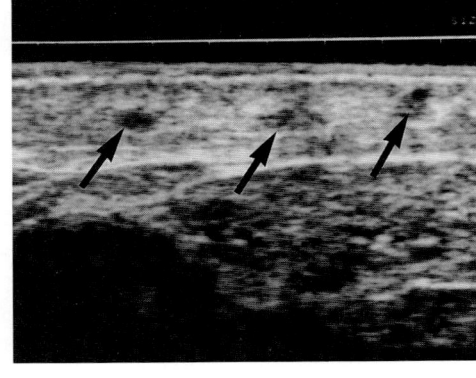

A   B

Fig. 7-5 Normal skin of the calf examined at 30 MHz. **(A)** Sonogram shows the echogenic skin (*open arrows*). Oblique hypoechoic striae (*arrows*) represent hair follicles scanned longitudinally. **(B)** Sonogram obtained after the transducer was rotated 90 degrees. Transverse sections of the hair follicles appear as round hypoechoic structures (*arrows*). (Scans obtained with UX-02 scanner from Rion Co., Tokyo, Japan.)

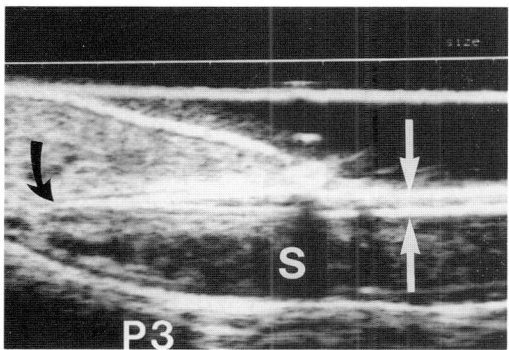

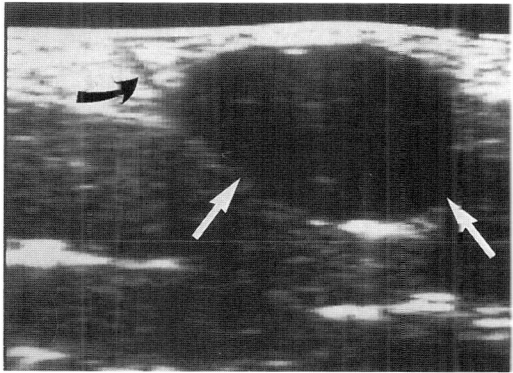

Fig. 7-6 Longitudinal sonogram of the base of the nail of the third finger obtained at 30 MHz. *Arrows* point to the 0.5-mm-thick nail, which appears as a double line. *Curved arrow* indicates the matrix of the nail. P3, dorsal aspect of distal phalanx; S, subungual space. (Scan obtained with UX-02 scanner from Rion Co., Tokyo, Japan.)

Fig. 7-7 Sebaceous cyst. A 20-MHz sonogram shows a 7-mm, well-defined, markedly hypoechoic lesion (*arrows*) with moderate sound-through transmission and regular margins. Note the associated hair follicle (*curved arrow*) traversing the normal skin.

(but not always) regular margins, are rarely totally anechoic, and often lack sound enhancement; thus, a solid mass cannot always be ruled out. On 20-MHz scans, keratin and lipid-rich debris may give rise to internal echoes that are occasionally associated with acoustic shadowing. Careful scanning of a sebaceous cyst should demonstrate the hair follicle from which the lesion arose as a hypoechoic band through the superficial dermis (Fig. 7-7). Despite their lower resolution, scans at 10 MHz may be useful in confirming that a superficial, palpable mass originates from within the skin.

NEVI

Most nevi are too thin to be resolved at 10 MHz. On 20-MHz scans, nevi appear as flat, markedly hypoechoic, well-defined, superficial masses, with an average thickness of 0.8 ± 0.6 mm (Fig. 7-8).[11] Because 20-MHz sonography cannot usually delineate the normal epidermis, it cannot distinguish between junctional (at the dermoepidermal junction), intradermal (in the dermis), and compound (both intraepidermal and intradermal) nevi, although very thin or very thick nevi are likely to be junctional or intradermal, respectively.

OTHER BENIGN TUMORS

Skin hemangiomas are very common, and their clinical diagnosis is straightforward. Their sonographic appearance at 20 MHz varies from a well-defined, superficial hy-

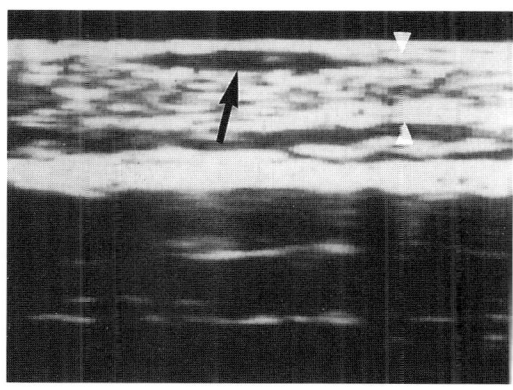

Fig. 7-8 Nevus. A 20-MHz sonogram shows a flat, well-defined, nearly anechoic lesion (*arrow*) at the surface of the skin. The nevus is 0.2 mm thick. *Arrowheads* delineate the normal skin

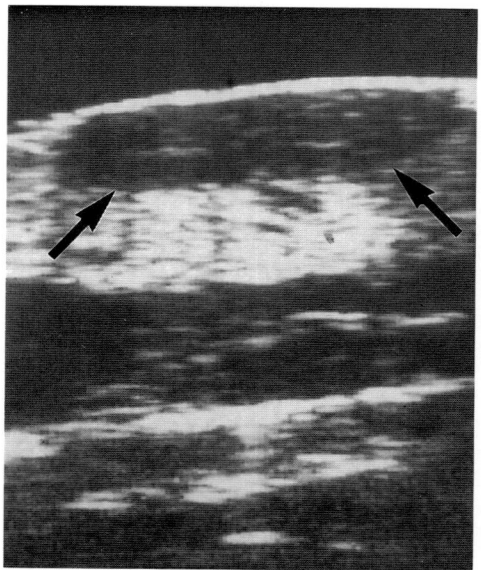

Fig. 7-9 Cutaneous hemangioma of the lower back. A 20-MHz sonogram shows a well-defined, hypoechoic, nonspecific mass (*arrows*) involving the superficial half of the skin.

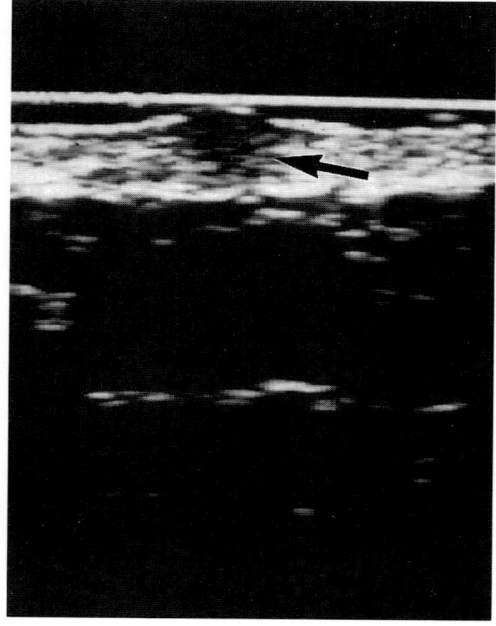

Fig. 7-10 Dermatofibroma of the forearm. A 20-MHz sonogram shows a hypoechoic mass with slightly irregular margins (*arrow*).

poechoic mass (Fig. 7-9) to a flat, intradermal, serpiginous mass.[11] Lymphangiomas have been reported to have a multicystic appearance.[9] With the use of continuous-wave Doppler sonography at 10 MHz, abnormal Doppler signals have been reported in 100 percent (8/8) of cutaneous angiomas.[13]

Dermatofibromas, also called fibrous histiocytomas, fibrous xanthomas, and nodular fibrosis, result from a proliferation of fibroblasts and histiocytes within the dermis. On 20-MHz scans, they appear as nonspecific hypoechoic masses, sometimes with irregular margins (Fig. 7-10). Dermatofibrosarcoma protuberans is a rare, benign, slow-growing but locally aggressive tumor of the dermis typically affecting the trunk; local recurrences following surgical excision are not uncommon. Scans performed at 7.5 MHz demonstrate the hypoechoic nodules impinging on the subcutaneous fat.[8]

Actinic keratosis, which occurs most commonly on the backs of the hands and on the face, is characterized by an atypical hyperplasia of keratinocytes (parakeratosis) combined with solar elastosis affecting the papillary dermis. Sonographically, actinic keratosis appears as a thin, superficial, irregular, hypoechoic band (Fig. 7-11).[11] This appearance is indistinguishable from that of in situ or even early invasive squamous cell carcinoma. In contrast, seborrheic keratosis is a more focal process, which develops superficially in the epidermis; sonograms obtained at 20 MHz demonstrate the integrity of the underlying dermis (Fig. 7-12).

INFLAMMATION

In inflammatory dermatoses such as eczema, the pathologic changes involve the epidermis and the papillary dermis, while the reticular dermis is preserved. On 20-MHz sonograms, this results in a thin, superficial, hypoechoic band (Fig. 7-13), which can serve as a marker on follow-up sonograms to evaluate response to anti-inflammatory drugs.

Sonographic quantification of the type IV reaction after intradermal application of recall antigens has proved to be accurate but not cost-effective.[14]

Marked edema of the skin can result from various causes such as cellulitis or external radiotherapy. The thickened skin can be readily measured using 7.5- to 10-MHz probes. The echogenicity of the dermis is slightly decreased, but more important, the boundary between the dermis and the subcutaneous fat becomes blurred and sometimes disappears. Color Doppler studies can demonstrate increased vascularity with low-resistance flow in the dermis (Plate 7-1).

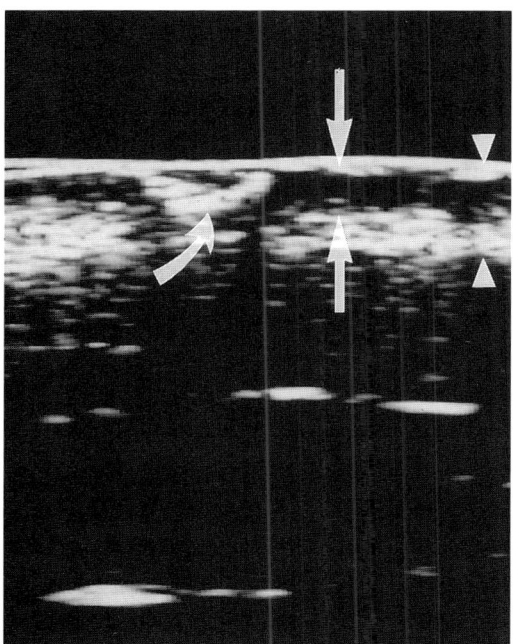

Fig. 7-11 Actinic keratosis. A 20-MHz sonogram shows a hypoechoic, irregular, superficial band (*arrows*). The *curved arrow* points to the site of a punch biopsy. *Arrowheads* delineate the skin.

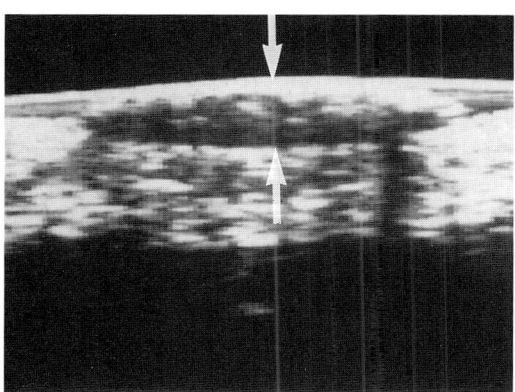

Fig. 7-12 Seborrheic keratosis. A 20-MHz sonogram shows a hypoechoic, well-defined, meniscus-shaped mass (arrows). The mass is sharply demarcated from the underlying dermis.

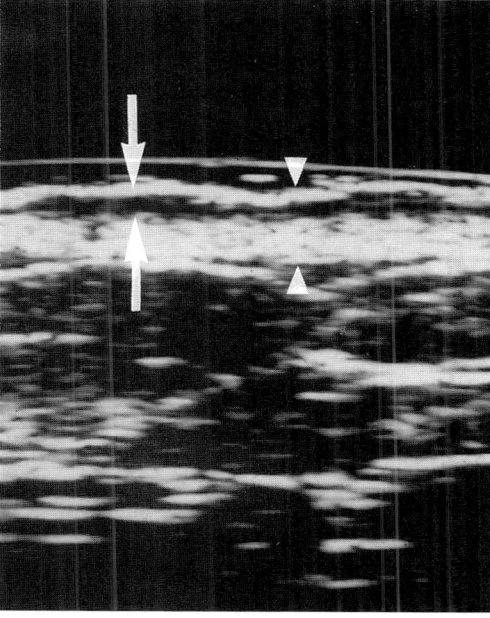

Fig. 7-13 Eczematous dermatitis. A 20-MHz sonogram shows a thin, superficial, hypoechoic band (*arrows*). *Arrowheads* delineate the skin.

PSORIASIS

Psoriasis is a widespread chronic disease of unclear etiology that affects more than 1 percent of the population in the United States. The epidermis is thickened and shows both hyperkeratosis and parakeratosis, with marked elongation of the rete ridges. Lymphocytic infiltrate is present in the underlying papillary dermis, with the reticular dermis usually being spared. Both the elongation of the rete ridges and the inflammatory infiltrate in the papillary dermis are thought to be responsible for the superficial, hypoechoic band seen on sonograms. The thickness of this layer is a marker of disease activity that can be used to evaluate response to treatment.[15]

MISCELLANEOUS

High-frequency sonography has been used to evaluate the degree of burns,[16,17] the skin atrophy induced by corticosteroids,[18] and wound healing.[19]

Malignant Tumors

Malignant skin tumors, regardless of their histologic type, show decreased or occasionally mixed echogenicity, and the majority of them exhibit irregular margins on 20-MHz scans.[11] However, because of overlap in the appearance of malignant and benign masses (many of which also have decreased echogenicity and irregular margins), sonography does not eliminate the need for biopsy. On the other hand, sonography's role in assessing tumor depth may be critical for staging of certain tumors.

MELANOMA

Malignant melanoma represents about 3 percent of all cancers in the United States and is by far the most aggressive cutaneous malignancy. The depth of the primary tumor is the single most important prognostic factor.

Clark's and Breslow's classification systems are used to classify melanoma on the basis of the level of dermal invasion and the actual thickness of the tumor (with break points at 0.75, 1.50, 2.25, and 3.00 mm), respectively, as determined from the histologic specimen (Table 7-1).[20-22] Preoperative knowledge of a tumor's thickness would help in planning treatment strategy, and sonography holds promise as an accurate, noninvasive means of assessing thickness.

Sonographically, melanoma appears as a markedly hypoechoic or anechoic mass developing from the surface of the skin inward. The deep margin of the tumor, which is regular or slightly irregular, contrasts sharply with the underlying echogenic dermis, thus allowing accurate measurement of the tumor.[11] Although 7.5- or 10-MHz scans can depict large melanomas, 20-MHz scans are more informative; for example, they can demonstrate a thin superficial spread undetected at a lower frequency (Fig. 7-14). Although a good correlation between preoperative sonographic measurements of tumor thickness and histopathologic findings was observed initially with 10-MHz scans,[5] it has been reported more recently that sonograms obtained at 20 MHz tend to overestimate tumor thickness, possibly because of the inclu-

Table 7-1 Clark's Classification of Melanoma

Level I	In situ melanoma remains above the basement membrane
Level II	Melanoma invades the papillary dermis
Level III	Melanoma reaches the junction of the papillary and reticular layers but does not invade the reticular dermis
Level IV	Melanoma invades the reticular dermis
Level V	Melanoma invades the subcutaneous fat

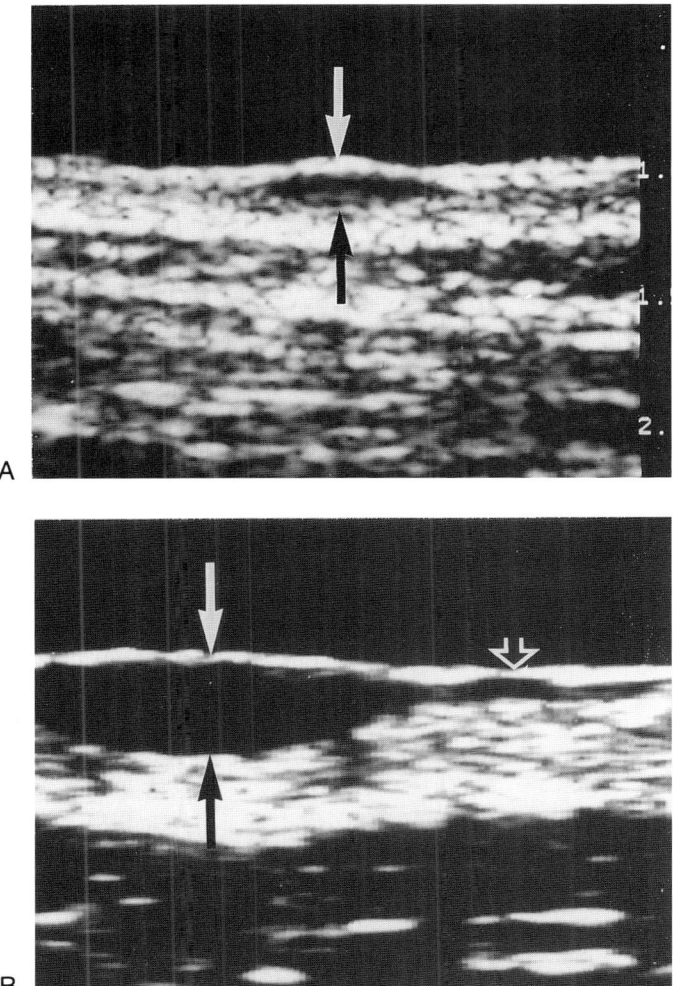

Fig. 7-14 Primary malignant melanoma of the skin. **(A)** A 10-MHz sonogram shows a hypoechoic nodule (*arrows*) at the surface of the skin. **(B)** A 20-MHz sonogram shows the margins of the tumor more clearly, allowing more accurate measurement of the thickness (*arrows*). This scan also shows the superficial spread of the tumor (*open arrow*), which was not apparent on the 10-MHz scan.

sion of subtumoral lymphocytic infiltrate in the measurement.[23]

The rich vascularity of primary melanoma has been documented using continuous-wave 10-MHz Doppler sonography, with 97 percent of tumors thicker than 0.8 mm showing detectable Doppler flow signals.[24] Color Doppler studies of thick melanomas using 7.5- to 10-MHz probes demonstrate the increased vascularity inside the tumor, especially at the deep margin of the tumor. Spectral analysis confirms the low-resistance flow pattern (Plate 7-2).

OTHER MALIGNANT TUMORS

Basal cell carcinomas, squamous cell carcinomas, and Kaposi's sarcomas appear sonographically as nonspecific hypoechoic masses; their margins are usually less regular than those of melanomas. Mycosis fungoides, a primary malignant T-cell lymphoma of the skin, has a different appearance. In its eczemalike or plaquelike form, mycosis fungoides appears as a markedly hypoechoic band involving the upper dermis, with a variable thickness depending on the degree of dermal invasion (Fig. 7-15).[11]

Metastases to the skin are relatively rare and derive most commonly from melanomas and carcinomas of the breast, lung, colon, and kidney. Skin metastases tend to have a more focal appearance than primary skin malignancies and an oval-to-round shape (Fig. 7-16). Metastases characteristically arise from within the dermis and extend toward the surface of the skin and the subcutaneous fat, whereas melanomas or squamous cell carcinomas develop from the skin surface and involve the dermis secondarily. Skin involvement is not infrequent in patients with lymphoma or leukemia and appears as a more diffuse process, occasionally with poor delineation between the dermis and subcutaneous fat.

Diffuse malignant infiltration of the skin of the breast is a hallmark of inflammatory breast cancer. The skin's thickness may exceed 1 cm; echogenicity may remain elevated, although most often it will be decreased. Color Doppler studies demonstrate significant low-resistance hypervascularity in and underneath the involved skin, indistinguishable from the findings observed in benign inflammatory processes.

PATHOLOGIC CONDITIONS OF THE SUBCUTANEOUS TISSUES

Sonography is highly sensitive in demonstrating diffuse and focal pathologic changes in the subcutaneous tissues.[8]

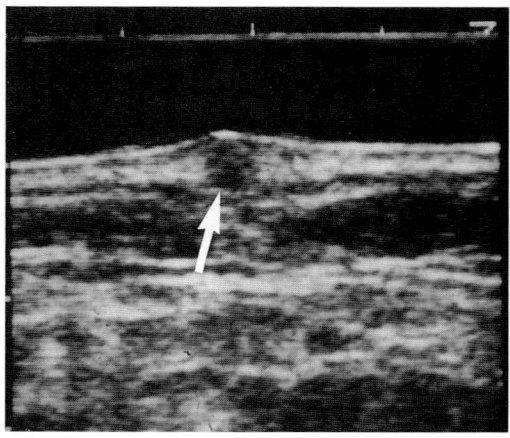

Fig. 7-16 Skin metastasis from a breast carcinoma. Sonogram obtained at 7.5 MHz shows a round, hypoechoic mass (*arrow*) about 4 mm in diameter, involving the full thickness of the skin.

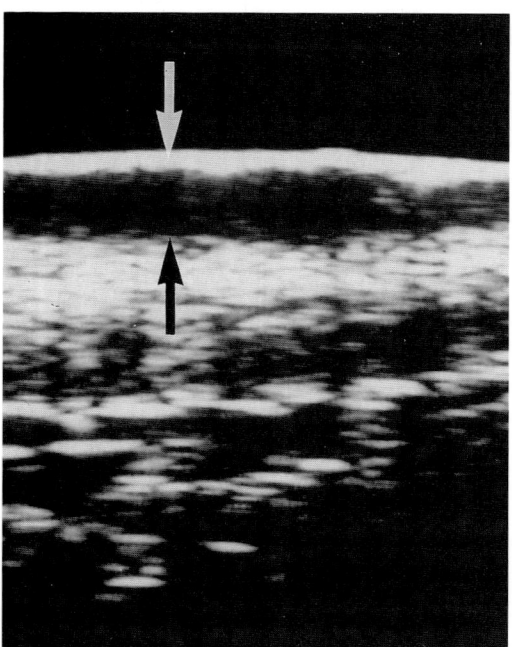

Fig. 7-15 Mycosis fungoides. A 20-MHz sonogram of a plaquelike lesion shows a hypoechoic band (*arrows*) involving the superficial half of the dermis.

Trauma

Trauma to subcutaneous tissues occurs frequently, though sonographers are rarely called upon to assess it. Subcutaneous hematomas have a wide spectrum of sonographic appearances[8,25,26]: a recent hemorrhage generally appears as a focal hyperechoic area (Fig. 7-17); as the hematoma liquefies and becomes organized, a complex fluid collection can develop, leading on rare occasions to a purely cystic collection.

The subcutaneous tissues are a common site of foreign bodies. Real-time sonography has proved to be highly sensitive in the detection of foreign bodies, including those that are not radiopaque (see Ch. 8).[27] Foreign bodies appear as brightly echogenic reflectors associated with shadowing, a trail of reverberation echoes ("comet-tail" artifact), both, or neither, depending on the object's physical nature; the comet-tail artifact is virtually pathognomonic of metallic objects. When present, inflammatory reaction to the foreign body appears as a hypoechoic area surrounding the echogenic foreign body. Sonography provides accurate preoperative three-dimensional localization of foreign bodies.[28]

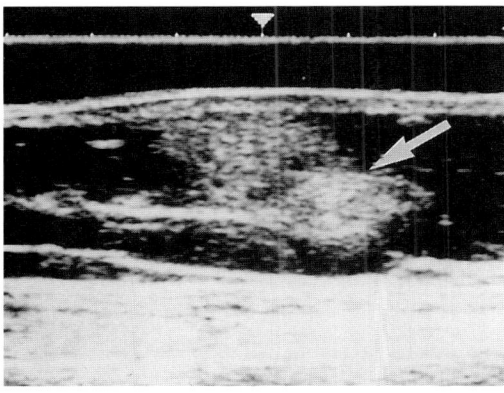

Fig. 7-17 Subcutaneous hematoma in the thigh after a dog bite. Sonogram shows an echogenic hematoma (*arrow*) contrasting with the hypoechoic subcutaneous fat.

Inflammatory Conditions

A common problem in evaluating inflamed subcutaneous tissues is differentiating clinically between cellulitis and abscess. Sonography is useful in this situation. In cellulitis, sonograms show a diffuse increase in the echogenicity of the subcutaneous fat, which often becomes indistinguishable from the skin. Color Doppler examination identifies increased blood flow (Plate 7-3). In contrast, abscesses appear sonographically as fluid-filled collections, often with internal echoes or a complex appearance, irregular walls, and increased vascularity at the periphery of the lesions on color Doppler scans. If an abscess is small or not easily defined by palpation, sonography can be used to guide a confirmatory fine-needle aspiration and subsequent percutaneous drainage.

Fat necrosis (panniculitis) can develop following trauma; it is also seen in patients with systemic diseases such as scleroderma or lupus erythematosus. On sonograms, the involved fat has a dirty appearance with an ill-defined focal area of increased echogenicity.[8,25]

Benign Tumors

The most common benign subcutaneous tumors are lipomas and hemangiomas. Both types of lesions display a wide spectrum of echogenicity (see Ch. 3). Lipomas may be hypoechoic, hyperechoic, or isoechoic relative to the surrounding fat (Fig. 7-18).[29] Because an isoechoic lipoma may be difficult to identify sonographically, it is good practice to use palpation under real-time sonoscopy, which allows localization of the mass on the sonogram with certainty. Hemangiomas are usually—but not always—hypoechoic. Bright echogenic foci indicating the presence of phleboliths are an important clue in the diagnosis of hemangiomas. Color Doppler examination inconsistently demonstrates significant blood flow within hemangiomas (see Ch. 3).

96 MUSCULOSKELETAL ULTRASOUND

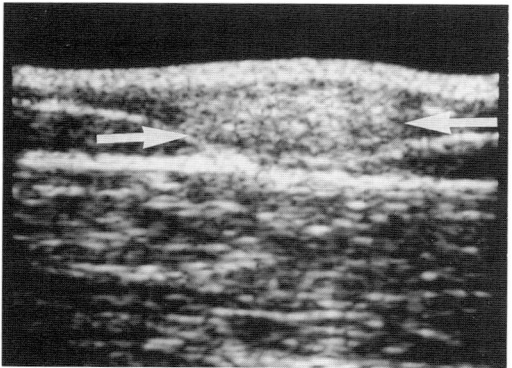

Fig. 7-18 Subcutaneous lipoma. Sonogram shows an elongated, echogenic mass (*arrows*) in the subcutaneous fat.

Malignant Tumors

Relatively few primary malignant tumors develop in the subcutaneous fat. However, this area is not infrequently the site of recurrent soft-tissue sarcomas and metastases from various types of primary tumors. In a patient with a history of malignant melanoma, the development of a hypoechoic mass in the subcutaneous tissues near the site of the primary tumor or between it and the draining nodal basin should be considered a local recurrence or in-transit metastasis until proven otherwise. Sonograms obtained at 7.5 or 10 MHz can demonstrate lesions that are only a few millimeters in diameter (Fig. 7-19).[30] Metastases from melanoma are highly vascularized and usually exhibit significant flow on color Doppler studies.

Sonography has been shown to be as accurate as MRI in the detection of early recurrence in patients treated for soft-tissue sarcoma.[31] Ultrasound-guided fine-needle aspiration can be performed within minutes to document the malignant nature of a suspicious lesion (Fig. 7-19).

Miscellaneous

Sonography has been used to measure the thickness of subcutaneous fat in various groups[32,33] and to evaluate its changes in patients with various clinical conditions, including malnutrition in children.[34] Sonog-

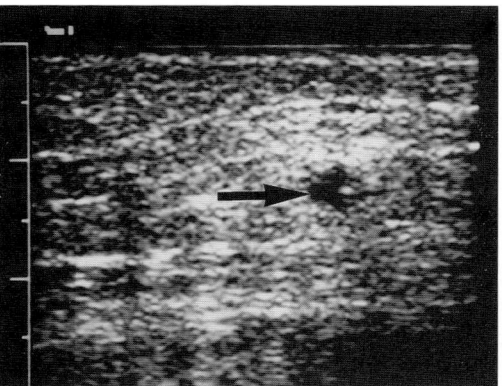

Fig. 7-19 Ultrasound-guided fine-needle aspiration biopsy of an in-transit melanoma metastasis in the subcutaneous fat of the thigh. Sonogram shows a 4-mm hypoechoic mass (*arrow*). The bright echo represents the tip of the fine needle. A single pass yielded an adequate specimen and allowed confirmation of the metastatic melanoma.

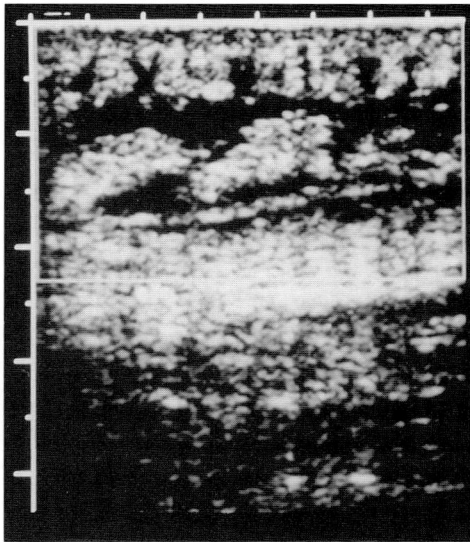

Fig. 7-20 Chronic lymphedema of the lower extremity. Sonogram shows fluid-distended lymphatics dissecting the subcutaneous fat and the deep dermis.

raphy has also proved useful for assessing changes during and after skin expansion in plastic surgery.[35] Some investigators have used sonography to measure the heel and sole pad thickness in patients with acromegaly or diabetes and in patients treated with phenytoin.[36] Sonography has been used recently to detect onchocercomas in patients with onchocerciasis.[37] Sonography has been proposed for guiding venipuncture for venography in patients with marked edema.[38] In chronic lymphedema, sonography shows dilated lymphatics dissecting the subcutaneous fat and the deep dermis (Fig. 7-20), with color Doppler study being negative.

CONCLUSIONS

Sonography of the skin cannot change the diagnostic approach to tumors of the skin, which usually require a biopsy. However, because it provides easy and accurate in vivo cross-sectional imaging of the skin, 20-MHz sonography is a powerful research tool, which has a place in any research project requiring noninvasive in vivo evaluation of the thickness of the skin or skin lesions and of changes in thickness over time. Improvement in the resolution of skin sonograms is expected with 40- to 50-MHz transducers.

The subcutaneous tissues represent probably the best application of external transcutaneous sonography using currently existing high-frequency probes. The high lesion detection rate and diagnostic accuracy of sonography combined with its low cost, wide availability, and ease of use make it the first-line examination with which to image subcutaneous tissues.

REFERENCES

1. Alexander H, Miller DL: Determining skin thickness with pulsed ultrasound. J Invest Dermatol 72:17, 1979
2. Tan CY, Statham B, Marks R, Payne PA: Skin thickness measurement by pulsed ultrasound: its reproducibility, validation and variability. Br J Dermatol 106:657, 1982
3. Cole GW, Handler SJ, Burnett K: The ultrasonic evaluation of skin thickness in scleredema. J Clin Ultrasound 9:501, 1981
4. Miyauchi S, Miki Y: Normal human skin echogram. Arch Dermatol Res 275:345, 1983
5. Shafir R, Itzchak Y, Heyman Z et al: Preoperative ultrasonic measurements of the thickness of cutaneous malignant melanoma. J Ultrasound Med 3:205, 1984
6. Fornage BD, Deshayes JL: Ultrasound of the normal skin. J Clin Ultrasound 14:619, 1986
7. Fornage B, Duvic M: High-frequency sonography of the skin. J Eur Acad Dermatol Venereol 3:47, 1993
8. Fornage BD: Sonography of the Extremities. Vigot, Paris, 1991
9. Nessi R, Betti R, Bencini PL et al: Ultrasonography of nodular and infiltrative lesions of the skin and subcutaneous tissues. J Clin Ultrasound 18:103, 1990
10. Fornage BD, Touche DH, Rifkin MD: Small parts real-time sonography: a new "waterpath." J Ultrasound Med 3:355, 1984
11. Fornage BD, McGavran MH, Duvic M, Waldron CA: Imaging of the skin with 20-MHz US. Radiology 189:69, 1993
12. Stiller MJ, Driller J, Shupack JL et al: Three-dimensional imaging for diagnostic ultrasound in dermatology. J Am Acad Dermatol 29:171, 1993
13. Srivastava A, Hughes BR, Hughes LE, Woodcock JP: Doppler ultrasound as an adjunct to the differential diagnosis of pigmented skin lesions. Br J Surg 73:790, 1986
14. Hoffmann K, Feldmann S, Dirschka T et al: Sonographic quantification of the type IV reaction after intradermal application of recall antigens. Skin Pharmacol 7:291, 1994
15. Olsen LO, Serup J: High-frequency ultrasound scan for non-invasive cross-sectional imaging of psoriasis. Acta Derm Venereol (Stockh) 73:185, 1993
16. Brink JA, Sheets PW, Dines KA et al: Quantitative assessment of burn injury in porcine skin with high-frequency ultrasonic imaging. Invest Radiol 21:645, 1986
17. Bauer JA, Sauer T: Cutaneous 10 MHz ultrasound B scan allows the quantitative assess-

ment of burn depth. Burns Incl Therm Inj 15:49, 1989
18. Lévy J, Gassmüller J, Schröder G et al: Comparison of the effects of calcipotriol, prednicarbate and clobetasol 17-propionate on normal skin assessed by ultrasound measurement of skin thickness. Skin Pharmacol 7:231, 1994
19. Hoffmann K, Winkler K, el-Gammal S, Altmeyer P: A wound healing model with sonographic monitoring. Clin Exp Dermatol 18:217, 1993
20. Clark WH, Ainsworth AM, Bernardino EA et al: The developmental biology of primary human malignant melanomas. Semin Oncol 2:83, 1975
21. Breslow A: Thickness, cross-sectional areas and depth of invasion in the prognosis of cutaneous melanoma. Ann Surg 172:902, 1970
22. Day CL Jr, Lew RA, Mihm MC et al: The natural break points for primary-tumor thickness in clinical stage I melanoma. N Engl J Med 305:1155, 1981
23. Hoffmann K, Jung J, el-Gammal S, Altmeyer P: Malignant melanoma in 20-MHz B-scan sonography. Dermatology 185:49, 1992
24. Srivastava A, Hughes LE, Woodcock JP, Laidler P: Vascularity in cutaneous melanoma detected by Doppler sonography and histology: correlation with tumour behaviour. Br J Cancer 59:89, 1989
25. Fornage BD: Sonography of the skin and subcutaneous tissues. Radiol Med (Torino) 85(suppl 1):149, 1993
26. Wilson DJ: Ultrasonic imaging of soft tissues. Clin Radiol 40:341, 1989
27. Fornage BD, Schernberg FL: Sonographic diagnosis of foreign bodies of the distal extremities. AJR 147:567, 1986
28. Fornage BD, Schernberg FL: Sonographic preoperative localization of a foreign body in the hand. J Ultrasound Med 6:217, 1987
29. Fornage BD, Tassin GB: Sonographic appearances of superficial soft-tissue lipomas. J Clin Ultrasound 19:215, 1991
30. Fornage BD, Lorigan J: Sonographic detection and fine-needle aspiration biopsy of nonpalpable recurrent or metastatic melanoma in subcutaneous tissues. J Ultrasound Med 8:421, 1989
31. Choi H, Varma DGK, Fornage BD et al: Soft-tissue sarcoma: MR imaging vs sonography for detection of local recurrence after surgery. AJR 157:353, 1991
32. Heckmatt JZ, Pier N, Dubowitz V: Measurement of quadriceps muscle thickness and subcutaneous tissue thickness in normal children by real-time ultrasound imaging. J Clin Ultrasound 16:171, 1988
33. Maruyama Y, Iizuka S, Yoshida K: Ultrasonic observation on distribution of subcutaneous fat in Japanese young adults with reference to sexual difference. Ann Physiol Anthropol 10:61, 1991
34. Koskelo E-K, Kivisaari LM, Saarinen UM, Siimes MA: Quantitation of muscles and fat by ultrasonography: a useful method in the assessment of malnutrition in children. Acta Paediatr Scand 80:682, 1991
35. Reali UM, Chiarugi C, De Siena GM, Giannotti V: Sonographic evaluation of dermis and subcutaneous tissue during and after skin expansion. Plast Reconstr Surg 93:1050, 1994
36. Gooding GAW, Stess RM, Graf PM et al: Sonography of the sole of the foot. Evidence for loss of foot pad thickness in diabetes and its relationship to ulceration of the foot. Invest Radiol 21:45, 1986
37. Ares-Vidal J, Bru-Saumell C, Bianchi-Cardona L, Corachan-Cuyas M: Detection of subcutaneous nodules with ultrasound in onchocerciasis (letter). Med Clin 92:316, 1989
38. Johns CM, Sumkin JH: US-guided venipuncture for venography in the edematous leg. Radiology 180:573, 1991

8
Foreign Bodies
Gretchen A. W. Gooding

Radiography in the emergency room is a common first step in the detection of foreign bodies. Although the site and location of glass and metal are easily determined (96% of cases),[1,2] only 15 percent of wooden fragments are seen on radiographs, the visualization of some plastics is problematic, and thin, fine glass fragments may be obscured by overlying structures.[3] Radiographs also have limited usefulness in guiding the removal of a foreign body. In contrast, studies with ultrasound demonstrate high sensitivity in detecting a wide range of foreign objects in the soft tissues. Sonographic localization for extraction is another major advantage of the modality.[4]

TECHNIQUE

Sonography, because of its noninvasive nature, its fine near-field of view with transducers of 7 to 10 MHz, and the ease with which it can be used, is an appropriate study for detecting suspected superficial foreign bodies. Superficial foreign bodies are best seen with high-resolution linear-array transducers, whereas intra-abdominal foreign bodies are identified best with sector transducers of 3.5 to 5 MHz. Intraoperatively, small high-resolution linear-array transducers are ideal.

Because of its mobility, the ultrasound unit can be taken to an emergency room or operating suite as needed.[5,6] Foreign-body removal requires a sterile field, so the ultrasound transducer is covered with a sterile sheath after the transducer footprint is lubricated with acoustic gel and any bubbles from the transducer surface–sheath interface are expressed. For external localization, sterile gel serves as the coupling agent and is applied to the skin at the site of the suspected foreign body. For intraoperative localization, sterile saline is used, and the transducer is merely placed within the saline bath without touching the organs themselves.

SONOGRAPHIC APPEARANCE OF FOREIGN BODIES

Experimental Studies

Experimental studies using ultrasound to detect a variety of foreign bodies have been done.[7–11] One such experimental method involves using a chicken breast as a model for the superficial soft tissues; echoes in it are similar to those in the plantar aspect of the heel of the foot.[10] The chicken breast tissue is implanted with a wooden splinter, a tiny glass rod, and a small nail; each object produces a different pattern of echoes when examined with ultrasound.[10] A wooden splinter tends to absorb the sound, producing an acoustic shadow beyond the proximal interface. A tiny air-filled glass rod produces bright hyperechoic interfaces along its length. A metal nail causes a reverberating comet-tail artifact, typical of metal in the soft tissues. Collagen can also be injected into the chicken breast to simulate collagen injection in the subcutaneous tissues, which is used by some dermatologists for cosmetic effects; injected collagen produces a hyperechoic area (Fig. 8-1).

A blind study using a 5-MHz transducer on cadavers with implants of glass, plastic, and wood yielded a sensitivity of 89 percent and a specificity of 93 percent in 65 episodes.[12] Because of its high sensitivity, sonography has particular relevance in the search for foreign bodies that are not radiopaque.[13]

In Vivo Appearances

Foreign objects may be noted in the soft tissues soon after injury or may be present for months or years. Unrecognized foreign bodies can be a source of inflammation and infection in unsuspecting patients. In a series of 200 cases, the average length of time before a foreign body in the hand was removed was 7 months.[3]

The typical sonographic appearance of a foreign body is a hyperreflective focus with an acoustic shadow; however, the introduction of air during surgical manipulation may obscure surrounding tissues. Glass fragments, wooden splinters, thorns, and metal needles represent only a few of the many types of foreign bodies that can be detected in the superficial soft tissues using sonography (Figs. 8-2 to 8-4).[7,14] Foreign bodies also may be associated with a surrounding inflammatory mass appearing as a hypoechoic halo (Figs. 8-3 and 8-4).[15] Glass fragments produce well-defined bright echoes in the soft tissues (Figs. 8-4 and 8-5). Depending on its size, a wood fragment may be seen as a hyperechoic focus with or without an acoustic shadow resulting from absorption of the beam (Figs. 8-3 and 8-6).[16] Metallic objects, wherever they occur, tend to cause reverberating comet-tail artifacts.[17-19] Large metallic objects such as hip prostheses, bullets, or shrapnel fragments are best seen with radiographs. On sonograms, bullets and shrapnel, like most metallic objects, produce highly reflective interfaces with acoustic shadowing.[20] Bullets can be removed under ultrasound guidance, particularly when they are located superficially. Surgical clips produce tiny, focal, hyperechoic, triangular comet-tail artifacts, whether in the neck (Fig. 8-7), around a peripheral graft (Plate 8-1), or in the abdomen, where they are less well appreciated on sonograms.

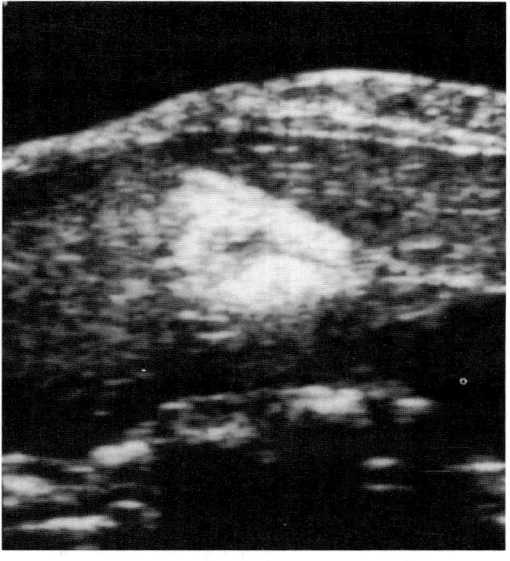

Fig. 8-1 Collagen injected into a chicken breast appears brightly echogenic. Note also the linear echoes from the needle.

FOREIGN BODIES **101**

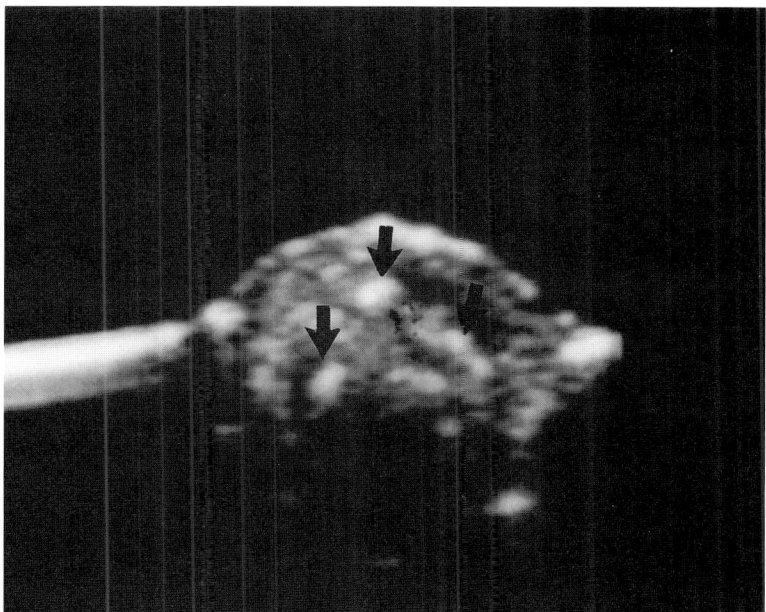

Fig. 8-2 Longitudinal sonogram of the fingertip shows bright echoes (*arrows*) from tiny glass fragments.

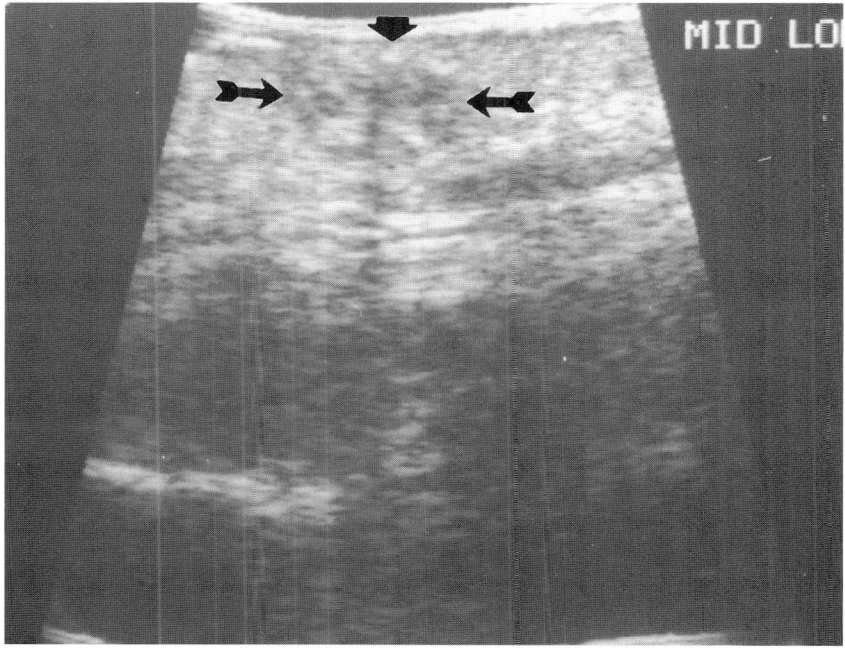

Fig. 8-3 Longitudinal sonogram of the sole of the foot shows in cross section a small wooden splinter (*short arrow*) that casts an acoustic shadow and is surrounded by a slightly hypoechoic inflammatory mass (*arrows*).

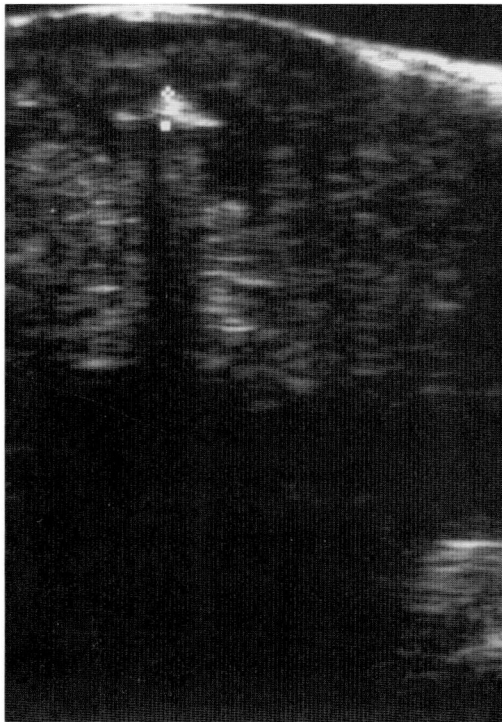

Fig. 8-4 Glass fragment in the soft tissues of the foot. Sonogram shows the foreign body as a bright hyperechoic mass (*cursors*) that casts an acoustic shadow.

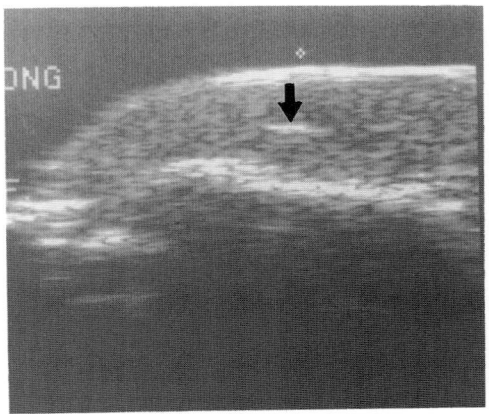
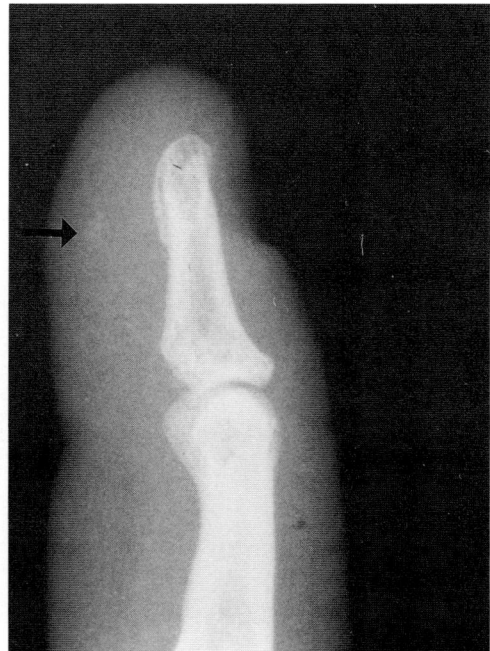

A B

Fig. 8-5 Small glass fragment in a finger from an injury years earlier. **(A)** Sonogram of the fingertip shows a thin linear echo (*arrow*). **(B)** Radiograph shows a faint density in the fingertip (*arrow*).

FOREIGN BODIES

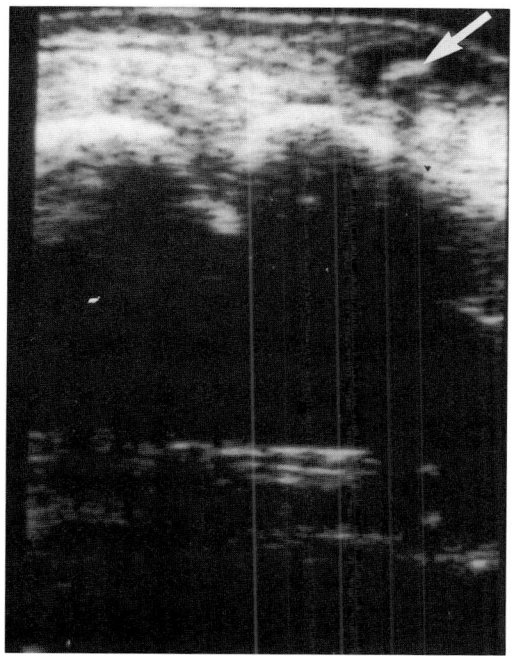

Fig. 8-6 Transverse sonogram of the sole of the foot posterior to the metatarsal heads shows a toothpick fragment as a hyperechoic focus (*arrow*) surrounded by a hypoechoic inflammatory mass.

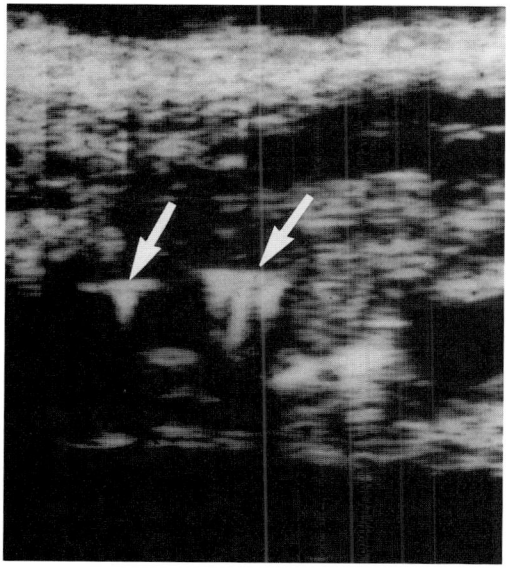

Fig. 8-7 Longitudinal sonogram of the neck shows surgical clips as bright reflectors (*arrows*) with posterior reverberations.

In foreign body detection, sonography is used to define an abnormal interface. Only the proximal surface is identified if the object absorbs sound and casts a shadow, and the entire foreign body may not be seen. If the correct line of interrogation is not achieved, sonography may foreshorten the foreign object (Fig. 8-8).[21,22]

A bright interface in the soft tissues even with acoustic shadowing does not always indicate a foreign body. Air in the soft tissues can also cause bright echoes that may cast acoustic shadows and be a source of confusion (Fig. 8-9). Bones cause marked acoustic shadowing as a result of both the absorption and reflection of the acoustic beam. Thin shadows may result from poor contact between the transducer and the surface of hairy skin and should not be confused with shadowing from a small, very superficial foreign body (Fig. 8-10).

COMMON SITES OF FOREIGN BODIES

Foot

The sole of the foot is one of the most common sites for foreign bodies. Pain in the foot may be the precipitating complaint when a sonogram is ordered to rule out a foreign body.

On sonographic examination, the sole of the foot produces echoes of medium intensity in the soft tissues with a bright interface at the skin surface and another strong interface at the bone (Fig. 8-11).[23,24] Various lesions, including keratotic skin lesions or even sesamoid bones, can have an appearance similar to that of a foreign body.[10] Sesamoid bones cast acoustic shadows with highly reflective echoes at the soft tissue–bone interface. Bursal calcifications adjacent to the calcaneus and elsewhere will also cause significant acoustic shadowing (Fig. 8-12). Iatrogenic

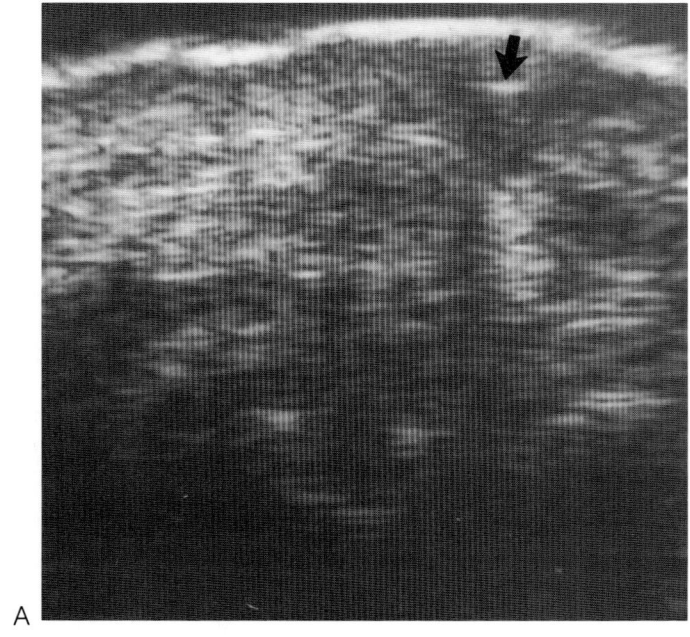
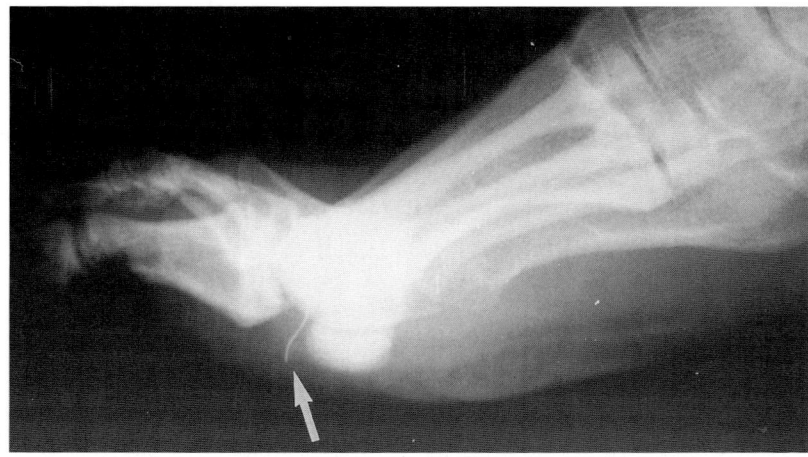

Fig. 8-8 Metallic wire in the sole of the foot. **(A)** Transverse sonogram of the sole of the foot shows only a minute hyperechoic focus (*arrow*) with an acoustic shadow representing the cross section of the wire. There is a surrounding inflammatory area of decreased echogenicity. **(B)** Lateral radiograph of the foot shows the wire at the ball of the foot (*arrow*).

FOREIGN BODIES **105**

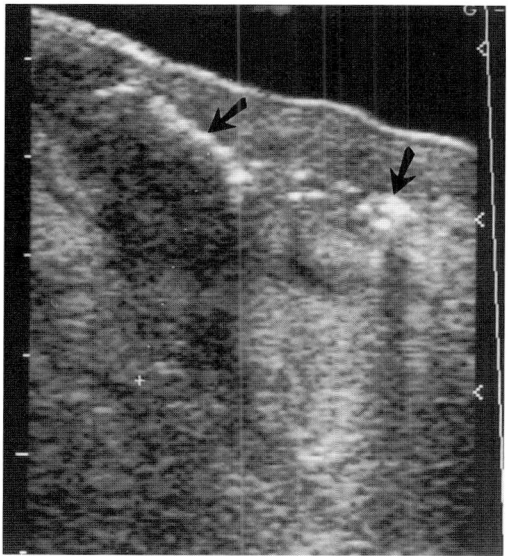

Fig. 8-9 Inflammatory mass of the thigh with subcutaneous air. Sonogram shows subcutaneous air bubbles as bright echoes (*arrows*).

calcifications occasionally develop in the foot at the site of local corticosteroid injections, which are used to reduce symptoms from conditions such as heel spurs.[25,26] Soft-tissue granulomas, whether superficial or deep, frequently calcify, and soft-tissue tumors may calcify as well, with resulting acoustic shadowing.[26] In instances of soft-tissue inflammation or neoplasms, an actual mass can be identified in association with the calcifications (Fig. 8-13).

On occasion, a patient will suspect that he or she has a foreign body when that is not the case. For instance, a carpenter may suspect a splinter in the hand but have a traumatic inclusion cyst instead, seen on sonograms as a small cystic lesion (Fig. 8-14). A Morton's neuroma appears as a small hypoechoic nodule (see Ch. 14).[27]

Neck

Surgical clips are commonly noted in the neck as bright hyperechoic foci in the soft tissues (Fig. 8-7). A color Doppler artifact may result from an encircling metallic ca-

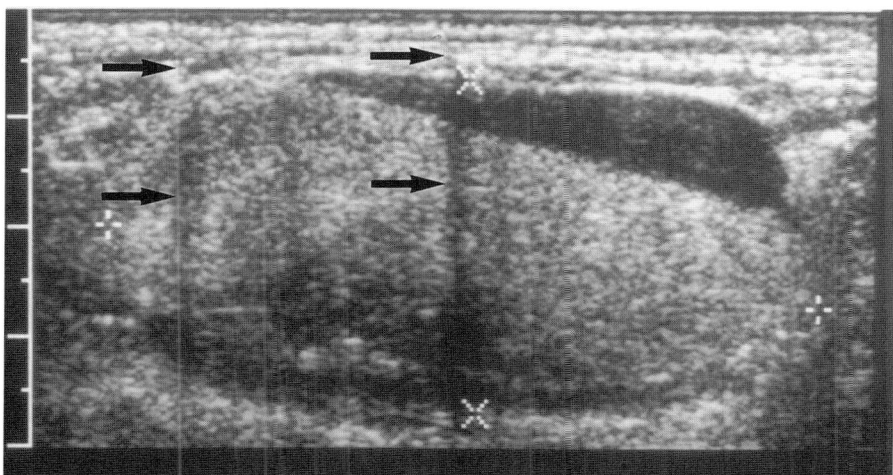

Fig. 8-10 Longitudinal sonogram of the elbow shows a large hematoma with a fluid–fluid level. Vertical shadows (*arrows*) are not due to foreign bodies but are related to a hairy arm and loss of good contact between skin and transducer.

106 MUSCULOSKELETAL ULTRASOUND

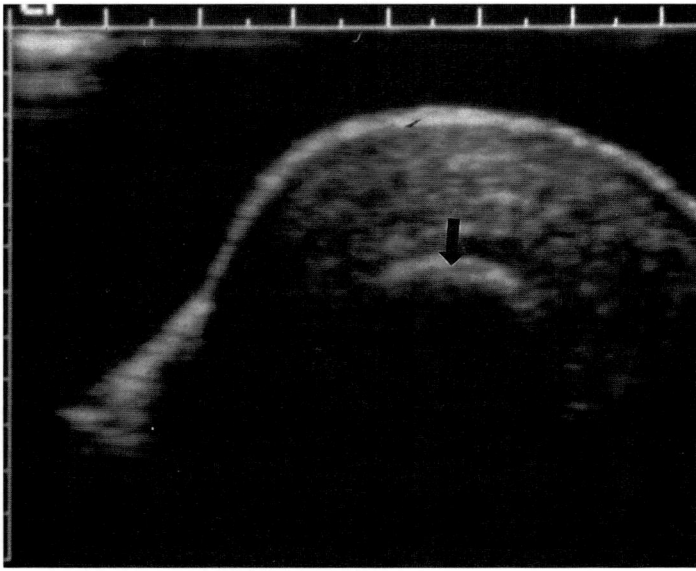

Fig. 8-11 Transverse sonogram shows the normal soft tissues of the heel. The arrow indicates the calcaneal bone interface.

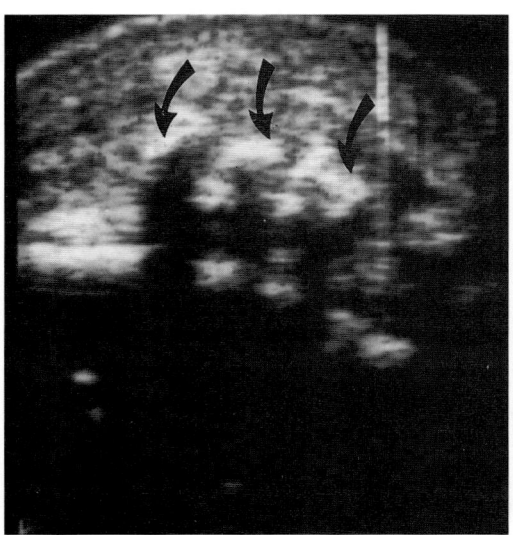

Fig. 8-12 Transverse sonogram of the heel shows calcifications in the calcaneal bursa (*arrows*).

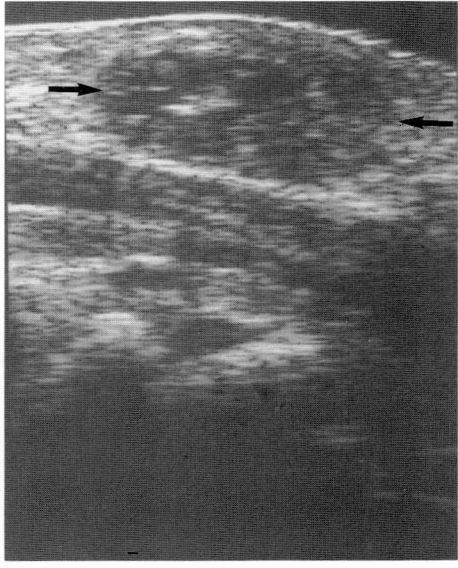

Fig. 8-13 Coccidioidomycosis. Longitudinal sonogram of the wrist shows a soft-tissue mass (*arrows*). The bright central echo represents calcification, not a foreign body.

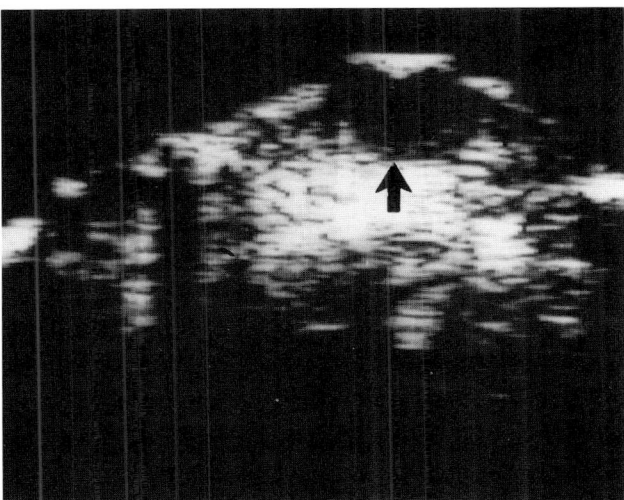

Fig. 8-14 Longitudinal sonogram of the index finger shows an inclusion cyst (*arrow*) in a carpenter who thought he had a wooden splinter.

rotid clamp placed to reduce flow to the brain, such as for a cranial arteriovenous malformation. The clamp produces a metallic comet-tail artifact in the soft tissues (Fig. 8-15), but the color Doppler signals reverse, causing a blue, rather than a red, focal flow abnormality in the carotid artery from a mirror artifact misregistration (Plate 8-2).[23] Jugular and subclavian vein catheterizations also may result in retained wires or catheter frag-

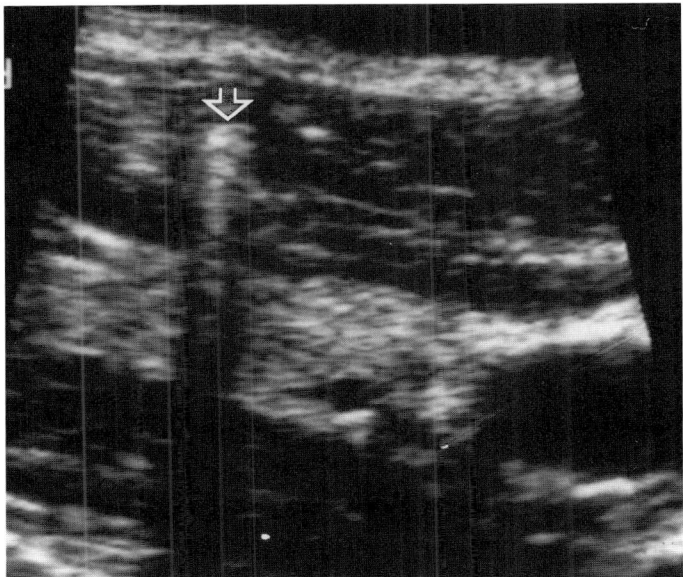

Fig. 8-15 Carotid clamp. Longitudinal sonogram of the neck shows a metallic foreign body—a circumferential carotid clamp (*arrow*)—causing a reverberating comet-tail artifact in the soft tissues (see Color Plate 8-2).

ments in the soft tissues or within the vessels (Fig. 8-16).

Esophageal foreign bodies such as fish bones or tracheobronchial foreign bodies such as coins are not visualized by sonography.

Blood Vessels

Catheters and pacemaker leads in blood vessels can be identified by ultrasound (Fig. 8-16),[29] and vascular grafts of prosthetic material can be recognized by their walls' sonographic characteristics, which depend on the corrugations of the material used (Figs. 8-17 and 8-18).

LOCALIZING A FOREIGN BODY

Preoperative localization of a foreign body is helpful to the surgeon concerned with removing a small object.[22,30-33]

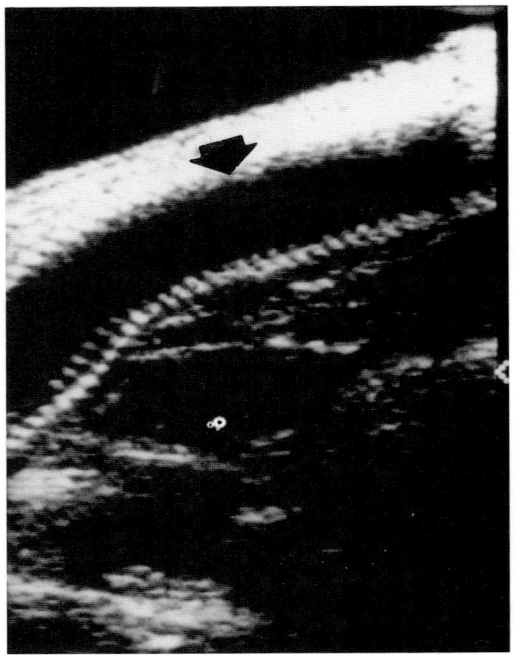

Fig. 8-17 Longitudinal sonogram shows multiple corrugations (*arrow*) from the prosthetic graft material used in this femoral-femoral graft.

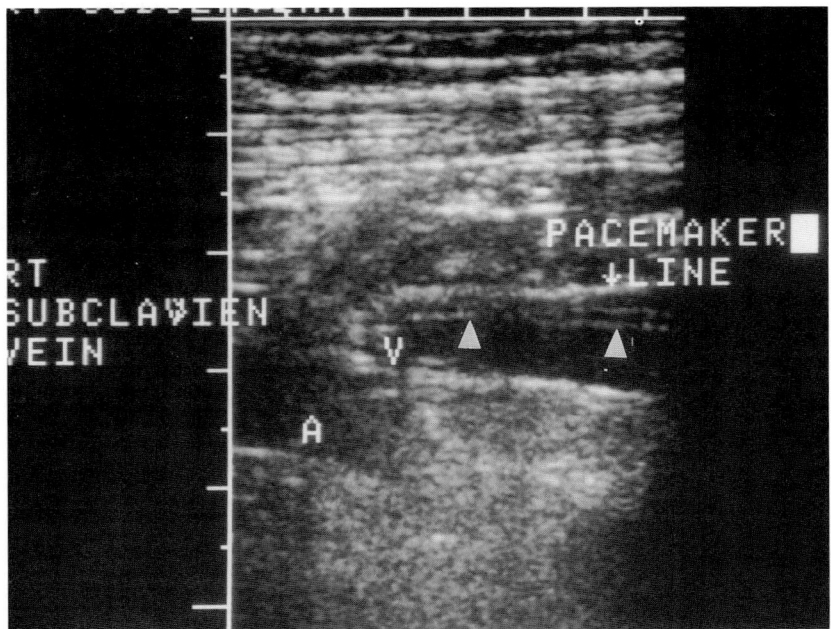

Fig. 8-16 Longitudinal sonogram of the right subclavian vein (V) shows a pacemaker line (*arrowheads*).

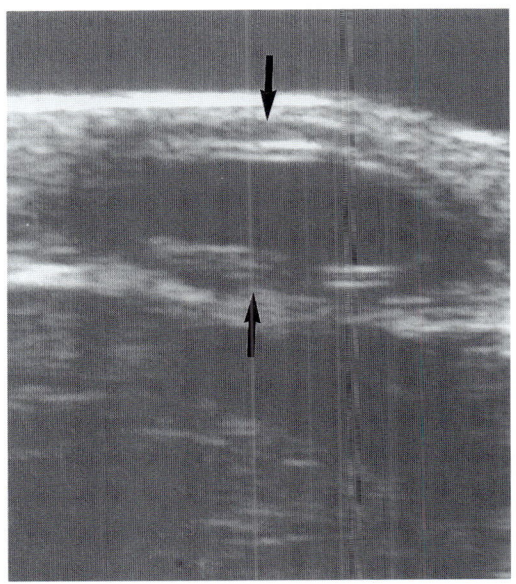

Fig. 8-18 Dialysis fistula in the forearm. Longitudinal sonogram shows the fistula with a rim of fluid around it (*arrows*), suggesting the possibility of infection.

If the foreign body may be in the superficial soft tissues, a high-resolution transducer of 7 to 10 MHz should be used. The smaller the object sought, the higher the resolution of the transducer should be. Detection of tiny, superficial objects requires a meticulous technique. A linear-array transducer should be used because of its superior near-field of view. A standoff pad can be helpful. Once the object is identified, scans should be obtained in various planes to determine its length and width. The depth from the surface should be measured and communicated to the surgeon. Sonography may be unable to determine the diameter of the object if the object casts a shadow that obscures all but the interface with the surrounding tissue. The object should be scanned in at least two planes perpendicular to each other. If color Doppler sonography is available, it is helpful in defining the relationship of the foreign body to adjacent vessels. The skin lying directly over the object to be removed should be marked so that the surgeon knows where to make the incision and how deep to go.[22] Sonography is used after surgery to observe whether any residual foreign object exists

Transducers with biopsy guides can be used to follow a localizing needle to the site of a foreign body under direct real-time observation. At times, the foreign body can be removed percutaneously without open surgery, as has been reported using radiographic guidance.[34]

If desired, sonography can be performed during surgical removal by covering the transducer (linear-array or sector) in a sterile sheath, using acoustic gel as a couplant between the transducer footprint and the sheath. The ensheathed transducer can then be placed in sterile saline in the wound (as an adequate acoustic window) to visualize the site for precise localization.

CONCLUSIONS

The ability to detect a foreign body in the soft tissues with ultrasound depends on the object's size, the character of the echoes produced, and the matrix in which the foreign body resides. The smaller the suspected object, the more important it is to use the highest frequency transducer available, preferably 10 MHz for the superficial soft tissues. Sonography is effective in identifying both radiopaque and radiotransparent foreign bodies (e.g., wood, plastic, glass, and metal fragments). A hyperechoic foreign body will be easy to identify in a homogeneous hypoechoic background, such as the sole of the foot, but may be difficult to see in a highly heterogeneous hyperechoic background, although its presence may be suspected if it produces a comet-tail artifact or an acoustic shadow.

Pre- or intraoperative sonographic localization has the potential to shorten the search for a foreign body, and it enhances the ability of the surgeon to recognize whether the object has moved with manipulation. Thus, tissue damage can be reduced.

REFERENCES

1. Fisher MS, Felman AH: The radiographic detection of glass in soft tissue. Radiology 92:1529, 1969
2. Tandberg D: Glass in the hand and foot. Will an x-ray film show it? JAMA 248:1872, 1982
3. Anderson MA, Newmeyer WL III, Kilgore ES, Jr: Diagnosis and treatment of retained foreign bodies in the hand. Am J Surg 44:63, 1982
4. Crawford R, Matheson AB: Clinical value of ultrasonography in the detection and removal of radiolucent foreign bodies. Injury 20:341, 1989
5. Shaikh K, Cilley J, O'Connor W, Del Rossi AJ: Intra-operative echocardiography: a useful tool in the localization of small intracardiac foreign bodies. J Cardiovasc Surg 30:42, 1989
6. de Campo JF, Myers D, Klug GL: Ultrasound guided removal of an intracranial bullet: case report. Australas Radiol 30:6, 1986
7. De Flaviis L, Scaglione P, Del Bo P, Nessi R: Detection of foreign bodies in soft tissues: experimental comparison of ultrasonography and xeroradiography. J Trauma 28:400, 1988
8. Suramo I, Pamilo M: Ultrasound examination of foreign bodies. An in vitro investigation. Acta Radiol 27:463, 1986
9. Little CM, Parker MG, Callowich MC, Sartori JC: The ultrasonic detection of soft tissue foreign bodies. Invest Radiol 21:275, 1986
10. Gooding GAW, Hardiman T, Sumers M et al: Sonography of the hand and foot in foreign body detection. J Ultrasound Med 6:441, 1987
11. Torfing KF, Teisen HG, Skodt T: Computed tomography, ultrasonography and plain radiography in the detection of foreign bodies in pork muscle tissue. ROFO Fortschr Geb Rontgenstr Nuklearmed 149:60, 1988
12. Blyme PJ, Lind T, Schantz K, Lavard P: Ultrasonographic detection of foreign bodies in soft tissue. A human cadaver study. Arch Orthop Trauma Surg 110:24, 1990
13. Gilbert FJ, Campbell RSD, Bayliss AP: The role of ultrasound in the detection of nonradiopaque foreign bodies. Clin Radiol 41:109, 1990
14. Fornage BD, Schernberg FL: Sonographic diagnosis of foreign bodies of the distal extremities. AJR 147:567, 1986
15. Fornage BD, Schernberg FL, Rifkin MD: Ultrasound examination of the hand. Radiology 155:785, 1985
16. Serrin DA, Eberhardt H, Hirsch JH: Ultrasonic localization of wooden splinter in the foot. Med Ultrasound 6:83, 1982
17. Fornage BD: Preoperative sonographic localization of a migrated transosseous stabilizing wire in the hand. J Ultrasound Med 6:471, 1987
18. Ziskin MC, Thickman DI, Jacobs-Goldenberg N et al: The comet tail artifact. J Ultrasound Med 1:1, 1982
19. Wendell BA, Athey PA: Ultrasonic appearance of metallic foreign bodies in parenchymal organs. J Clin Ultrasound 9:133, 1981
20. Yiengpruksawan A, Mariadason J, Ganepola GA, Freeman HP: Localization and retrieval of bullets under ultrasound guidance. Arch Surg 122:1082, 1987
21. Howard CB, Nyska M, Mellor I et al: Sonographic underestimation of the size of a foreign body. J Clin Ultrasound 20:412, 1992
22. Shiels WE II, Babcock DS, Wilson JL, Burch RA: Localization and guided removal of soft tissue foreign bodies with sonography. AJR 155:1277, 1990
23. Gooding GAW, Stess RM, Graf PM, Grunfeld C: Heel pad thickness: determination by high-resolution ultrasonography. J Ultrasound Med 4:173, 1985
24. Gooding GAW, Stess RM, Graf PM et al: Sonography of the sole of the foot. Evidence for loss of foot pad thickness in diabetes and its relationship to ulceration of the foot. Invest Radiol 21:45, 1986
25. Conti RJ, Shinder M: Soft tissue calcifications induced by local corticosteroid injection. J Foot Surg 30:34, 1991
26. Kaplan PA, Anderson JC, Norris MA, Matamoros A, Jr: Ultrasonography of post-traumatic soft-tissue lesions. Radiol Clin North Am 27:973, 1989
27. Redd RA, Peters VJ, Emery SF et al: Morton neuroma: sonographic evaluation. Radiology 171:415, 1989

28. Gooding GAW, Saloner D, Eisert W, Nagarkar S: Color Doppler artifact from metallic carotid clamp. J Ultrasound Med 10:691, 1991
29. Gooding GAW, Bank WO: Ultrasound visualization of 5-French catheter. Radiology 144:647, 1982
30. Banerjee B, Das RK: Sonographic detection of foreign bodies of the extremities. Br J Radiol 64:107, 1991
31. Fornage BD, Schernberg FL: Sonographic preoperative localization of a foreign body in the hand. J Ultrasound Med 6:217, 1987
32. Gordon D: Nonmetallic foreign bodies [letter]. Br J Radiol 58:574, 1985
33. Donaldson JS: Radiographic imaging of foreign bodies in the hand. Hand Clin 7:125, 1991
34. Nosher JL, Siegel R: Percutaneous retrieval of nonvascular foreign bodies. Radiology 187:649, 1993

9
Rotator Cuff

Laurence A. Mack
Frederick A. Matsen III

Shoulder pain and weakness on elevation of the arm are common clinical problems. In patients over 40 years of age, rotator cuff disease is a frequent etiology. Older and disused rotator cuffs and those of smokers fail more easily than others. In contrast, in healthy subjects under the age of 40, a major injury is required to disrupt the tendons of the rotator cuff, and the bone of the greater tuberosity may fracture before the cuff tears.[1]

Cuff fibers may fail a few at a time, giving rise to a clinical presentation often misinterpreted as bursitis or tendinitis. Failure of large groups of cuff fibers leads to sudden weakness on elevation and external rotation. Cuff disease may lead to symptomatic abrasion against the undersurface of the coracoacromial arch with painful crepitus.[1]

Contrast arthrography has served for many years as the primary radiologic examination used to investigate the integrity of the rotator cuff tendons.[2] Although this technique is accurate for the diagnosis of full-thickness rotator cuff tears, it is less helpful in assessing the size of tears and in detecting partial-thickness tears.[3] Magnetic resonance imaging (MRI) has also gained wide acceptance as a technique for evaluation of the rotator cuff. This technique is accurate in the diagnosis of full-thickness tears, but recent reports have questioned whether MRI can reliably distinguish full- or partial-thickness cuff tears from degeneration of the cuff.[3]

An initial report by Seltzer et al found that articulated-arm ultrasound scanners could visualize the rotator cuff.[4] Farrar et al, in 1983, described the first technique of examination of the cuff tendons using real-time mechanical sector scanners.[5] A number of more recent reports using high-resolution linear-array transducers have demonstrated that real-time sonography may serve as an alternative means of examining the rotator cuff.[6-9]

TECHNICAL CONSIDERATIONS

Instrumentation

Initial studies of the real-time evaluation of the rotator cuff used mechanical sector transducers with frequencies of 5 to 10 MHz.[10] Although such transducers produce high-quality images, they are limited by suboptimal superficial resolution secondary to near-field artifacts and a narrow superficial image field. In addition, they are prone to artifacts. Since only small portions of parallel tendon fibers in the center of the image have the specular reflection geometry, an artifactually heterogeneous appearance of the tendons may be observed.

High-resolution linear-array transducers are best for examining the shoulder and other musculoskeletal areas. Transducers of 7 to 10 MHz are preferable to those of lower frequency. Recent introduction of wide-aperture, broad-bandwidth technology with improved electronic focusing has further improved spatial and contrast resolution. In patients with excessive subcutaneous fat or fatty muscular infiltration, lower-frequency (3.5–5 MHz) transducers may be required for adequate tissue penetration. An acoustic standoff material such as Kitecko (3 M, Minneapolis, MN) may be helpful in improving the near-field image when these lower-frequency devices are used. Compared with mechanical transducers, linear-array transducers demonstrate superior near-field resolution. In addition, their broad superficial field of view is helpful in evaluating superficial abnormalities, and a greater portion of the parallel tendon fibers will fulfill the specular reflection condition that best delineates tendon pathology and avoids artifacts.

Technique

The sonographic anatomy of the rotator cuff is complex in three dimensions, and the study is made more difficult by adjacent bony structures that limit ultrasound access to underlying soft-tissue structures. Several excellent discussions of normal anatomy have been published.[11,12] In addition, further study of anatomic relationships in the dissection room or surgical suite is helpful.

Several different techniques for examination of the rotator cuff have been described.[6,10,11] All share the underlying principle of attempting to examine the tendons in two orthogonal planes whenever possible. The following technique for the evaluation of the rotator cuff was developed at the University of Washington.[13]

The patient is scanned while seated on a revolving stool. This permits easy positioning during the scanning of both shoulders. The examiner is also seated on a stool, which has wheels to enhance mobility. Both of the patient's shoulders are examined, starting with the less symptomatic side. The complex curved anatomy and acoustic couplant make transducer positioning difficult. Increased transducer stability is provided by holding the transducer so that the examiner's hand is resting on the skin surface.[11] Recorded images are oriented on film as if one were facing the patient. For example, while imaging the right shoulder, the left side of the image represents the lateral region and the right side the medial region. Use of paired images of the same structure may be helpful in demonstrating abnormalities.

NORMAL ULTRASOUND ANATOMY

Biceps Tendon and Bicipital Groove

TRANSVERSE VIEW (POSITION 1)

When viewed transversely, the bicipital groove, which contains the tendon of the long head of the biceps brachii muscle, appears as a concavity in the bony surface of the humerus (Fig. 9-1). This tendon functions as an anatomic landmark that differentiates the medially located subscapularis tendon from the supraspinatus tendon. With the arm in neutral position, the biceps tendon is visualized as a hyperechoic oval structure within the bicipital groove. The biceps tendon is imaged throughout its length, and im-

ROTATOR CUFF

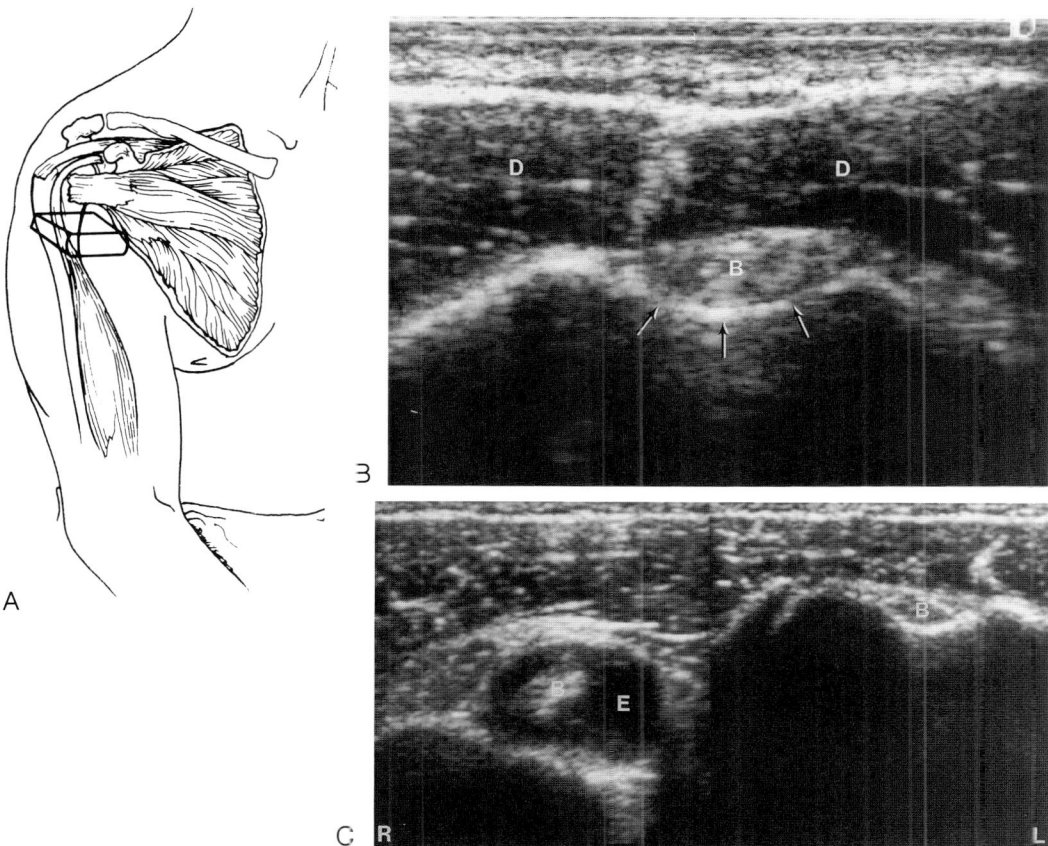

Fig. 9-1 Biceps tendon. Transverse views. **(A)** Anatomic drawing showing the position of the transducer. **(B)** Sonogram perpendicular to the axis of the tendon demonstrates the bicipital groove (*arrows*) containing the echogenic oval biceps tendon (B). The deltoid muscle (D) overlies the tendon. **(C)** Paired images of the biceps tendon (B) demonstrate an edematous right biceps tendon, which is surrounded by a large effusion (E). R, right side; L, left side.

ages are recorded at several levels (musculotendinous junction, midtendon, and rotator interval). Middleton et al demonstrated that small amounts of intra-articular fluid surrounding the biceps tendon can be visualized on scans done in this position.[8] Care should be taken that excessive transducer pressure does not compress a fluid-filled tendon sheath.

LONGITUDINAL VIEW (POSITION 2)

The biceps tendon is viewed parallel to its axis by scanning at 90 degrees to position 1. The tendon will appear as hyperechoic parallel lines (Fig. 9-2). If the transducer is not maintained parallel to the tendon, portions of the tendon may appear artifactually hypoechoic.[14] Often, the transducer must be rocked slightly to ensure complete visualization.

Subscapularis Tendon

LONGITUDINAL VIEW (POSITION 3)

Scanning proximally along the humeral shaft from position 1 will visualize the subscapu-

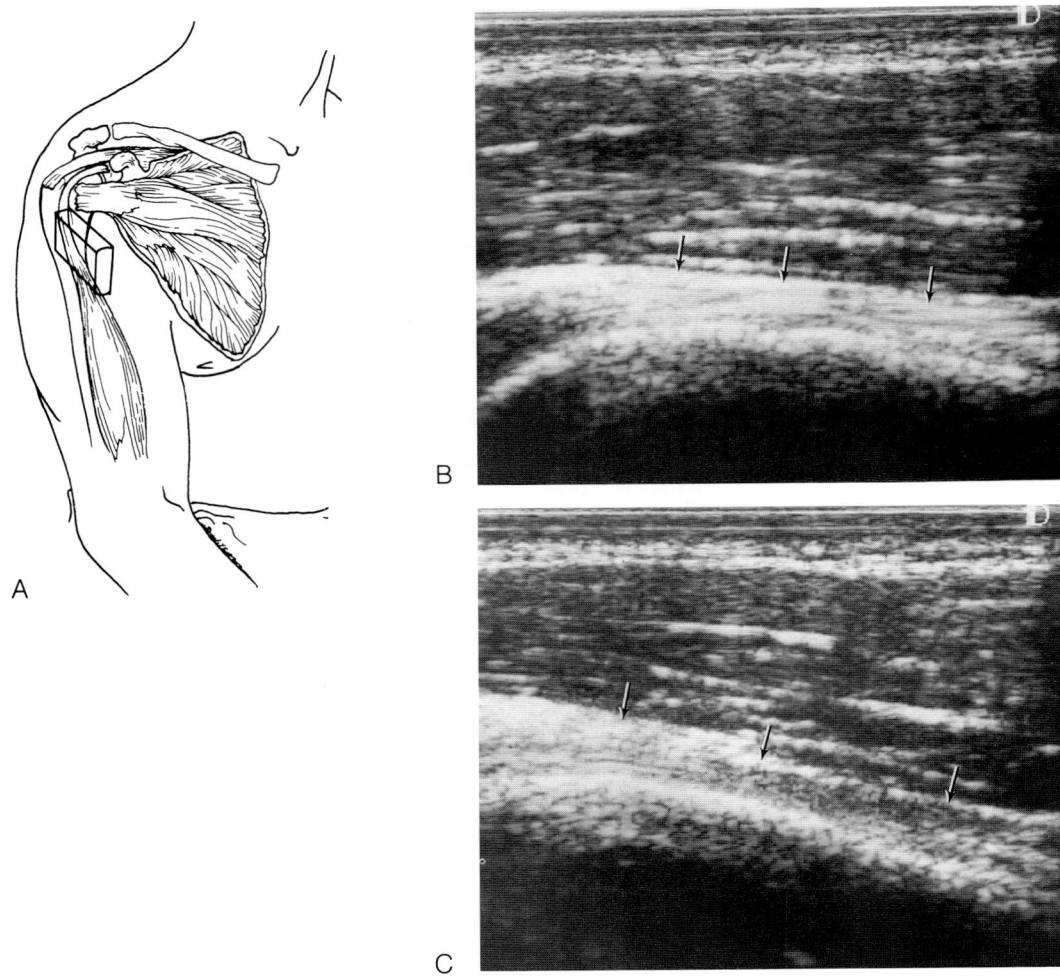

Fig. 9-2 Biceps tendon. Longitudinal views. **(A)** Anatomic drawing showing the position of the transducer. **(B)** Sonogram parallel to the axis of the biceps tendon shows the tendon (*arrows*) as a hyperechoic structure with multiple linear internal septa. The soft tissue overlying this structure is the deltoid muscle. **(C)** When the transducer is tilted to place the tendon (*arrows*) out of the specular condition, the lower half of the tendon is not well visualized.

laris tendon in a plane that is longitudinal (parallel) to its axis (Fig. 9-3). This tendon appears as a band of medium-level echoes deep to the thin, convex echogenic line representing the subdeltoid bursa. Scanning during passive internal and external rotation with the arm adducted may be helpful in assessing the integrity of the subscapularis tendon.

TRANSVERSE VIEW (POSITION 4)

The subscapularis tendon is visualized perpendicular to its fibers by turning the transducer 90 degrees to position 3 (Fig. 9-4). In this view, the tendon appears oval. The transverse view may be helpful in detecting tendinous attrition in patients with chronic anterior dislocation.

ROTATOR CUFF 117

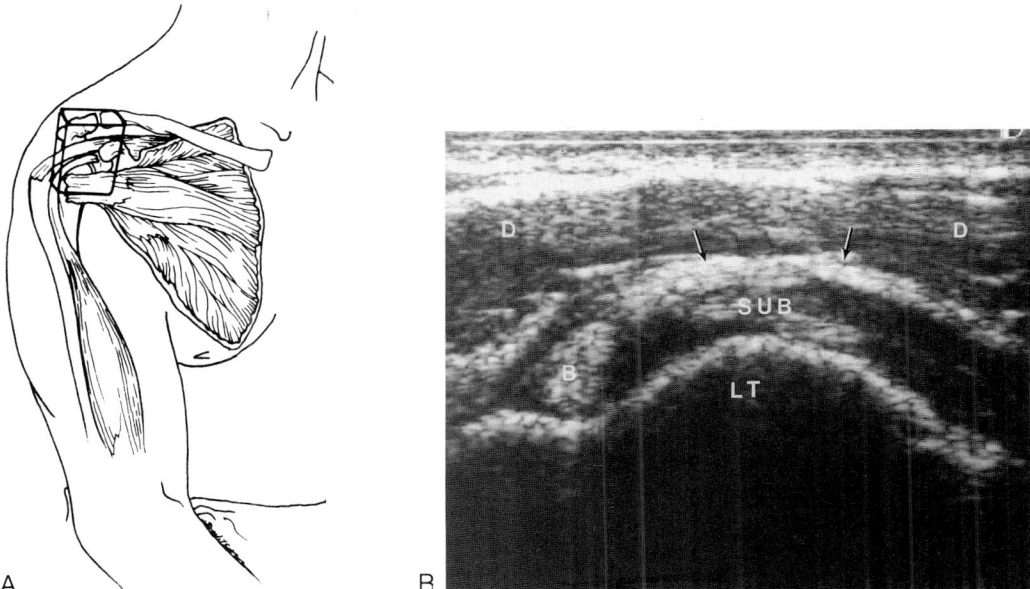

Fig. 9-3 Subscapularis tendon, longitudinal view. **(A)** Anatomic drawing showing the position of the transducer. **(B)** Sonogram parallel to the axis of the subscapularis tendon demonstrates the tendon (SUB) as a band of soft tissue deep to the subdeltoid bursa (*arrows*) and medial to the biceps tendon (B). The bony contour of the lesser tuberosity (LT) is also noted. D, deltoid muscle.

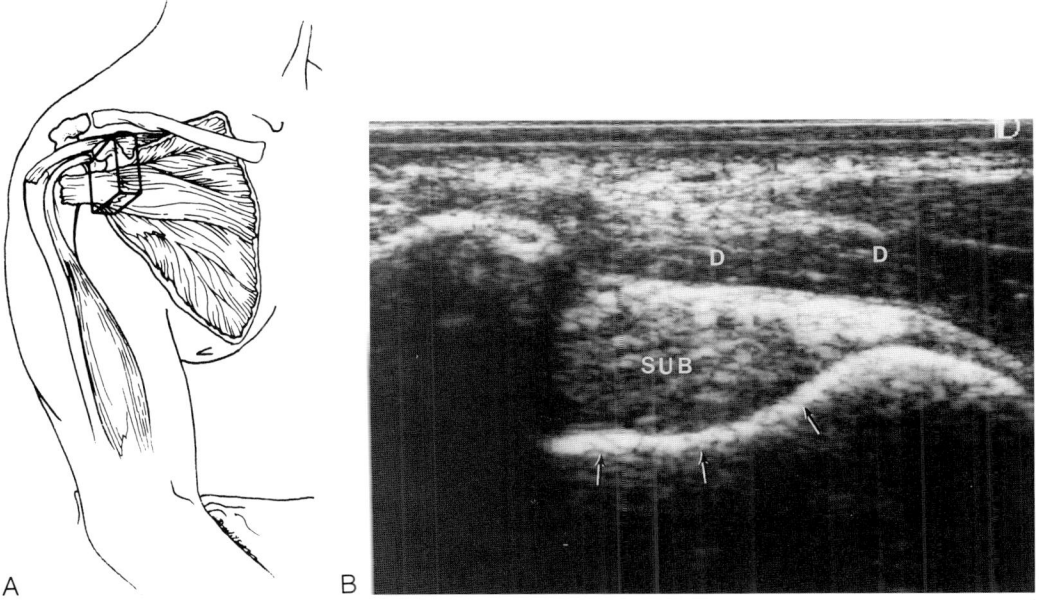

Fig. 9-4 Subscapularis tendon, transverse view. **(A)** Anatomic drawing showing the position of the transducer. **(B)** Sonogram perpendicular to the axis of the subscapularis tendon demonstrates the tendon as an oval soft-tissue structure (SUB) superficial to the humerus (*arrows*) and deep to the deltoid muscle (D).

Supraspinatus Tendon

TRANSVERSE VIEW (NEUTRAL POSITION) (POSITION 5)

Moving the transducer laterally and posteriorly from position 3 will visualize the supraspinatus tendon as a band of medium-level echoes deep to the subdeltoid bursa and superficial to the bright echoes originating from the bony surface of the greater tuberosity (Fig. 9-5). The articular cartilage of the humeral head may be visualized as a hypoechoic band just superficial to the bony margin. The sonographic window for visualization of this structure is limited by the acromion. Therefore, careful transducer positioning is essential for successful imaging. Images that demonstrate the critical zone, the portion of the tendon that begins approximately 1 cm posterolateral to the biceps tendon and that is most susceptible to injury, are especially important because failure to adequately visualize this area can result in false-negative findings. Imaging lateral to the apex of the greater tuberosity can result in false-positive findings.

LONGITUDINAL VIEW (NEUTRAL POSITION) (POSITION 6)

When the supraspinatus tendon is viewed parallel (longitudinal) to its axis by turning the transducer 90 degrees to position 5, it appears as a beak-shaped soft-tissue structure extending from under the acromion, which casts an acoustic shadow, to its attachment along the greater tuberosity (Fig. 9-6). The bright linear echoes from the subdeltoid bursa identify the superficial margin of the supraspinatus tendon. Passive abduction and adduction with the patient's palm parallel to the trunk is often very helpful in assessing the integrity of the supraspinatus tendon. It is important to scan lateral to the greater tuberosity so that small effusions in the subdeltoid bursa can be visualized. Care must be taken not to use excessive transducer pressure as small amounts of fluid may be compressed and therefore not visualized.

VIEWS IN EXTENSION AND INTERNAL ROTATION OF THE ARM (POSITIONS 7 AND 8)

After the supraspinatus tendon is scanned in neutral position, it should then be scanned with the arm in extension and internal rotation.[10] This position is best achieved by placing the patient's arm behind his or her back. It is important to remember that when the arm is in this position, the greater tuberosity (and thus the supraspinatus attachment) is more anteriorly located than when the arm is in the neutral position. The supraspinatus tendon is scanned perpendicular (position 7) and parallel (position 8) to its fibers (Fig. 9-7). Often, tendons that were obscured by a laterally placed acromion can be more fully visualized. Small tears and effusions are often accentuated with this maneuver.

Infraspinatus Tendon (Position 9)

The infraspinatus tendon is visualized by moving the transducer posteriorly from position 5 to a plane parallel to the scapular spine (Fig. 9-8). A common error is to use a more vertically oriented transducer position, which results in confusing images. The infraspinatus tendon is visualized as a beak-shaped soft-tissue structure that progressively thins as it approaches its attachment to the posterior aspect of the greater tuberosity. Passive internal and external rotation may be helpful in confirming the integrity of the infraspinatus tendon. Additional structures visualized in this scan position include the posterior glenoid labrum, seen as a hyperechoic, triangular structure deep to the tendon, and the articular cartilage of the humeral head, imaged as a thin, hypoechoic layer superficial to the high-level echoes originating from the bony surface.

ROTATOR CUFF 119

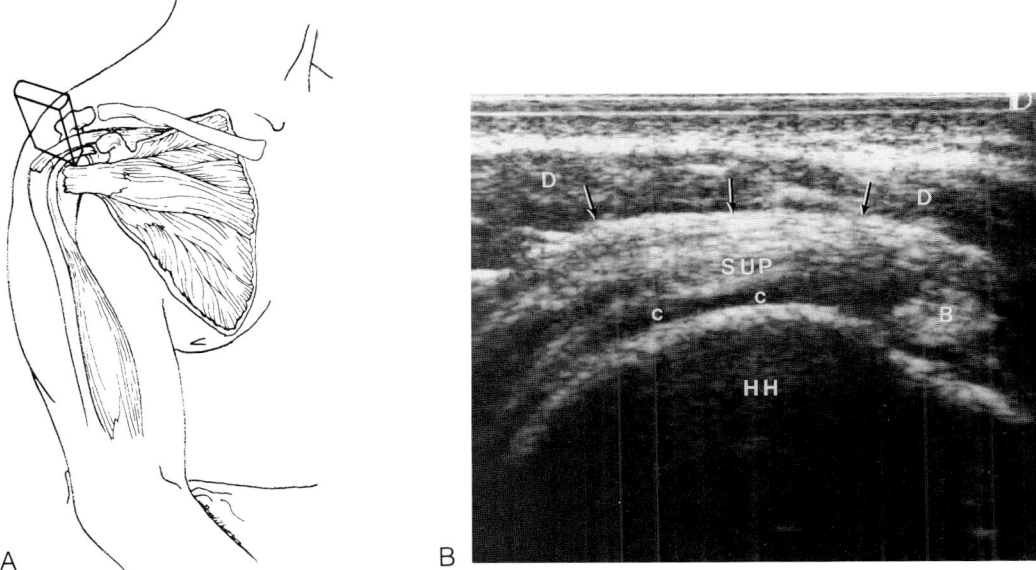

Fig. 9-5 Supraspinatus tendon, transverse view. **(A)** Anatomic drawing showing the position of the transducer. **(B)** Sonogram demonstrates the supraspinatus tendon (SUP) as a band of soft tissue deep to the subdeltoid bursal complex (*arrows*) and superficial to the cartilage of the humeral head (c). B, biceps tendon; D, deltoid; HH, humeral head.

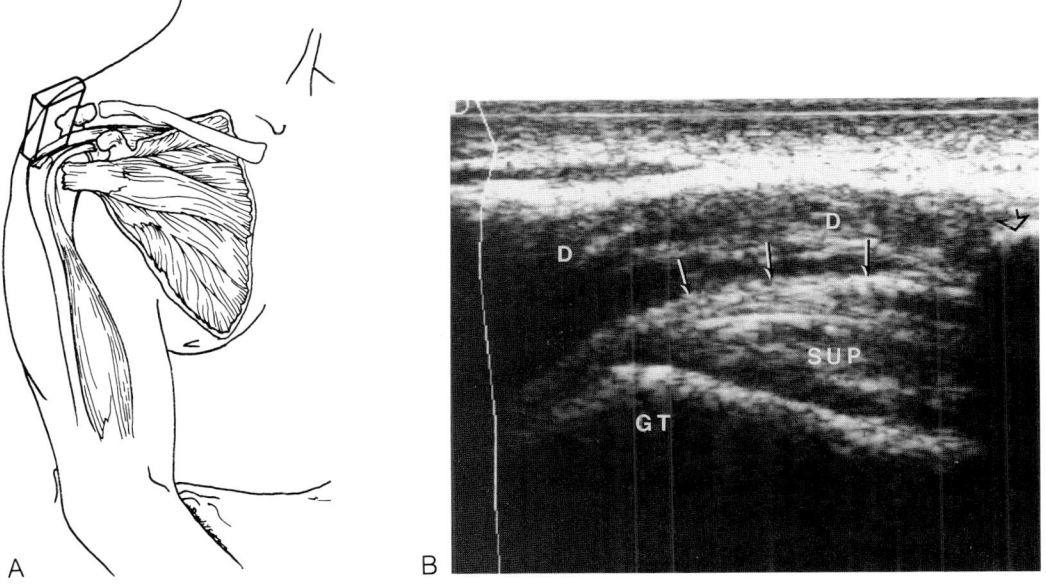

Fig. 9-6 Supraspinatus tendon, longitudinal view. **(A)** Anatomic drawing showing the position of the transducer. **(B)** Sonogram obtained at 90 degrees to view in Figure 5B demonstrates the supraspinatus tendon (SUP) as a beak-shaped soft-tissue structure extending out from under the acoustic shadow cast by the acromion (*open arrow*) and attaching to the greater tuberosity (GT). *Arrows*, subdeltoid bursal complex; D, deltoid.

120 MUSCULOSKELETAL ULTRASOUND

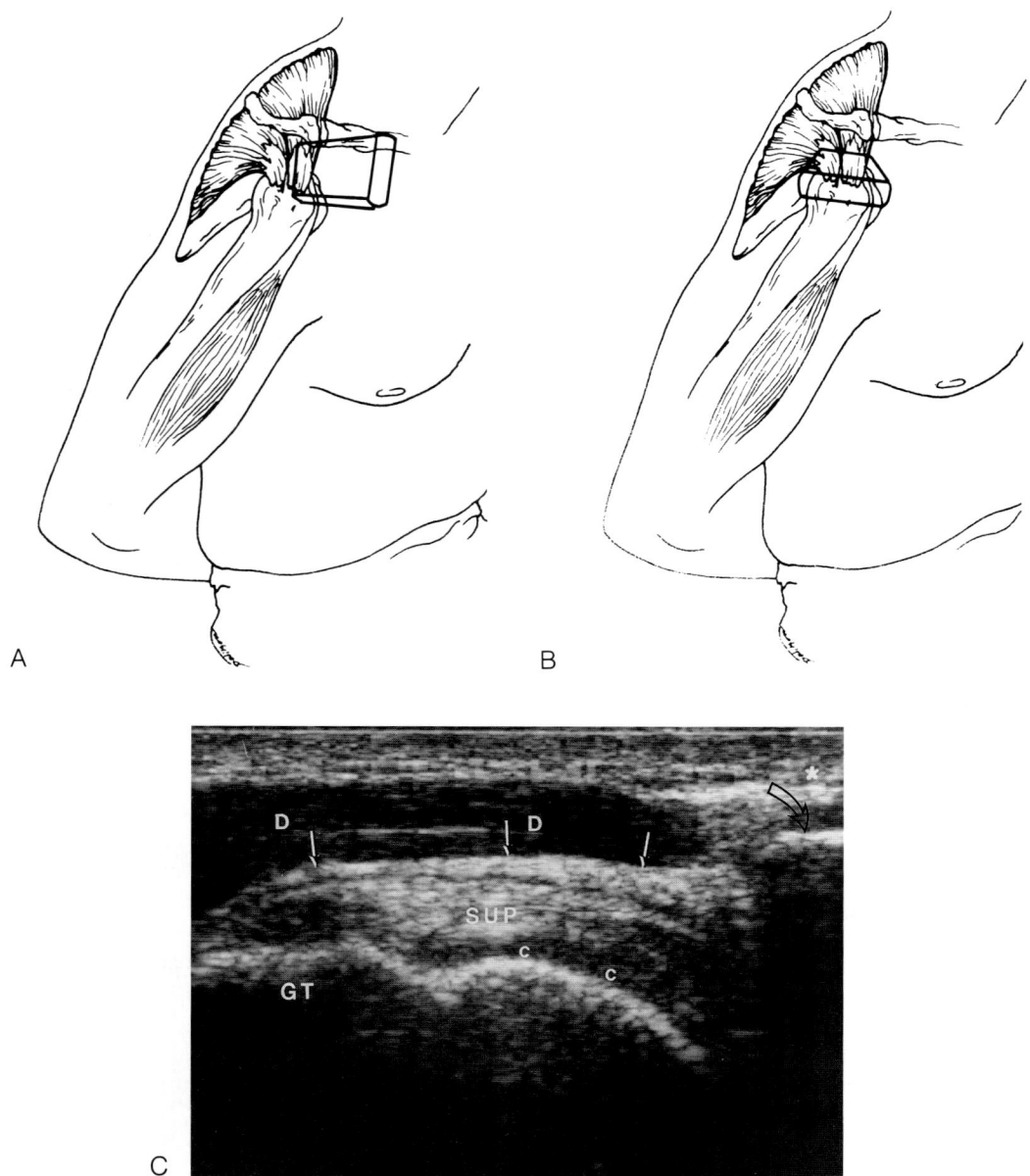

Fig. 9-7 Supraspinatus tendon. Views in extension and internal rotation of the arm. **(A, B)** Anatomic drawings showing the position of the transducer for **(A)** longitudinal and **(B)** transverse sonograms of the tendon. **(C)** Sonogram demonstrates improved visibility of the supraspinatus tendon (SUP) in the same subject scanned in Figure 9-6B. More of the tendon is seen. Views perpendicular to the axis of the tendon (not shown) are similar to those in neutral position (Figure 9-5B). *Curved open arrow,* acromion; *straight arrows,* subdeltoid bursal complex; c, humeral head cartilage; D, deltoid muscle; GT, greater tuberosity.

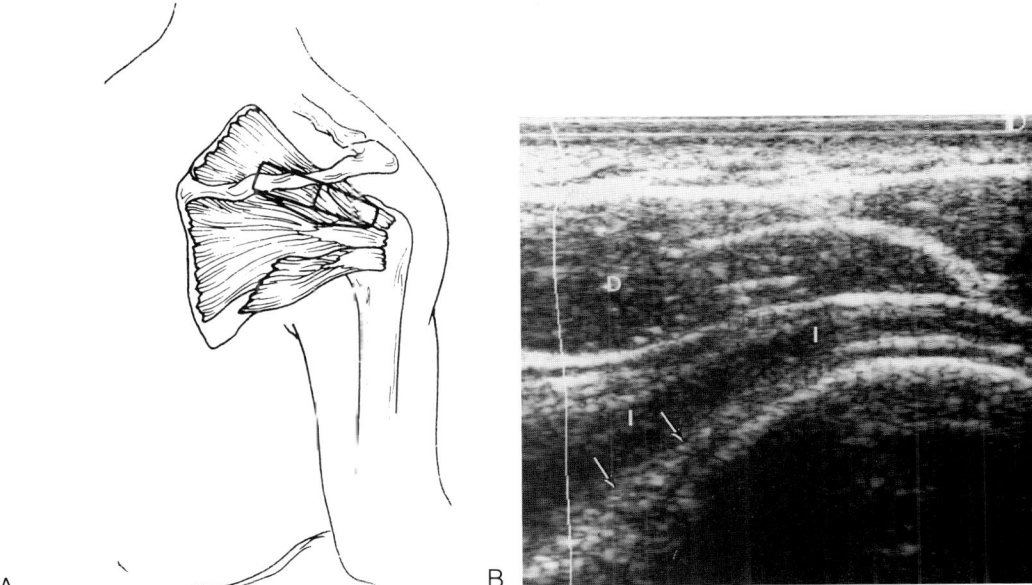

Fig. 9-8 Infraspinatus tendon, longitudinal view. **(A)** Anatomic drawing showing the position of the transducer. **(B)** Sonogram demonstrates the infraspinatus tendon (I) as a beak-shaped structure coursing over the humeral head. *Arrows,* posterior glenoid labrum; D, deltoid muscle.

Teres Minor Tendon (Position 10)

Moving the transducer distally along the humerus from position 9 and keeping it parallel to the scapular spine will visualize the teres minor tendon as a trapezoidal soft-tissue structure. Its oblique internal echoes differentiate the teres minor from the infraspinatus tendon, with its more horizontal internal echoes. Recent reports have suggested that very small intra-articular effusions may be best visualized at this level.[15] Although tears of the teres minor tendon are very rare, its visualization ensures that the entire infraspinatus tendon has been scanned.

ROTATOR CUFF TEAR

Diagnostic Criteria

As suggested by Middleton, previously published criteria for the diagnosis of rotator cuff tears can be categorized into four groups: (1) nonvisualization of the cuff, (2) localized absence or focal nonvisualization, (3) discontinuity, and (4) focal abnormal echogenicity.[16] The efficacy of these criteria is supported by a recent report of Weiner and Seitz.[17] The following discussion of the sonographic appearances of rotator cuff pathologies will be based on these groups.

In patients with large or massive rotator cuff tears, no cuff tendon will be visualized. The absence of cuff tendon allows the subdeltoid bursa to directly approximate the surface of the humeral head or humeral articular cartilage. In the longitudinal supraspinatus view, such tears will result in a characteristic concave configuration of the subdeltoid bursa (Fig. 9-9). The subdeltoid bursa may be quite thickened in this context, measuring up to 5 mm in width. Passive humeral movement (adduction and abduction) is often helpful in confirming the absence of cuff ten-

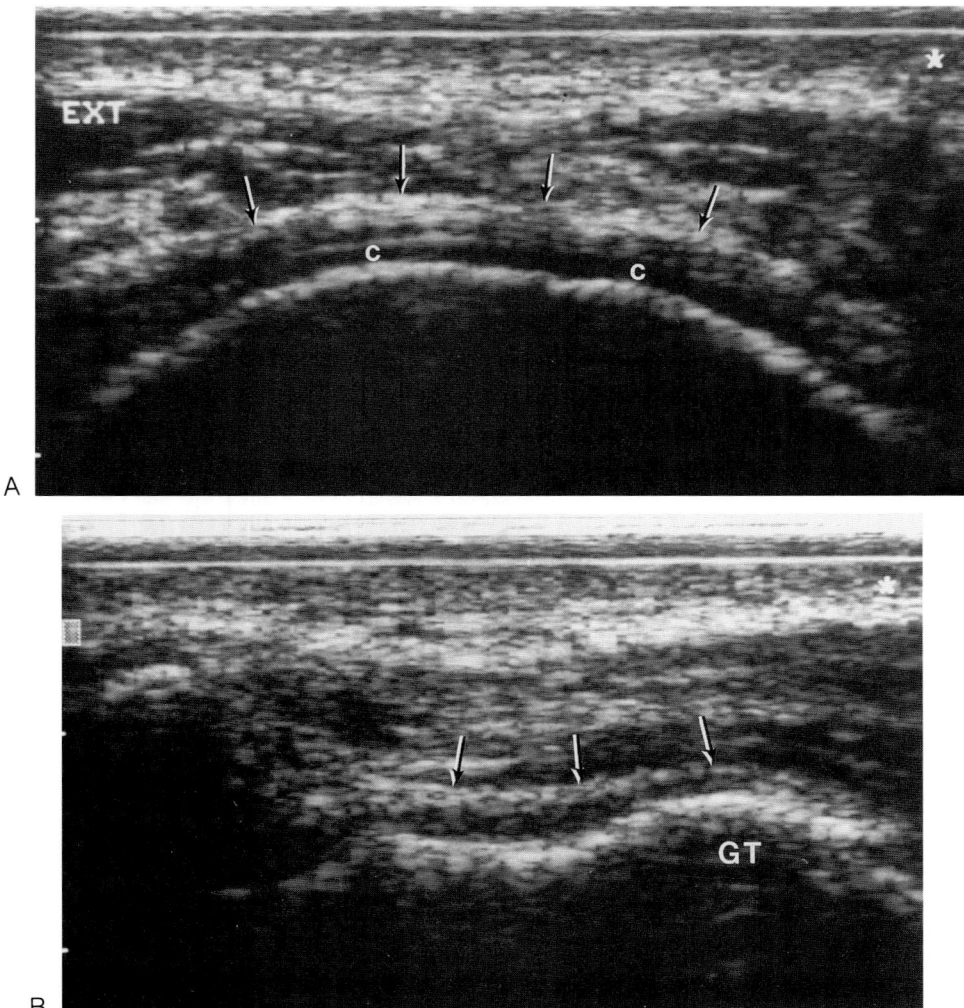

Fig. 9-9 Massive rotator cuff tear. **(A)** Transverse and **(B)** longitudinal sonograms in the region of the supraspinatus tendon indicate total absence of the tendon. The subdeltoid bursa (*arrows*) lies directly upon the cartilage of the humeral head (c). In the longitudinal plane, the subdeltoid bursa (*arrows*) assumes a characteristic concave configuration. GT, greater tuberosity.

don. Normal movement of the greater tuberosity will not be accompanied by motion of the supraspinatus tendon. Characteristic contour changes of the bursa will also be seen during such maneuvers as the subdeltoid bursa is elevated by the greater tuberosity. Joint and bursal effusions commonly accompany large tears and will be seen along the biceps tendon and lateral to the greater tuberosity, respectively. Large tears of the supraspinatus tendon are often accompanied by disruption of other cuff tendons. Such tears may extend posteriorly to involve the infraspinatus tendon or anteriorly to involve the biceps tendon and the subscapularis tendon.

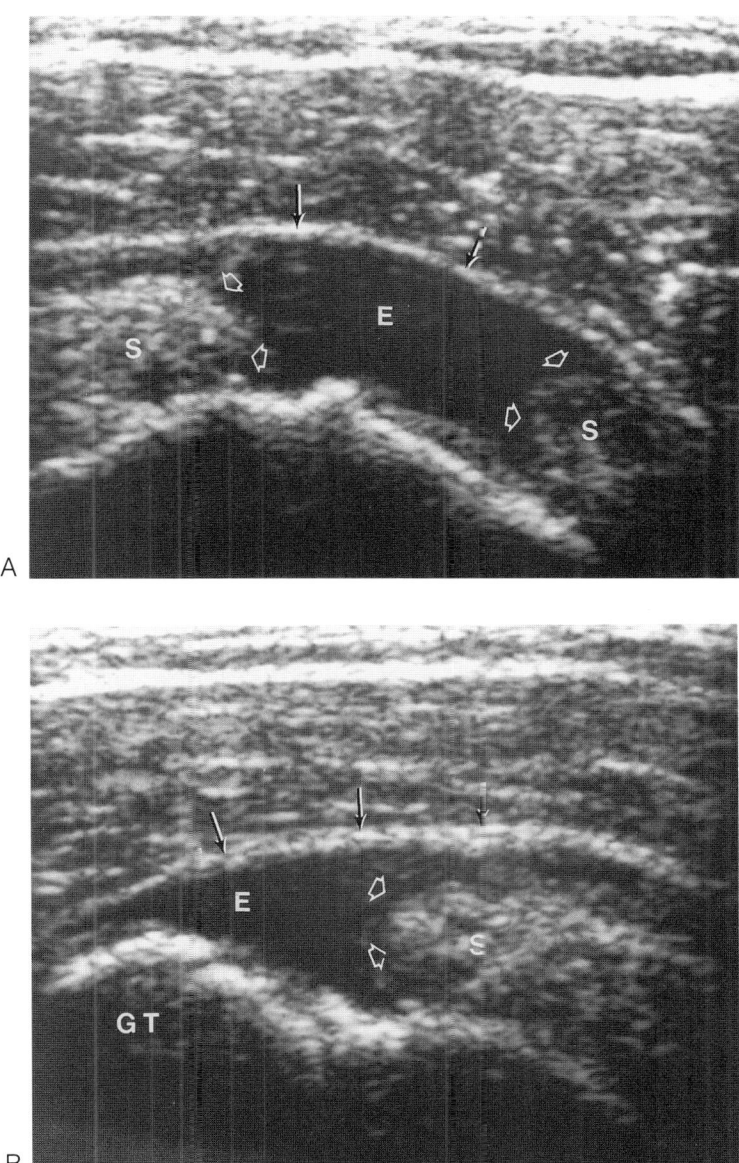

Fig. 9-10 Medium-sized rotator cuff tear. **(A)** Transverse and **(B)** longitudinal sonograms show a medium-sized tear filled with joint fluid (E). Note the echogenic margins of the residual cuff (*open arrowheads*). GT, greater tuberosity; S, residual supraspinatus tendon; *arrows,* subdeltoid bursal complex.

A smaller full-thickness tear will appear as a localized absence in the cuff, with the subdeltoid bursa touching the humeral surface (Figs. 9-10 and 9-11). This results in loss of the normal anterior arc of the subdeltoid bursa and the development of focal concavity. The most common location of such smaller tears is the anterolateral supraspinatus tendon in the critical zone. Characteristically, these tears begin approximately 1 cm

124 MUSCULOSKELETAL ULTRASOUND

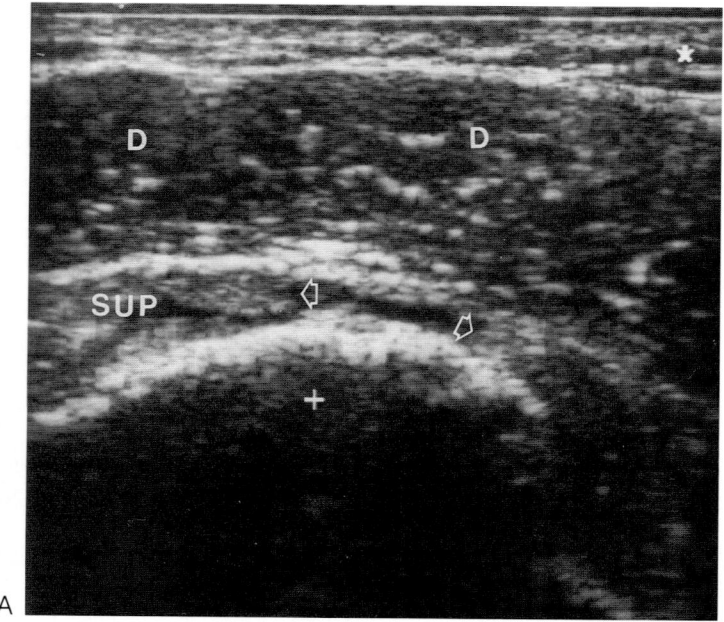

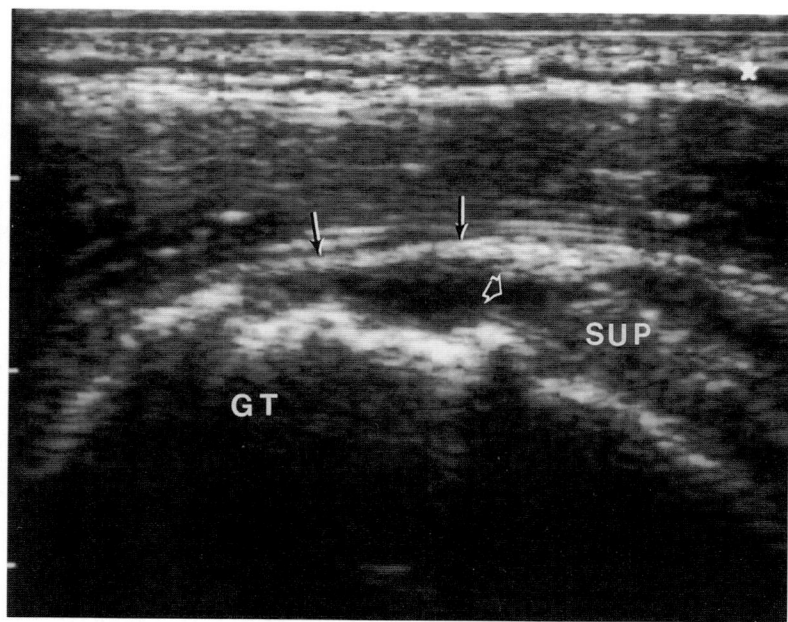

Fig. 9-11 Small rotator cuff tear. **(A)** Transverse and **(B)** longitudinal sonograms show a small tear. *Arrows,* subdeltoid bursal complex; *open arrows,* edges of cuff tear; D, deltoid; GT, greater tuberosity; SUP, supraspinatus tendon.

lateral to the biceps tendon. Because inadequate technique may result in images that mimic the appearance of small tears, such defects must be confirmed by visualization in two perpendicular scan planes. Tears will be sharply demarcated with abrupt transition from normal to abnormal cuff, in contrast to more diffuse cuff thinning, which is often seen in advanced fiber failure. Visualization of these smaller tears is often best with the arm in extension and internal rotation. In this position, tension on the cuff will accentuate the defect, and often small amounts of joint or bursal effusion will be displaced into the defect, further highlighting the tear.

Even smaller tears will appear as a discontinuity of the cuff filled with hypoechoic joint fluid or hyperechoic reactive tissue, which highlights the defect (Fig. 9-12). As noted above, small cuff lesions are often best visualized during extension and internal rotation. A small amount of bursal fluid is commonly present in these smaller tears and may be the only abnormal sonographic finding with the arm in neutral position.

Abnormalities of cuff echogenicity may be diffuse or focal. Diffuse abnormality of cuff echogenicity is an unreliable sonographic sign of cuff tear but may indicate diffuse cuff inflammation or fibrosis. Use of paired images (Fig. 9-13) to confirm that a disparity is real and not secondary to geometric artifacts is important in avoiding false-positive results (see section on pitfalls). Although diffuse abnormality of cuff echogenicity is an unreliable indicator of full-thickness tears, significant disparity in cuff thickness should suggest cuff attrition. Significant degrees of cuff attrition are associated with partial-thickness tears and also suggest that progressive cuff changes are present.

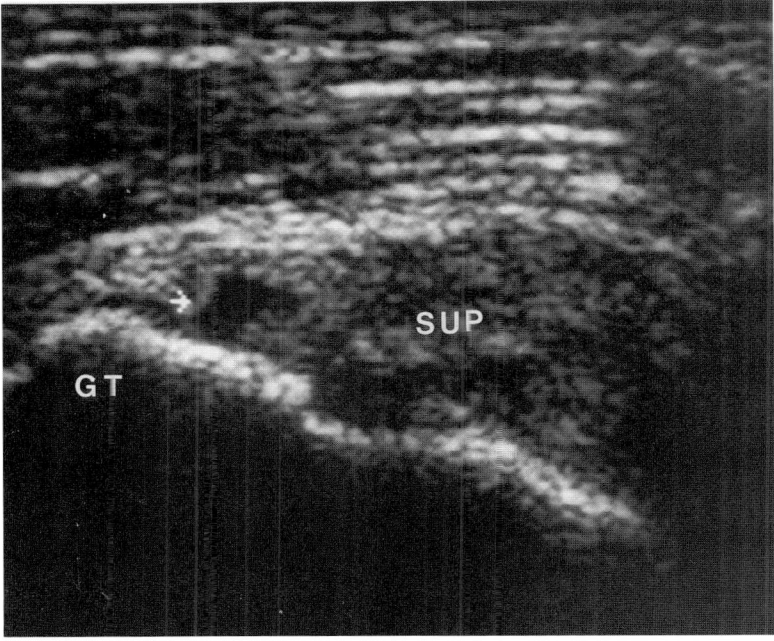

Fig. 9-12 Very small rotator cuff tear. Longitudinal sonogram of the supraspinatus tendon (SUP) demonstrates a 4-mm tear filled with joint fluid (*arrow*). GT, greater tuberosity.

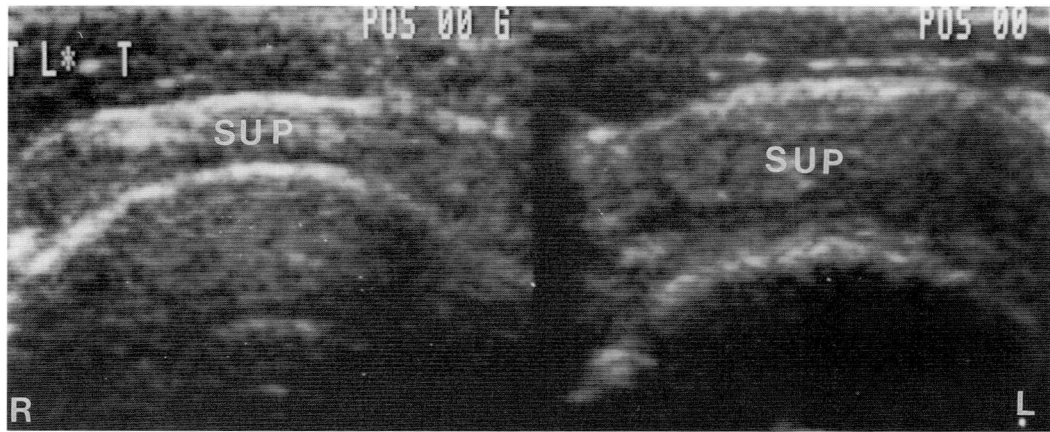

Fig. 9-13 Cuff attrition. Paired transverse sonograms of the supraspinatus tendon (SUP) demonstrate thinning of the right cuff (R) with increased echogenicity and irregularity of the superior margin. This thinned cuff became completely ruptured over the next 3 years. L, normal left cuff.

In contrast, focal abnormal echogenicity has been associated with small full- and partial-thickness tears. Crass et al reported that the area of increased echogenicity results from granulation tissue, hypertrophied synovium, and hemorrhage.[9] More recently, the visualization of either an intratendinous hypoechoic focus or a dominant echogenic focus has been suggested to be a criterion for diagnosis of partial-thickness tears.[17]

In addition to the major criteria described above, several minor sonographic findings have proven helpful in our experience. The most common and reliable of these secondary sonographic findings is the visualization of a subdeltoid bursal effusion (Fig. 9-14). The identification of a small effusion may be the only abnormal sonographic finding in patients with small tears. The importance of this finding increases when intra-articular fluid is also visualized.[8] Bursal fluid is most easily seen lateral to the greater tuberosity, especially with the arm in extension and internal rotation. Care must be taken not to compress or displace small amounts of fluid, as a false-negative scan could result. In the presence of an otherwise normal sonogram but appropriate clinical symptoms, visualization of an effusion should always be followed by additional evaluation using arthrography or MRI.

The shape of the supraspinatus tendon and its position relative to adjacent structures also may be helpful. In normal subjects, the bright linear echoes from the subdeltoid bursa are upwardly convex. In patients with partial attrition or full-thickness tears, concavity of the subdeltoid bursal contour may be noted.[11] In patients with large tears, the humeral head may be elevated relative to the acromion, obliterating the space normally occupied by the supraspinatus tendon. In such patients, the acromial contour also may be altered when compared with the normal side. Both of these sonographic findings correspond to similar plain film findings in patients with large, chronic tears.

Postoperative Appearances

Sonography also plays an important role in the evaluation of patients after acromioplasty and, especially, repair of full-thickness tears.[18,19] Symptoms, which are common in

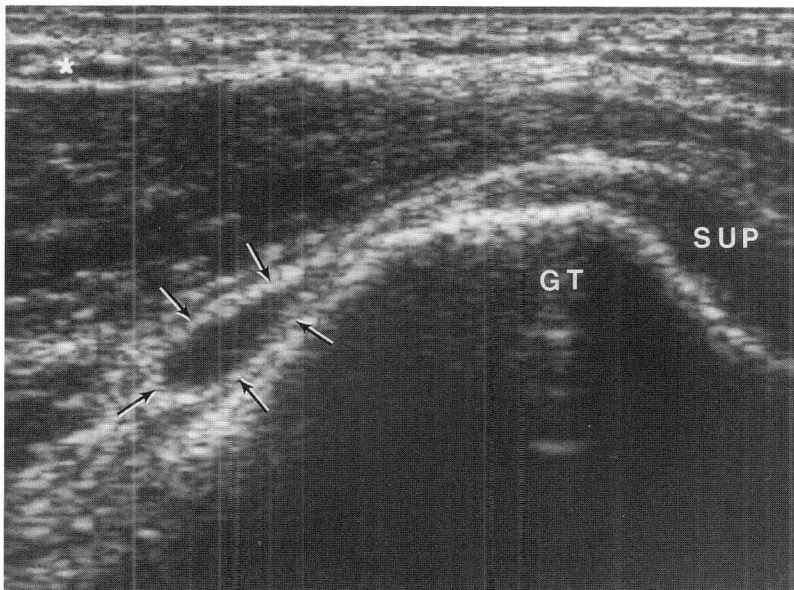

Fig. 9-14 Subdeltoid bursal effusion. Sonogram parallel to the axis of the supraspinatus tendon (SUP) and lateral to the greater tuberosity (GT) demonstrates a small amount of bursal fluid (arrows) in a characteristic teardrop shape.

this group of patients, can result from a variety of causes, including recurrent tendinitis or impingement. In addition, a recent report by Harryman et al demonstrated a high rate of recurrent rotator cuff tear, especially when the original tear was large.[20] Differentiating among these possibilities on the basis of clinical or physical findings is often difficult. Other imaging modalities, including arthrography and MRI, are less reliable after rotator cuff repair than preoperatively and may produce false-negative or false-positive results.[3,21] However, several studies have demonstrated that sonography may be helpful in this difficult group of patients.[18,19]

The postoperative patient is far more difficult to evaluate sonographically than the preoperative patient because surgery distorts important sonographic landmarks. For this reason, the surgical procedures used in acromioplasty and cuff repair and the resulting sonographic appearances must be understood if imaging is to be successful.

In decompression acromioplasty, the surgeon removes the anterior, inferior aspect of the acromion to provide additional space for the underlying supraspinatus tendon. This surgery results in a more pointed, irregular configuration than the rounded, smooth contour of the normal acromion. Postoperative changes may mimic the acromial findings in chronic, large rotator cuff tears with a high-riding humerus. Because the inferior aspect of the acromion is removed, more of the supraspinatus tendon may be visualized.

During the repair of a full-thickness cuff tear, the cuff tendons are reimplanted into a trough made perpendicular to the axis of the supraspinatus tendon on the greater tuberosity, rather than being attached directly to the surface of the greater tuberosity.[1] The reimplantation trough is placed in a location that provides optimal tension for the remaining tendons; therefore, it may be medial to a variable degree, reflecting the amount of cuff tendon that is preserved. The trough appears

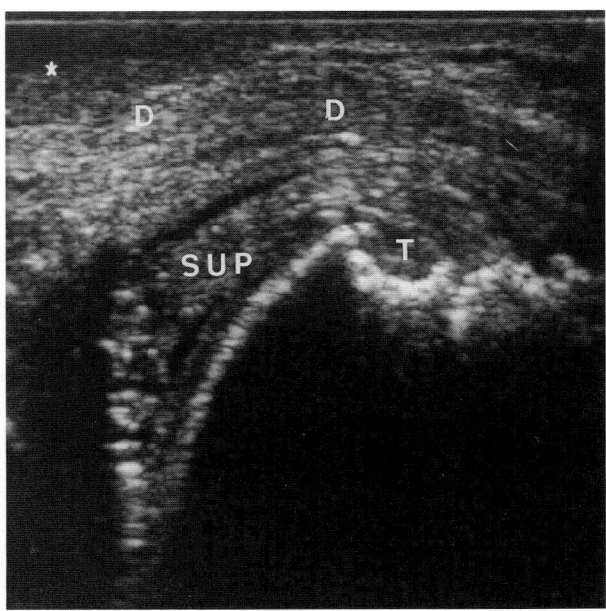

Fig. 9-15 Postoperative evaluation of rotator cuff repair. Postoperative sonogram parallel to the axis of the supraspinatus tendon in a patient with a repaired tear of the left rotator cuff demonstrates the reimplantation trough (T). Note the absence of the subdeltoid bursal complex. The supraspinatus tendon (SUP) is nearly isoechoic compared with the overlying deltoid muscle (D). In this instance, dynamic scans are helpful in confirming recurrent cuff tears.

sonographically as a rounded defect in the humeral contour when the supraspinatus tendon is viewed longitudinally (Fig. 9-15). The nonresorbable sutures used to secure the tendon are often visualized as specular echoes deep in the trough. Because of the trough's medial location following the repair of large tears, scanning with the arm in extension and internal rotation may be necessary to visualize the site of tendon reimplantation. Failure to scan in this position can lead to a false-positive diagnosis. Such a maneuver should be used with care, however, especially in the immediate postoperative period, to avoid re-injury to the friable newly reimplanted tendons.

Sonographic appearances of the cuff tendons are abnormal in the postoperative patient, especially after cuff repair. These tendons, especially the supraspinatus, are often echogenic and thinned compared with the contralateral normal shoulder.[18,19] The subdeltoid bursa is usually resected during the repair of a cuff tear, thereby removing an important sonographic landmark separating deltoid from supraspinatus. Dynamic scanning with passive abduction and adduction motion is therefore important in identifying a thin but intact cuff, which may be isoechoic compared with adjacent deltoid muscle. Joint effusions are common postoperatively and best visualized along the biceps tendon. Although extra-articular effusions may occasionally be identified, a restricted site for fluid collection is disrupted by bursal resection, making visualization more difficult.

Recurrent cuff tear is common, occurring in 20 percent of patients in whom a defect limited to the supraspinatus tendon was repaired and in more than 50 percent of patients who

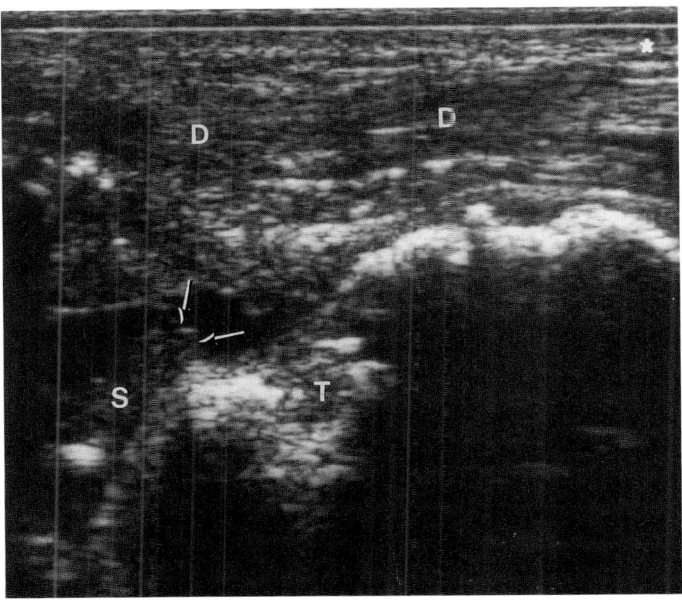

Fig. 9-16 Recurrent rotator cuff tear. Sonogram parallel to the axis of the left supraspinatus tendon demonstrates disruption of the supraspinatus tendon (S). The tip of the tendon (*arrows*) is distant from the reimplantation trough (T). A small amount of residual tendon is noted within the trough. D, deltoid.

preoperatively had large tears involving more than the supraspinatus.[20] Because recurrent tears are often large and multiple tendons are often involved, such tears most often appear as total absence of the cuff (Fig. 9-16). It is difficult to differentiate small recurrent tears from the appearance created when only a small amount of cuff tendon remains to be reattached unless baseline scans performed in the immediate postoperative period are available.

Cuff arthropathy is commonly seen in patients with recurrent cuff tear. The term was first used by Neer et al. to describe degenerative changes of humeral cartilage caused by repetitive minor trauma and disruption of normal nutrient pathways in patients with chronic cuff disruption.[22] Cuff arthropathy appears sonographically as irregularity of the bony surface of the humerus and loss of normal hypoechoic cartilage (Fig. 9-17).

TRAUMA

In patients with acute injury and weakness on abduction, it may be difficult to distinguish clinically between an acute greater tuberosity fracture and an acute rotator cuff tear. The distinction is important because the treatments differ vastly. Bassett and Cofield demonstrated that early repair of a cuff tear produces better postoperative results.[23] Therefore, diagnosis of a tear in the acute phase is important.

Diagnosis of fracture in the adult humerus is a new application for ultrasound. The bony contour is evaluated sonographically for abnormal shape or discontinuity. Sonographic diagnosis has proven useful in the shoulder, where radiographic evaluation of the region of the greater tuberosity may be limited because of the often irregular surface and the difficulty of obtaining tangential radiographs

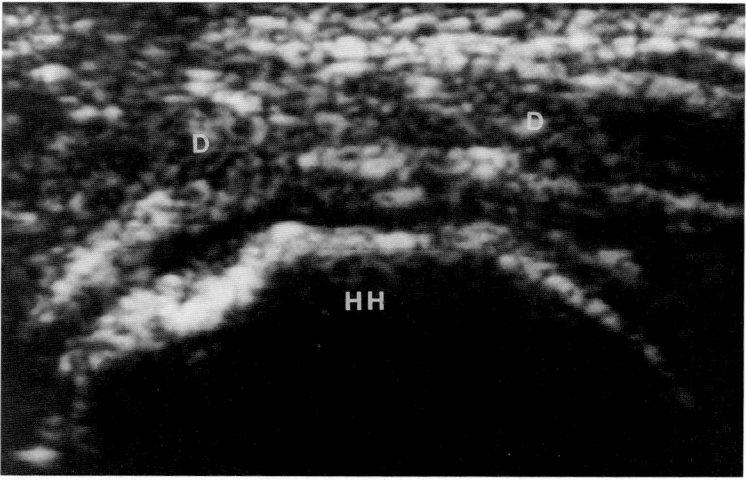

Fig. 9-17 Cuff arthropathy. Sonogram perpendicular to the axis of the supraspinatus tendon demonstrates irregularity of the humeral surface with absence of cartilage. Cuff arthropathy was noted at surgery. D, deltoid; HH, humeral head.

to demonstrate fracture. For this reason, fluoroscopy or computed tomography has traditionally been necessary to confirm such lesions. A greater tuberosity fracture appears sonographically as a discontinuity in the normally smooth bony surface (Fig. 9-18). The overlying supraspinatus tendon may be thickened secondary to edema or hemorrhage. Fat may be seen within the joint as a highly echogenic joint effusion that exhibits a characteristic swirling appearance on joint motion. Such findings correspond to a fat-fluid level that may be seen with other imaging techniques.

Patten et al demonstrated that such fractures could be readily seen on sonograms, whereas in 42 percent of cases they were not visible on initial plain films.[24] In a limited number of cases, the appearance of degenerative changes and calcific tendinitis mimicked characteristic findings of fracture. Therefore, these authors recommend that all positive sonograms be followed by further radiologic evaluation.

PITFALLS

Inadequate transducer positioning is the most common technical error in sonography of the rotator cuff. Most often, this reflects a failure to understand the complex anatomy of the shoulder. Positioning errors can lead to both false-positive and false-negative results. For example, scanning the supraspinatus tendon transversely with the transducer placed laterally may artifactually mimic a rotator cuff tear. More medial placement will demonstrate an intact supraspinatus. Similarly, longitudinal images of the supraspinatus tendon obtained with excess anterior angulation may also mimic a cuff tear, whereas correctly angled scans will demonstrate the intact tendon. For this reason, cuff tears, especially of the supraspinatus, should be viewed in two orthogonal planes whenever possible.

Failure to maintain proper transducer orientation with respect to tendon fibers causes an artifactually heterogeneous tendon appear-

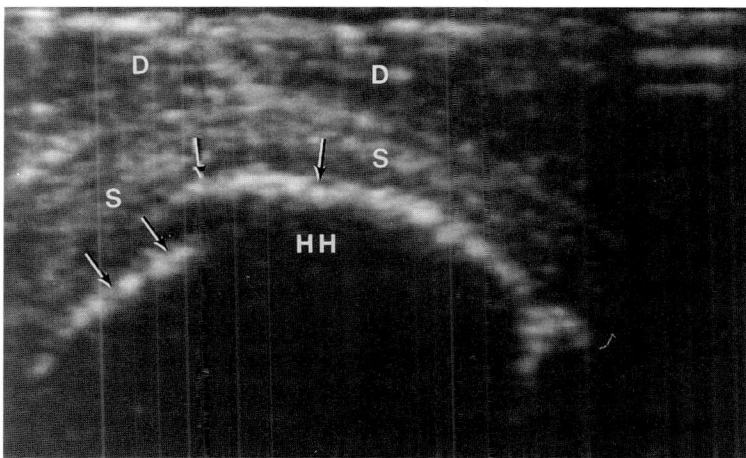

Fig. 9-18 Humeral head fracture. Sonogram demonstrates discontinuity of the greater tuberosity surface (*arrows*) deep to the supraspinatus tendon (S). D, deltoid; HH, humeral head.

ance. As demonstrated by Crass et al[25] and Fornage,[14] failure to orient the transducer parallel to the fibers of the tendon may result in artifactual areas of decreased echogenicity. When only a small area of the tendon is parallel to the transducer, a focal area of increased echogenicity may be produced, mimicking a small partial- or full-thickness tear. This artifact is especially pronounced with sector transducers.

The use of electronic transducers with multiple electronic focal zones leads to another possible technical error: multiple focal zones may hide characteristic acoustic shadows created by areas of calcific tendinitis. When such areas are encountered, it is important to use a single focal zone to distinguish between calcium in the tendon and focal linear areas of increased echogenicity. Even when this precaution is taken, the size of the calcifications may be sufficiently small to preclude shadowing.

Patient-related factors, including truncal obesity and large subdeltoid bursal effusions, may limit diagnostic information available from sonograms. In such patients, the lateral displacement of the transducer by excess soft tissue significantly narrows the sonographic window. Excess soft tissues also may require the use of lower-frequency transducers that will in turn limit the accuracy of the examination.

RESULTS

A number of authors have reported sensitivity and specificity above 90 percent in the sonographic evaluation of preoperative patients. However, other groups have had less success with this technique. The sonographic criteria used, the patient population studied, the diagnostic information desired, and operator experience and interest are crucial factors in explaining the wide variation in published results.[6–9,13,16,26–30]

Sonographic Criteria

A wide variety of sonographic criteria have been used in published reports. Failure to visualize the cuff is the least controversial criterion and has a high predictive value in the hands of most investigators. Focal absence of the cuff with approximation of bursa and humerus is nearly as reliable a criterion for full-thickness cuff tear. Cuff discontinuity is a less reliable criterion, with both false-negative and false-positive findings reported. Focally increased echogenicity is the most controversial criterion. This sonographic finding may have several causes and should be used only when clear asymmetry exists between shoulders.

Patient Population

The patient population at a given center also may influence results. Patients in a referral center for shoulder surgery can be expected to have a higher prevalence of pathology and more advanced lesions. Imaging in such patient populations, which were studied in the earlier investigations, can be expected to produce better results than in a center with younger patients who have a higher percentage of small full-thickness tears and partial-thickness tears.

Expectations

The type of information desired from rotator cuff sonography can vary widely. As is the case with MRI and arthrography, sonography cannot realistically be expected to diagnose all lesions of the cuff. The statistics from some of the less favorable reports suffer from overly broad expectations. This is especially true in the diagnosis of small partial-thickness cuff tears. Clinicians and sonologists must have realistic expectations regarding the utility of this (or any) imaging modality.[3] In a given patient, it may be necessary to employ a combination of techniques to define precisely the state of the rotator cuff.

Operator

Many authors have commented on the difficulty of performing rotator cuff sonography.[11-13] A relatively long learning curve exists as well. While this factor is difficult to quantitate, operator skill and experience undoubtedly play an important role in explaining the differences in outcomes reported in various publications. Similar disparities are also introduced by differences in scanners, transducers, and operating software used in various studies. Rapid improvements in high-resolution transducers and signal processing have increased the diagnostic capabilities of this application, making comparisons more difficult.

In postoperative patients, sonography is highly reliable. This may reflect the large size of recurrent tears. We previously reported that sonography was able to correctly diagnose all instances of recurrent cuff tear verified at surgery and to confirm 10 of 11 intact tendons.[18] Crass et al reported similar results when sonography was compared with surgical findings.[19]

CONCLUSIONS

Rotator cuff sonography has the potential to serve as a screening examination for patients with shoulder pain. It will be most useful in patients over 50 years of age who might be expected to have larger lesions. In younger patients with persistent symptoms, negative sonography should be followed by additional examinations. In the difficult group of postoperative patients, sonography is the best examination.

REFERENCES

1. Matsen FA III, Arntz CT: Subacromial impingement. In Rockwood CA, Matsen FA (eds): The Shoulder. Vol. 2. WB Saunders, Philadelphia, 1990

2. Resnick D: Shoulder arthrography. Radiol Clin North Am 19:243, 1981
3. Stiles RG, Otte MT: Imaging of the shoulder. Radiology 188:603, 1993
4. Seltzer SE, Finberg HJ, Weissman BN: Arthrosonography. Technique, sonographic anatomy and pathology. Invest Radiol 15:19, 1980
5. Farrar IL, Matsen FA III, Rogers JV et al: Dynamic sonographic study of lesion of the rotator cuff, abstracted. American Academy of Orthopaedic Surgeons 50th Annual Meeting, Anaheim, CA, March 10–15, 1983
6. Mack LA, Matsen FA III, Kilcoyne JF et al: Ultrasound evaluation of the rotator cuff. Radiology 157:205, 1985
7. Mack LA, Gannon MK, Kilcoyne JF, Matsen FA III: Sonographic evaluation of the rotator cuff. Accuracy in patients without prior surgery. Clin Orthop 234:21, 1988
8. Middleton WD, Reinus WR, Totty WF et al: Ultrasonographic evaluation of the rotator cuff and biceps tendon. J Bone Joint Surg 68A:440, 1986
9. Crass JR, Craig EV, Feinberg SB: Ultrasonography of rotator cuff tears: a review of 500 diagnostic studies. J Clin Ultrasound 16:313, 1988
10. Crass JR, Craig EV, Feinberg SB: The hyperextended internal rotation view in rotator cuff ultrasound. J Clin Ultrasound 15:416, 1987
11. Middleton WD: Ultrasonography of the shoulder. Radiol Clin North Am 30:927, 1992
12. Katthagen B-D: Anatomy as revealed by sonography. p. 45. In Ultrasonography of the Shoulder. Technique, Anatomy, Pathology. Thieme Medical Publishers, New York, 1990
13. Mack LA, Nyberg DA, Matsen FA III: Sonographic evaluation of the rotator cuff. Radiol Clin North Am 26:161, 1988
14. Fornage BD: The hypoechoic normal tendon. A pitfall. J Ultrasound Med 6:19, 1987
15. van Holsbeeck M, Introcaso J, Hoogmartens M: Sonographic detection and evaluation of shoulder joint effusion, abstracted. Radiology 177(P):214, 1990
16. Middleton WD: Status of rotator cuff sonography. Radiology 173:307, 1989.
17. Weiner SN, Seitz WH: Sonography of the shoulder in patients with tears of the rotator cuff: accuracy and value for selecting surgical options. AJR 160:103, 1993
18. Mack LA, Nyberg DA, Matsen FA III et al: Sonography of the postoperative shoulder. AJR 150:1089, 1988
19. Crass JR, Craig EV, Feinberg SB: Sonography of the postoperative rotator cuff. AJR 146:561, 1986
20. Harryman DDT II, Mack LA, Wang KY et al: Rotator cuff repair: correlation of functional results with cuff integrity. J Bone Joint Surg 73A:982, 1991
21. Calvert PT, Packer WP, Stoker DJ et al: Arthrography of the shoulder after operative repair at the torn rotator cuff. J Bone Joint Surg 68B:147, 1986
22. Neer CS II, Craig EV, Fukuda H: Cuff tear arthropathy. J Bone Joint Surg 65A:1232, 1983
23. Bassett RW, Cofield RH: Acute tears of the rotator cuff: the timing of surgical repair. Clin Orthop 175:18, 1983
24. Patten RM, Mack LA, Wang KY, Lingel J: Nondisplaced fractures of the greater tuberosity of the humerus: sonographic detection. Radiology 182:201, 1992
25. Crass JB, Van de Vegte GL, Harkavy LA Tendon echogenicity: ex vivo study. Radiology 167:499, 1988
26. Hodler J, Fretz CJ, Terrier F, Gerber C: Rotator cuff tears: correlation of sonographic and surgical findings. Radiology 169:791, 1988
27. Brandt TD, Cardone BW, Grant TH et al: Rotator cuff sonography: a reassessment. Radiology 169:791, 1989
28. Furtschegge A, Resch H: Value of ultrasonography in preoperative diagnosis of rotator cuff tears and postoperative follow-up. Eur J Radiol 8:69, 1988
29. Miller CL, Karasick D, Kurtz AB, Fenlin JM: Limited sensitivity of ultrasound for the detection of rotator cuff tear. Skeletal Radiol 18:179, 1989
30. Soble MG, Kaye AD, Guay RC: Rotator cuff tear: clinical experience with sonographic detection. Radiology 173:319, 1989

10
Elbow

Lori L. Barr

ANATOMY

The elbow is a large synovial hinge joint that allows the proximal radius and ulna to articulate with the distal humerus. In proximity is the radioulnar joint, which enables pronation and supination of the hand. The proximal radius and ulna are attached by the annular ligament of the radius. The elbow and radioulnar joints share a common fibrous capsule and synovial cavity. Skeletal maturation at the elbow is more rapid in girls than in boys: the capitellum is visible at 4 to 6 months in girls compared with 18 months in boys, and complete skeletal maturation occurs by 15 years in girls compared with 16 years in boys.[1] Ossifying epiphyses begin as spherical, ovoid, or irregular sites of calcification within the cartilaginous preformed bone ends. The articular surfaces of the elbow include the trochlea and capitellum of the humerus, the trochlear notch of the ulna, and the radial head. The trochlea of the humerus articulates with the trochlear notch of the ulna; the concave upper surface of the radial head articulates with the capitellum; and the raised margin of the radial head articulates with the capitellotrochlear groove. The articular surfaces are most fully in contact when the forearm is halfway between full pronation and supination; i.e., with the elbow flexed at 90 degrees.[2]

The relatively weak fibrous capsule of the elbow is fortified laterally by the radial and ulnar collateral ligaments. Anteriorly, the fibrous capsule attaches to the humerus above the radial and coronoid fossae, along the coronoid process of the ulna, and to the anterior portions of the annular ligament of the radius. Deep fibers from the brachialis muscle insert onto the superficial surface of the anterior capsule. Posteriorly, the capsule attaches to the olecranon fossa of the humerus at the margin of the olecranon. Laterally and medially, the posterior capsule blends with the corresponding collateral ligaments. The radial collateral ligament lies deep to the common extensor tendon of the forearm, while the ulnar collateral ligament helps to form the common forearm flexor tendon. The posterior portion of the capsule is reinforced by attachments to the overlying triceps brachii tendon.[2]

The fat pads occupying the coronoid, radial, and olecranon fossae are intracapsular but extrasynovial. The synovium covers the deep surface of the capsule, which pouches down along the neck of the radius past the inferior margin of the annular ligament. The bursae around the elbow are the radiohumeral bursa, the interosseous bursa, the bicipitoradial bursa, and the olecranon bursa.[2]

The muscles and tendons surrounding the elbow include the anconeus muscle and tendon of the triceps brachii posteriorly, the biceps brachii tendon and brachioradialis muscles anteriorly, the forearm flexors medially, and the forearm extensors laterally. The ulnar nerve courses around the elbow in the medial olecranon groove.

TECHNIQUE

Sonography allows complete evaluation of the elbow and surrounding musculoskeletal structures. In children younger than 2 years, a 7.5-MHz linear-array transducer provides optimal visualization of the anatomy. In older patients, a 5-MHz or broadband linear-array transducer is best. In patients who cannot fully extend their arms, a curved-array transducer may be useful.

Examination of the anterolateral aspect of the elbow is performed in longitudinal and longitudinal oblique planes, with the patient's hand halfway between pronation and supination.[3] The radiohumeral articulation is best visualized in the longitudinal plane (Fig. 10-1). The longitudinal oblique plane allows optimal visualization of the attachment of the common extensor tendon on the lateral epicondyle (Fig. 10-2). A longitudinal view in the midline allows visualization of the anterior joint capsule and attached synovium and the anterior fat pad (Fig. 10-3). Transverse views at the level of the antecubital fossa and slightly below the elbow joint space can be used to demonstrate the vascular anatomy and its variations and the radioulnar articulation (Fig. 10-4).

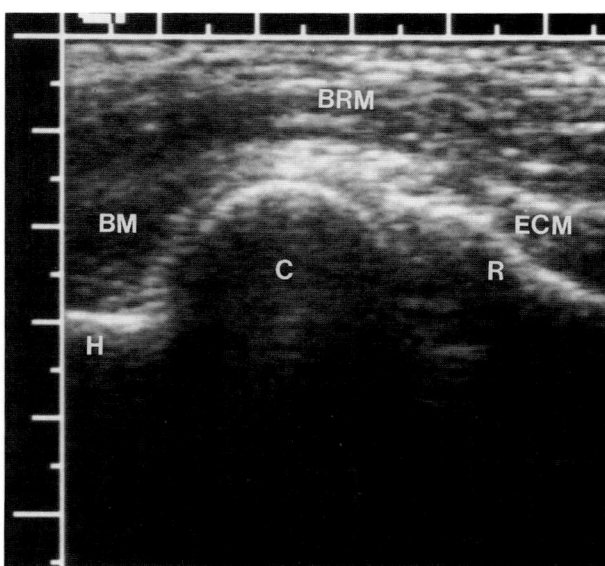

Fig. 10-1 Longitudinal sonogram along the anterolateral aspect of the extended elbow shows the radiohumeral joint. The radial head (R) articulates with the capitellum (C) of the humerus (H). Visible surrounding muscles are the brachioradialis muscle (BRM), the extensor carpi radialis muscle group (ECM), and the biceps brachii muscle (BM).

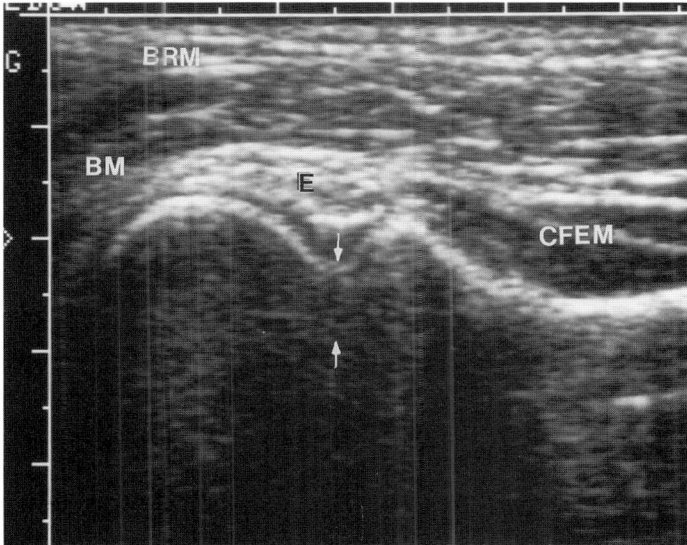

Fig. 10-2 Longitudinal oblique sonogram along the anterolateral aspect of the extended elbow shows the insertion of the common forearm extensor tendon (E) on the lateral epicondyle. Other visualized structures include the radiohumeral articulation (*arrows*), the common forearm extensor muscle group (CFEM), the biceps brachii muscle (BM), and the brachioradialis muscle (BRM).

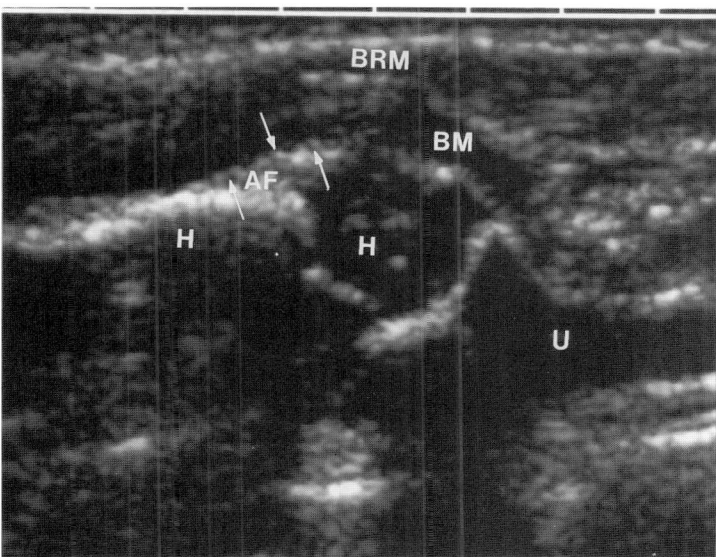

Fig. 10-3 Longitudinal midline sonogram of the anterior extended elbow. Note the echogenic anterior joint capsule (*arrows*) and the anterior fat pad (AF). Surrounding muscles include the brachioradialis muscle (BRM) and the biceps brachii muscle (BM). H, humerus; U, ulna.

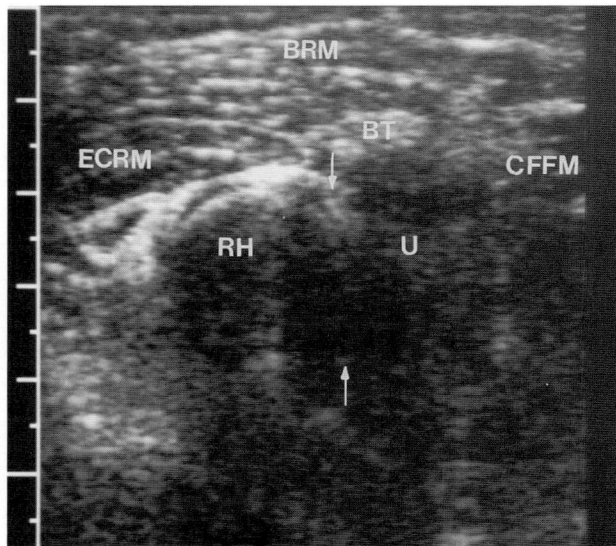

Fig. 10-4 Transverse sonogram slightly below the extended elbow at the radioulnar articulation (*arrows*). The radial head (RH) remains perpendicular to the ultrasound beam. Because its surface at this point is oblique to the beam, the ulna (U) is not clearly defined. Visible surrounding muscles and tendons include the extensor carpi radialis muscles (ECRM), the brachioradialis muscle (BRM), the biceps brachii tendon (BT), and the common forearm flexor muscles (CFFM).

The anteromedial aspect of the elbow is visualized using longitudinal and longitudinal oblique views when the hand is in supination. The longitudinal view demonstrates the ulnohumeral articulation (Fig. 10-5), whereas the longitudinal oblique view best demonstrates the common forearm flexor tendon (Fig. 10-6).

Posterior assessment of the elbow is easily performed with the arm in 180-degree elevation, 90-degree flexion, and slight abduction. This position is accomplished by having the patient rest his hand behind his head. The posterior longitudinal view allows the olecranon fossa to be visualized (Fig. 10-7). Posterior transverse views are useful for identifying posterior portions of the fibrous capsule and synovium and the posterior fat pad (Fig. 10-8). The ulnar nerve may be visualized on posterior views obtained with the arm in extension.[4] The loose, thick skin overlying the posterior aspect of the elbow may make acoustic coupling difficult; a thin standoff pad may provide superior contact in this region.

Common pathologic conditions of the elbow detectable by sonography are trauma, inflammation, masses, and infection. Using sonography in patients with unexplained elbow pain may lead to discovery of previously undetected occult fractures, cartilaginous avulsions, periostitis, and synovial abnormalities. Sonography may be limited if the patient's range of motion is inhibited because of contractures, previous surgery, or intense pain. In such cases, it may be necessary to perform the examination after sedation or in the operating room. Another limitation of sonography is its inability to detect subtle alterations in structure without comparison with a normal elbow. In many cases, the patient's opposite elbow will be normal,

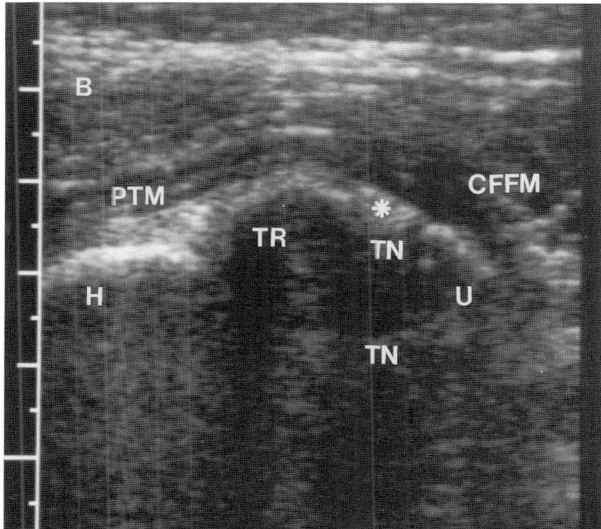

Fig. 10-5 Longitudinal sonogram along the anteromedial aspect of the extended elbow with the hand in supination. This view allows identification of the trochlea (TR) of the humerus (H) as it articulates with the trochlear notch (TN) of the ulna (U). The coronoid process (*) of the ulna appears echogenic after ossification, as in this adult. In children, it may appear hypoechoic, depending on the amount of cartilage not ossified at the time of sonography. Surrounding muscular structures include the brachialis muscle (B), the pronator teres muscle (PTM), and the common forearm flexor muscles (CFFM).

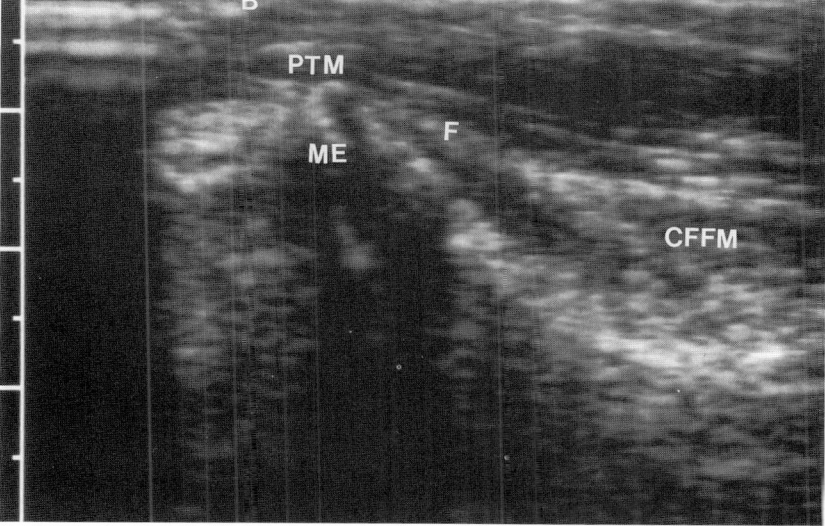

Fig. 10-6 Longitudinal oblique sonogram along the anteromedial surface of the extended elbow shows the common forearm flexor tendon (F) as it attaches to the medial epicondyle (ME). Visualized muscular structures are the brachialis muscle (B), the pronator teres muscle (PTM), and the common forearm flexor muscles (CFFM).

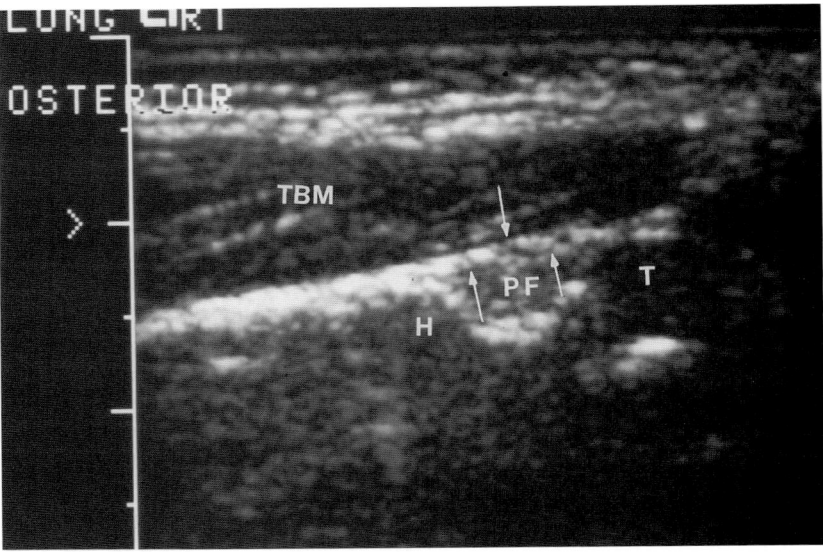

Fig. 10-7 Midline longitudinal sonogram along the posterior aspect of the flexed elbow. The olecranon fossa of the humerus (H) contains the posterior fat pad (PF). The trochlea (T) forms the distal bony landmark. The triceps brachii muscle (TBM) is visible superficial to the bony structures. The fibrous capsule of the joint (*arrows*) forms an echogenic line superficial to the posterior fat pad.

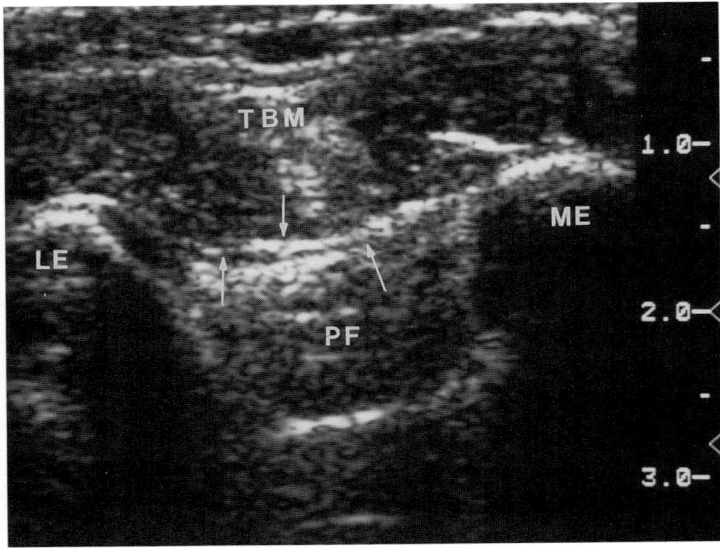

Fig. 10-8 Transverse sonogram along the posterior aspect of the flexed elbow at the level of the olecranon fossa. The olecranon fossa is bounded by the lateral epicondyle (LE) and medial epicondyle (ME). Deep to the triceps brachii muscle (TBM) is the fibrous capsule of the joint (*arrows*). The posterior fat pad (PF) is again noted in the fossa.

and scanning it allows immediate comparison with the area in question.

BONE ABNORMALITIES

Because growing bones are more elastic than fully formed bones, careful attention should be paid in children to the contour of the cortex in order to detect acute buckle fractures. The entire surface of the bone should be scanned in the area of pain or soft-tissue swelling. At all ages, a frank breach in the cortex is detectable with ultrasound (Fig. 10-9). Fractures may be accompanied by subperiosteal hematoma and soft-tissue swelling. Because sonography also reveals cortical disruptions[5] and fluid collections ad-

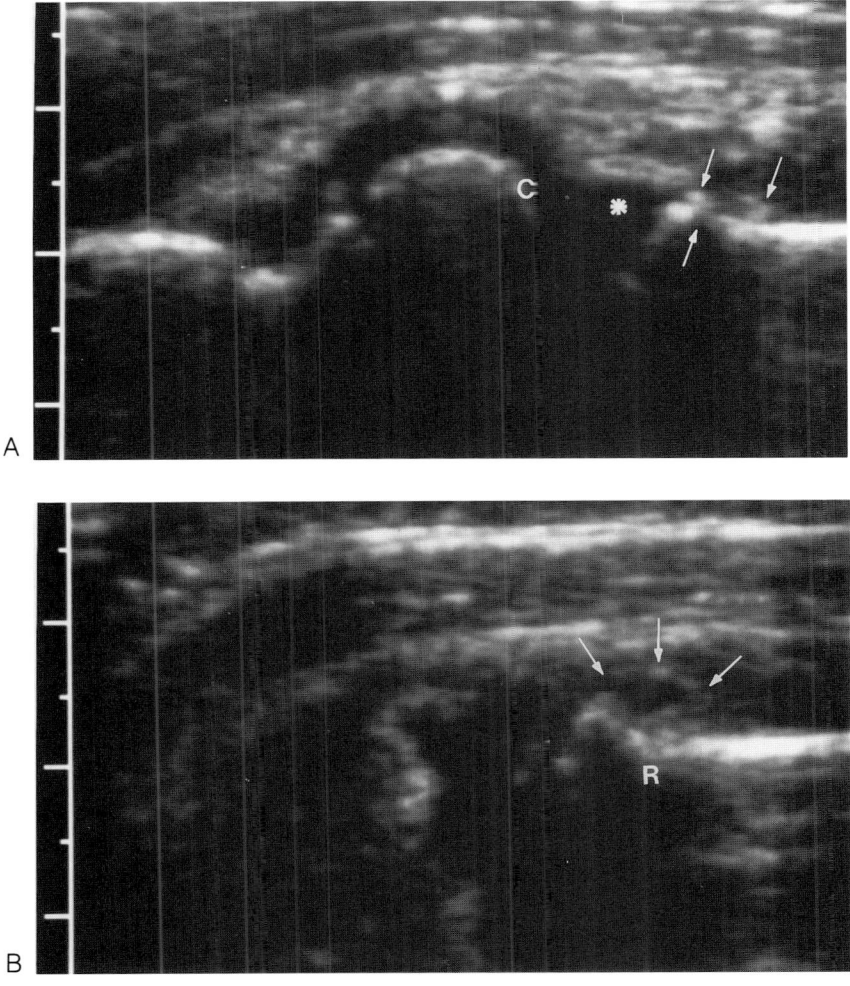

Fig. 10-9 Subacute radial head fracture. **(A)** Longitudinal sonogram along the anterolateral aspect of the extended elbow in a 13-year-old child shows increased echogenicity and irregularity around the radial metaphysis (*arrows*). Neither the capitellum (C) nor the radial head (*) has ossified yet. **(B)** Oblique sonogram shows a hypoechoic periosteal fluid collection (*arrows*) around the neck of the radius (R).

jacent to bone[6] in cases of osteomyelitis, careful correlation with the patient's history is essential to accurate assessment.

Periosteal elevation may be seen in subacute fractures and when adjacent inflammation increases surrounding blood supply to the elbow. Old fractures and surgical intervention lead to remodeling of the bone architecture, which may be difficult to sort out without comparison with a normal elbow (Fig. 10-10).

Children with suspected dislocations at the elbow may benefit in particular from sonography because the unossified cartilage is visible.[7] Evaluation of patients with suspected elbow dislocation often requires sedation. Displacement of bony structures outside normal scan planes may be accompanied by hemarthrosis[8] or joint effusion. Relocation is easily confirmed by repeating the sonogram. In cases of combined fracture and dislocation, epiphyseal and/or physeal fractures, and failure of a fracture to remain satisfactorily reduced, sonography demonstrates unacceptable motion at the injury site.[8,9] Intraoperative scanning may be used to optimize reduction without the need for arthrography.

CARTILAGINOUS ABNORMALITIES

Cartilage is visualized as varying amounts of hypoechoic tissue along the articular surfaces and at sites of skeletal immaturity. Sonography may be used to define cartilaginous avulsions as partial or complete. These injuries are often accompanied by joint effusion. The site of the loose body formed by complete avulsion also may be documented, obviating arthrography. Scanning in multiple directions and planes at the site of soft-tissue swelling will maximize detection of cartilaginous injuries.

JOINT ABNORMALITIES

Fluid in the joint may accompany trauma, infection, or inflammation. In children without a history of trauma, effusions are considered signs of septic arthritis until proven otherwise by needle aspiration. Joint effusions are easily identified by sonography using either an anterior or a posterior approach. Displacement of the posterior fat pad (Fig. 10-11) corresponds to the well-known radiographic finding of joint fluid. Uncomplicated effusions are anechoic. Septic arthritis effusions may appear as anechoic fluid or echogenic pus. Sonography can be used to guide needle localization and subsequent joint aspiration. Other causes of increased echogenicity in a distended joint include hemarthrosis[8] and pannus formation (Fig. 10-12).

The thickness of the normal elbow joint capsule increases with age (Table 10-1).[10] Normal synovial thickness is not measurable with current clinical ultrasound equipment. Capsular measurements exceeding those listed for the age group may indicate synovitis.[11] Septic arthritis, juvenile rheumatoid arthritis, and hemophilia are common causes of synovitis in childhood. In children with joint pain, sonography is often a more sensitive indicator of combined capsular and synovial thickening than is physical examination of the elbow.[11] Scanning in the posterior transverse plane allows easy measurement of the capsule thickness.[10] In adults, polyarthritides and rheumatoid arthritis also cause synovial thickening.

FIBROUS-TISSUE INJURIES

Lateral epicondylitis has been associated with a number of occupations (e.g., baggage handling) and sports (e.g., tennis and scuba diving).[12] Although less common than the lateral form, medial epicondylitis is also as-

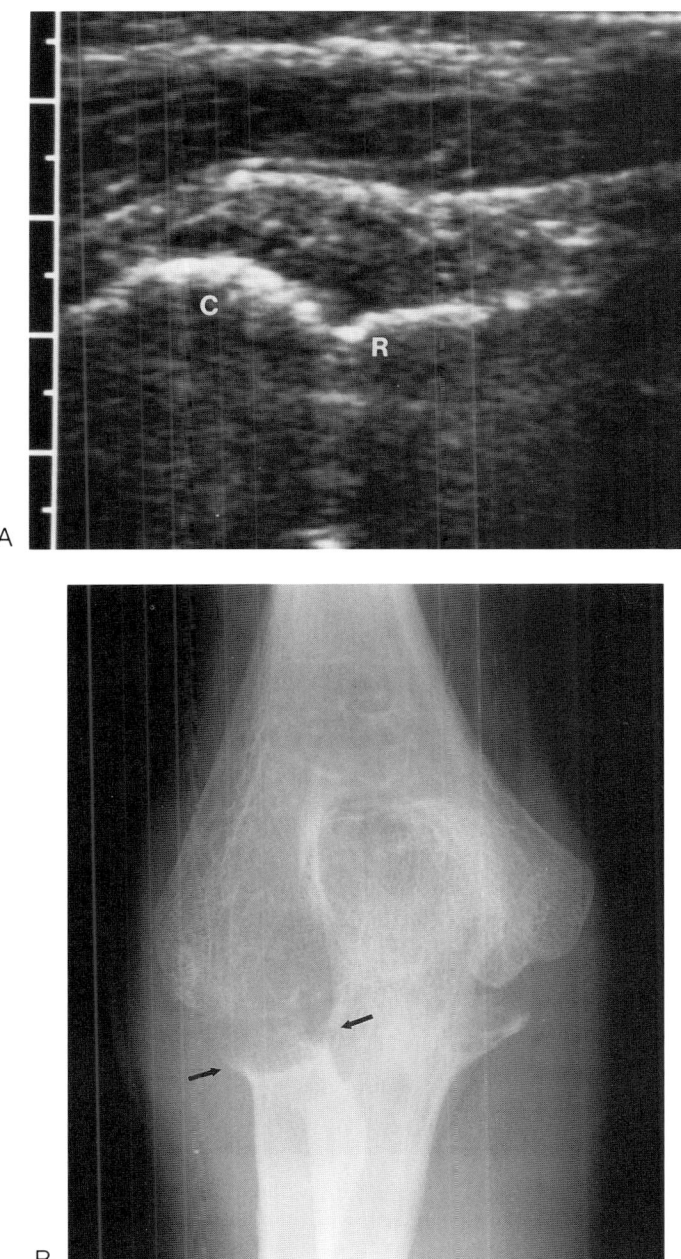

Fig. 10-10 Postoperative deformity in an 11-year-old hemophiliac. **(A)** Longitudinal sonogram along the anterolateral aspect of the extended elbow shows an absence of the normal configuration of the radial head (R) as it articulates with the capitellum (C) (compare to Fig. 10-1). **(B)** Anteroposterior radiograph confirms that the radial head was resected (*arrows*).

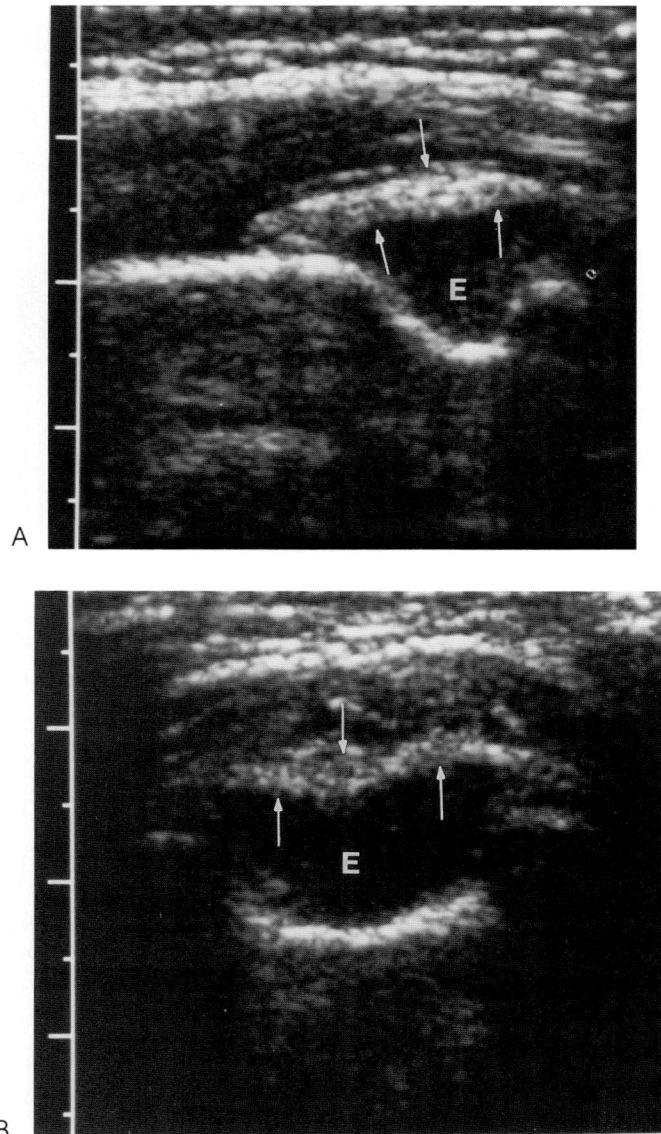

Fig. 10-11 Septic arthritis in a 3-year-old child. **(A)** Longitudinal sonogram along the posterior midline aspect of the flexed elbow shows distention of the joint capsule and synovium with displacement of the posterior fat pad (*arrows*). The olecranon fossa is filled with complex fluid (E). **(B)** Transverse sonogram shows capsular and synovial thickening (*arrows*) and the joint exudate (E). Needle aspiration confirmed the diagnosis of septic arthritis.

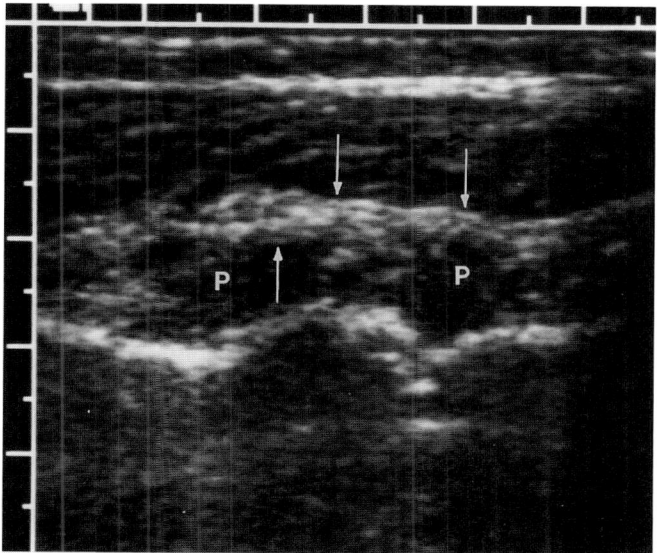

Fig. 10-12 Pannus formation in a 4-year-old child with juvenile rheumatoid arthritis. Midline longitudinal sonogram along the anterior aspect of the elbow shows thickening and irregularity of the joint capsule (*arrows*). The mildly echogenic material within the joint space proved to be pannus formation (P).

sociated with activities that exaggerate forearm flexion. In acute and chronic phases of injury, sonography can be used to document abnormalities by scanning in the appropriate oblique plane. Findings reported with ligamentous injuries include thickening of the tendon (Fig. 10-13), partial or complete avulsion from the bony origin, adjacent blood or fluid collections (Fig. 10-14), and inhomogeneous echogenicity of the tendon or ligament.[13] Calcification in or around the fibrous tissue indicates chronic injury. Postsurgical changes in the fibrous structures will have a similar sonographic appearance.

Communication with patients is therefore important for accurate interpretation.

SOFT-TISSUE ABNORMALITIES

Deep edema of the soft tissues is usually associated with bony trauma or osteomyelitis. In the absence of such conditions, increased echogenicity of the muscles and subcutaneous fat and effacement of the fibrous planes indicate cellulitis. Because lymphatic drainage of the forearm is limited to one or two epitrochlear nodes, identification of a

Table 10-1. Age-Related Differences in Elbow Joint Capsule Thickness

Age (yr)	Mean Thickness (mm)	Standard Deviation	Number of Cases
<2	0.74	0.13	13
2–15	1.04	0.21	29
>15	1.86	0.62	11

(From Hogan et al.[10] with permission.)

146 MUSCULOSKELETAL ULTRASOUND

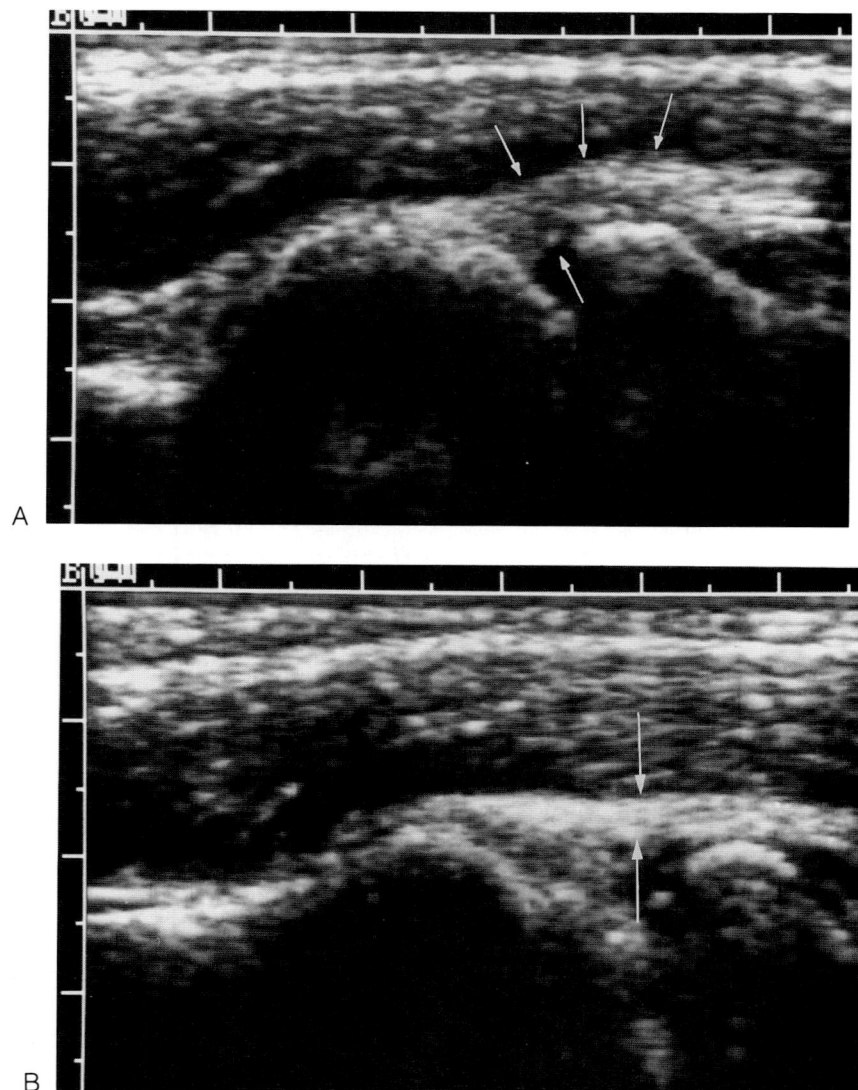

Fig. 10-13 Lateral epicondylitis. **(A)** Longitudinal oblique sonogram along the anterolateral aspect of the elbow in an adult with acute symptoms. Note the thickening of the common forearm extensor tendon and the indistinctness of its margins (*arrows*). **(B)** Longitudinal oblique sonogram of the asymptomatic contralateral elbow shows a normal common forearm extensor tendon (*arrows*).

hypoechoic, ovoid mass is indicative of acute epitrochlear lymphadenitis (Fig. 10-15).[14]

In many cases of neoplasms and congenital anomalies in the elbow, sonographic findings are nonspecific (Fig. 10-16). The role of the sonographer often is to determine the solid or cystic nature of the mass and to evaluate patterns of blood flow in the region with Doppler sonography.

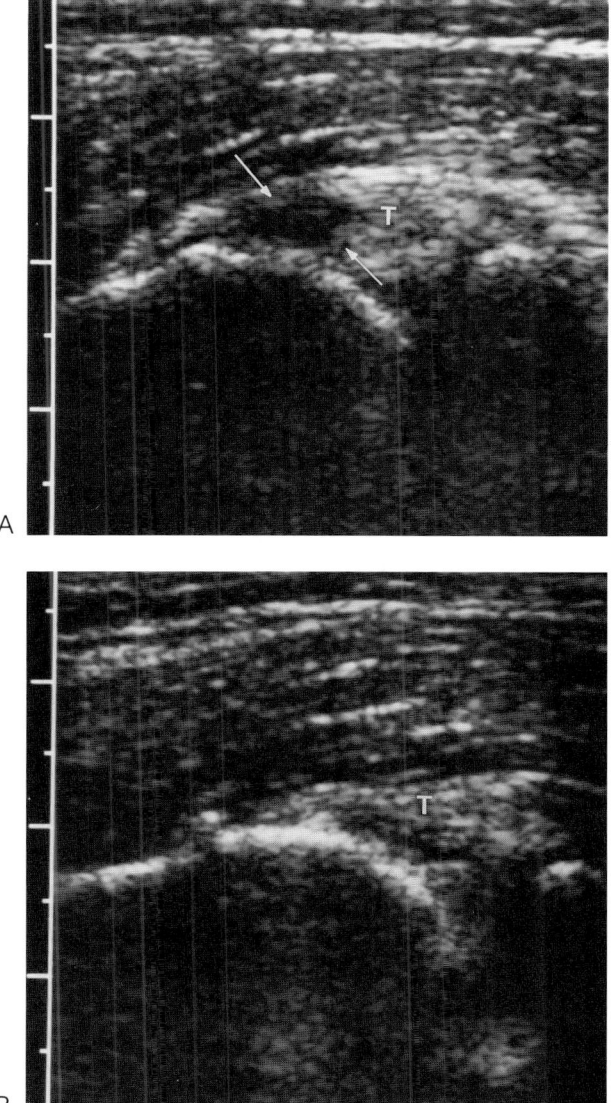

Fig. 10-14 Partial avulsion of the common forearm extensor tendon. **(A)** Longitudinal oblique sonogram along the anterolateral aspect of the elbow shows a hypoechoic fluid collection (*arrows*) at the insertion of the common forearm extensor tendon (T). **(B)** Longitudinal oblique sonogram of the normal contralateral common forearm extensor tendon (T), for comparison.

A common indication for Doppler study of the antecubital fossa is to detect an occluding thrombus at the origin or end of a hemodialysis shunt. Depending on age, thrombi may appear hyper- or hypoechoic. The occluded native vessels cannot be augmented. The spectral display will demonstrate an abnormal pattern. It is important to define the extent of a thrombus so that appropriate anticoagulation or surgical therapy can be

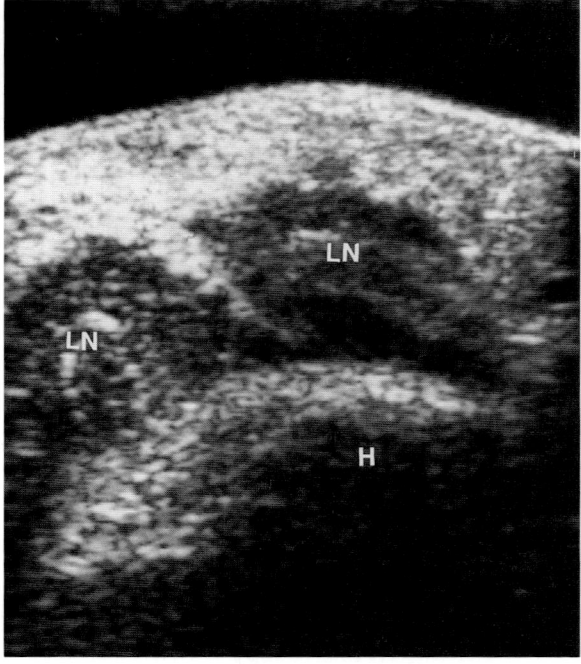

Fig. 10-15 Acute epitrochlear lymphadenitis in a 7-year-old patient. Transverse posterior sonogram at the distal humeral diaphysis (H). The inflamed lymph nodes (LN) appear as hypoechoic ovoid masses with central echoes, deep to the indurated subcutaneous tissues.

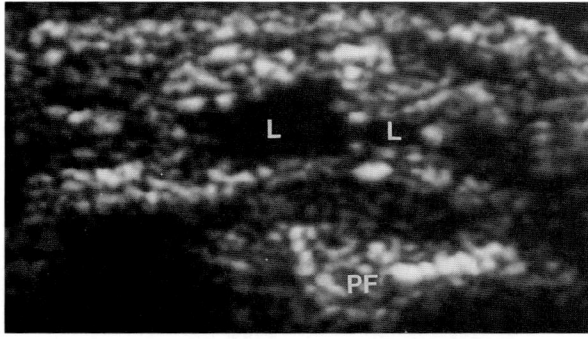

Fig. 10-16 Lymphangioma in a 9-year-old patient. Transverse sonogram along the posterior aspect of the elbow at the level of the posterior fat pad (PF). The irregular, anechoic mass (L) was compressible. Doppler study revealed no identifiable blood vessels around the lesion. The patient had no history of infection or trauma. At surgery, the diagnosis of lymphangioma was confirmed.

initiated promptly. Follow-up sonograms can be used to evaluate the shunt for subsequent patency.

CONCLUSIONS

The soft tissues and fibrous and bony structures of the elbow are readily delineated by sonography. A standardized approach and familiarity with common pathologic conditions are helpful for complete evaluation. Frequent comparison to the patient's opposite, normal elbow or an age-matched normal elbow aids in detecting subtle abnormalities.

REFERENCES

1. Brodeur AE, Silberstein MJ, Graviss ER: Radiology of the Pediatric Elbow. Hall Medical, Boston, 1981

2. Anson BJ, Maddock WG: Callander's Surgical Anatomy, 3rd Ed. WB Saunders, Philadelphia, 1952
3. Barr LL, Babcock DS: Sonography of the normal elbow. AJR 157:793, 1991
4. Hoogland PH: Ultrasonography of entrapment neuropathy of the median and ulnar nerves (abstract). Proc 29th Ann Meeting of the ASNR, Washington, DC 1991
5. Williamson SL, Seibert JJ, Glasier CM et al: Ultrasound in advanced pediatric osteomyelitis. Pediatr Radiol 21:288, 1991
6. Abiri MM, Kirpekar M, Ablow RC: Osteomyelitis: detection with US. Radiology 172:509, 1989
7. Lamont AC, Dias JJ: Ultrasonic diagnosis of dislocation of the radius in an infant with Down's syndrome. Br J Radiol 64:849, 1991
8. Markowitz RI, Davidson RS, Harty MP et al: Sonography of the elbow in infants and children. AJR 159:829, 1992
9. Dias JJ, Lamont AC, Jones JM: Ultrasonic diagnosis of neonatal separation of the distal humeral epiphysis. J Bone Joint Surg 70B:825, 1988
10. Hogan MJ, Rupich RC, Bruder JB, Barr LL: Age-related variability in elbow joint capsule thickness in asymptomatic children and adults. J Ultrasound Med 13:211, 1994
11. Barr LL, Passo MH, Stafford PA, Hahn TF: Sonographic findings in the synovial joints of symptomatic children (abstract). J Ultrasound Med 11(suppl):24, 1992
12. Barr LL, Martin LR: Tank carrier's lateral epicondylitis: case reports and a new cause for an old entity. South Pacific Undersea Medicine Journal 21:35, 1991
13. Giannini S, Lipparini M, Della Villa S et al: Ultrasonography in pathological conditions of muscles, tendons and joints. Ital J Orthop Traumatol 13:253, 1987
14. Barr LL, Kirks DR: Ultrasonography of acute epitrochlear lymphadenitis. Pediatr Radiol 23:72, 1993

11
Hand and Wrist

Luca De Flaviis
Maurizio Giulio Musso

Hand imaging began in 1895, when Wilhelm Conrad Roentgen discovered x-rays and took a radiograph of his wife's hand. For many years, no dramatic improvements in hand imaging occurred, even after the introduction of tomography. In the past 20 years, however, hand imaging has significantly improved. Special x-ray projections, such as dynamic, scaphoid, and carpal tunnel views, have been developed; and new techniques, such as arthrography, xeroradiography, sonography, computed tomography (CT), and, ultimately, magnetic resonance imaging (MRI), have been introduced.[1] These procedures are helpful in diagnosing not only bone lesions but also lesions in the soft tissues between the skin and the bone. An advantage of sonography in particular is its ability to visualize fine structures, such as tendons, nerves, and vessels, and to elucidate the nature of any kind of soft-tissue swelling, such as a cyst or synovial enlargement.[2] In addition, sonography is inexpensive, noninvasive, and devoid of side effects. Moreover, real-time sonography is uniquely able to provide a dynamic study of tendons gliding.[3]

TECHNICAL CONSIDERATIONS

Sonographic examination of the soft tissues in the hand and wrist has been made possible by the development of high-resolution transducers with frequencies of 7.5 to 15 MHz.[4-5] Although linear-array transducers are usually superior to mechanical sector probes for imaging very superficial structures, the smaller footprint and greater maneuverability of high-frequency mechanical scanners with built-in water paths may prove very useful for examining some areas of the hand and wrist, especially small joints. However, with sector probes, only the narrow segment of the tendon in the central area of the scan is perpendicular to the ultrasound beam; the result is an artifactual hypoechogenicity of the tendon laterally.[7] A thin standoff pad between the skin and the transducer's footprint ensures optimal focusing for very superficial structures, such as the median nerve or extensor tendons of the fingers. As usual, both longitudinal and transverse scans must be performed, especially when evaluating tendons.

NORMAL ANATOMY AND SONOGRAPHIC APPEARANCE

The complex anatomy of the hand and, especially, the wrist makes sonographic examination technically difficult. Therefore, an understanding of the normal anatomy of the hand and wrist is essential to mastering this technique.[8] The soft-tissue structures of the hand and wrist that can be seen with sonography include muscles, tendons and their sheaths, nerves, and blood vessels.

Muscles

The muscles that power the hand may be divided into two groups: intrinsic and extrinsic. The intrinsic muscles, which include the thenar, interosseous, lumbrical, and hypothenar muscle groups, have their origins and insertions within the hand. These muscles are markedly hypoechoic in relation to the adjacent tendons. Longitudinal sonograms demonstrate the striated pattern typical of skeletal muscles. The extrinsic muscles are located within the forearm and supply their motor force through the action of their tendons. They are subdivided into extensor and flexor muscles. The extensors overlie the dorsum of the forearm and extend the wrist and fingers. The flexors are on the volar aspect of the forearm and flex the wrist and fingers.

Tendons

Six synovial compartments are located on the dorsal aspect of the wrist beneath the extensor retinaculum; through these channels, the extensor tendons leave the forearm and pass through the dorsum of the hand. Knowledge of the location of these compartments is essential in the diagnosis of tendinous diseases of the dorsum of the hand and wrist.

Beginning from the radial aspect of the wrist, the first compartment contains the abductor pollicis longus and the extensor pollicis brevis tendons. There is often a second abductor pollicis longus tendon, lying in a separate compartment, which may cause a recurrence of de Quervain's disease if not released during surgical therapy for that condition (discussed later in chapter). The first compartment is palpable as the radial boundary of the so-called anatomic snuffbox.

In the second compartment lie the extensor carpi radialis longus and extensor carpi radialis brevis tendons. The dorsal tubercle of the radius separates the second compartment from the third compartment, which contains the extensor pollicis longus tendon. This represents the ulnar boundary of the aforementioned snuffbox. The extensor pollicis longus tendon abruptly changes its direction radially just below the radial tubercle; ruptures of this tendon frequently occur here.

The fourth compartment is the widest. This channel holds the four tendons of the extensor digitorum muscle and the extensor indicis tendon and is located in the midline of the dorsal aspect of the wrist. These tendons are often affected by tenosynovitis, which results in a swelling of the dorsum of the wrist that can mimic a ganglion.[9,10]

The fifth compartment contains the extensor digiti minimi tendon. Finally, the sixth compartment, whose bony floor is formed by the dorsal aspect of the ulnar head, contains the extensor carpi ulnaris.

The carpal tunnel is a fibro-osseous canal that lies in the palmar aspect of the wrist and through which pass a number of tendons and the median nerve. Its bony floor is composed of the carpal bones, and its roof is formed by the flexor retinaculum, which continues distally as the palmar fascia. The strong, thick sheet of the flexor retinaculum attaches to the osseous boundaries of the tunnel (the scaphoid tubercle and the ridge of the trapezium bone radially, the pisiform bone and the hook of the hamate bone medially). Many important structures are located in the carpal tunnel. Superficially and radially in the tunnel lies the median nerve. Lateral to it courses the flexor pollicis longus tendon in the radial bursa. The four flexor digitorum superficialis and four flexor digitorum profundus tendons are wrapped in the ulnar bursa.

At the ulnar aspect of the volar wrist lies the Guyon's canal, or ulnar tunnel, a fibro-osseous passage through which the ulnar neurovascular bundle enters the palm. This canal is limited by the palmar carpal ligament, which forms its roof, the flexor retinaculum, which forms its floor, and the pisiform bone and the hook of the hamate bone, medially.

Distal to the carpal tunnel, the superficial and deep flexor tendons of the second, third, fourth, and fifth fingers diverge and course through the palm underneath the palmar fascia. At the level of the metacarpophalangeal (MP) joints, each tendon, covered by a synovial sheath, enters a digital fibrous sheath. This fibrous sheath is composed of a series of annular ligaments, or pulleys, that, together with the bones, form a semicylindric fibro-osseous digital canal, which keeps the tendons in contact with the bones and prevents them from "bowstringing" during flexion. At the first phalanx, the flexor digitorum superficialis tendon begins its division and winds around the flexor digitorum profundus tendon so that distal to the proximal digital crease, the latter becomes the more superficial of the two. Repair of tendon injuries occurring between the first annular pulley and the distal insertion of the flexor digitorum profundus tendon on the base of the distal phalanx is often difficult.

The sonographic appearance of tendons is similar throughout the body. In the absence of technical artifacts, tendons are highly echogenic, with a characteristic fibrillar

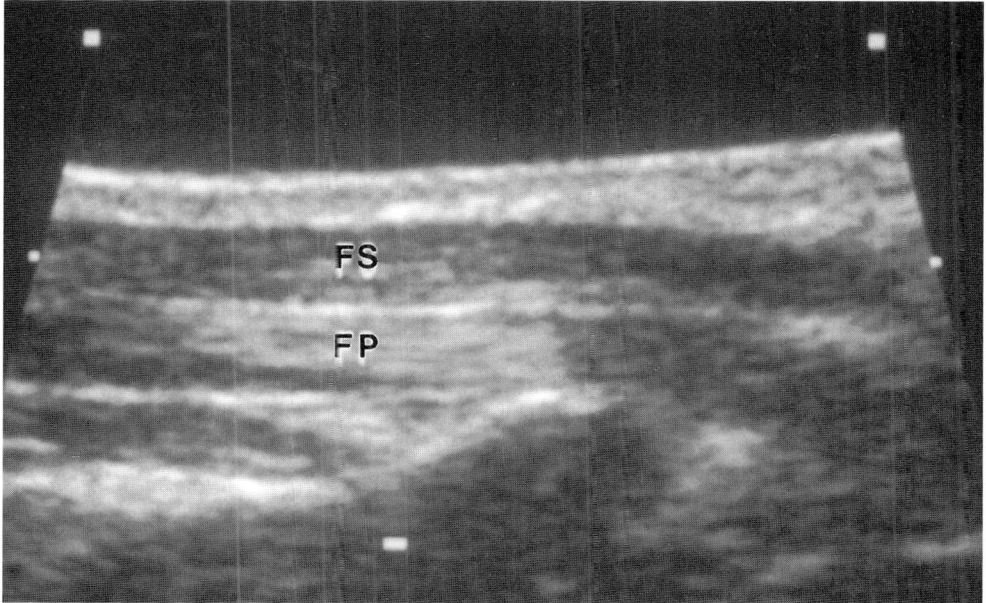

Fig. 11-1 Longitudinal sonogram of a normal wrist shows the normal appearance of the superficial (FS) and deep (FP) flexor tendons of a finger.

echotexture (Fig. 11-1). Tendon sheaths, when present, may appear on longitudinal sonograms as a very thin hypoechoic lining of the echogenic tendons.

Nerves and Blood Vessels

Nerves are slightly less echogenic than tendons. On longitudinal sonograms, the normal median nerve appears in the wrist as a superficial, echogenic, 3- to 5-mm thick, ribbonlike structure that lies anterior to the flexor tendons (Fig. 11-2). On transverse sonograms, the median nerve appears as a rounded echogenic area with a dotted texture. Sonographic identification of the median nerve is highly dependent on the operator. The most common pitfall is confusing the median nerve with surrounding echogenic flexor tendons. However, it is easy to identify the nerve by its immobility during flexion and extension of the fingers on real-time examination. It is more difficult to visualize the ulnar nerve in the Guyon's canal.

In the wrist, the radial and ulnar arteries appear as anechoic tubular structures that can be seen pulsating on real-time examination. The common palmar digital arteries also can be seen, distal to the superficial palmar arch.

DISEASES OF THE TENDONS

Tenosynovitis

Tendons affected by exudative or hypertrophic tenosynovitis are swollen and appear on sonograms as hyperechoic ribbons in a fluid-filled sheath (Fig. 11-3). Sonography allows the detection of even small synovial fluid collections. Internal echoes may be recognized in chronic, rheumatoid, or infected effusions. The flexor and extensor tendons at

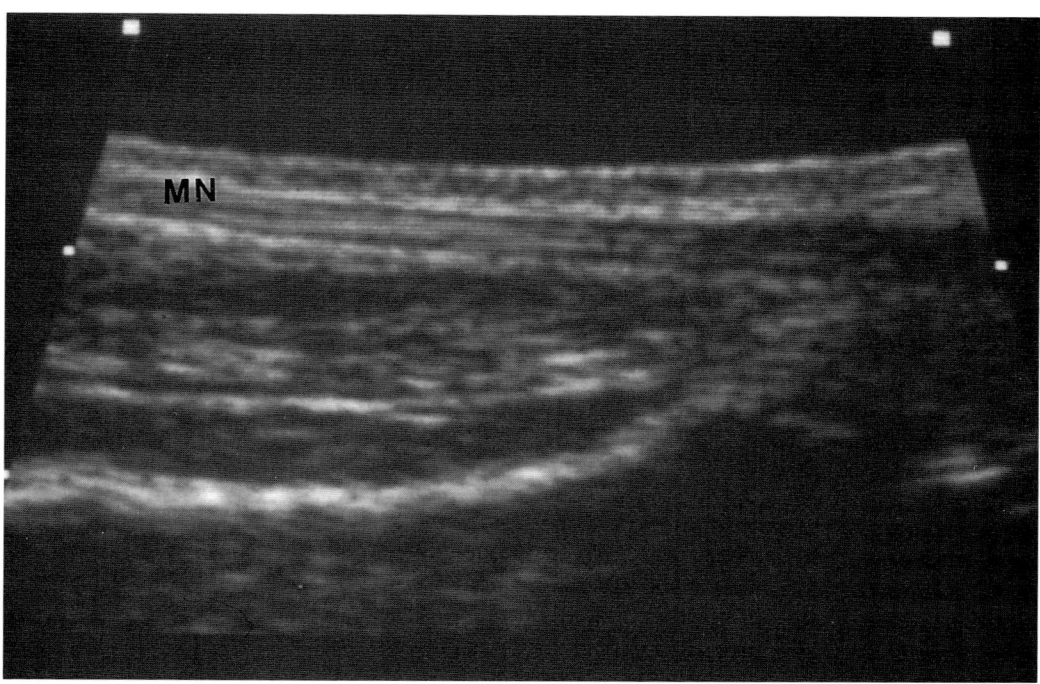

Fig. 11-2 Longitudinal sonogram of a normal median nerve (MN) in the wrist.

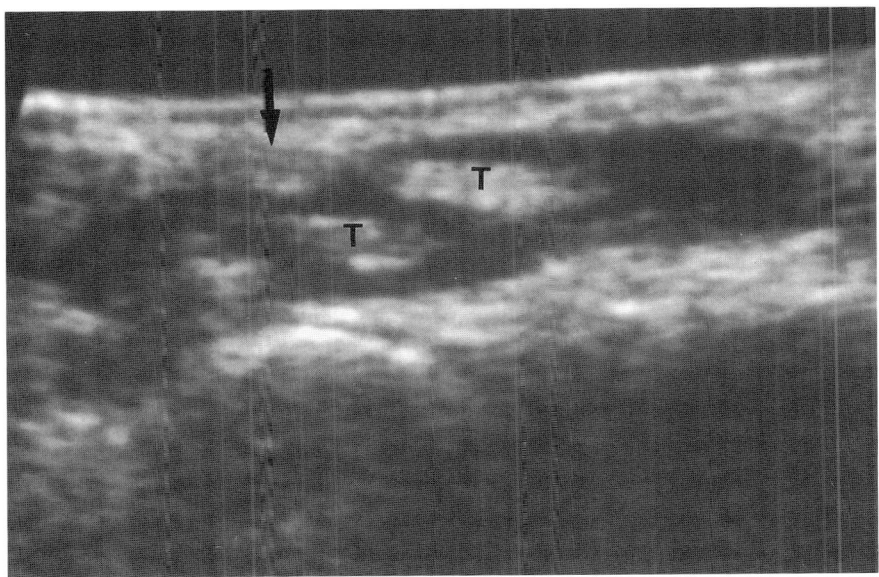
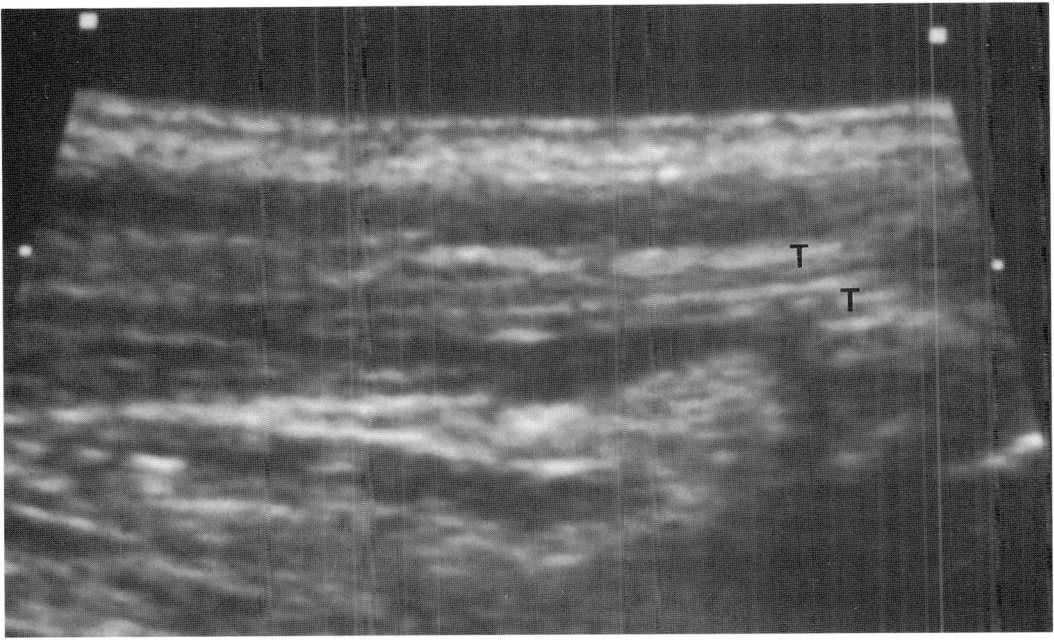

Fig. 11-3 Tenosynovitis. **(A)** Transverse and **(B)** longitudinal sonograms of the volar aspect of the wrist. The ulnar bursa is markedly distended by hypoechoic fluid. The tendons (T) can be appreciated as hyperechoic structures inside the sheath. *Arrow*, median nerve.

the wrist are primarily affected. Sagittal sonograms are most informative, but in cases of diffuse involvement, transverse sonograms are also useful for identifying the affected tendons. In advanced cases, the fluid collection is associated with degenerative changes in the tendons themselves, which appear inhomogeneous, with irregular margins. A complication of tenosynovitis, especially in rheumatoid disease, is pathologic rupture, which is caused by ischemic necrosis and the mechanical damage from friction against osseous structures.[11,12]

Stenosing Tenosynovitis

In stenosing tenosynovitis, the affected tendon is trapped within a fibro-osseous canal or beneath a ligament, resulting in stenosis of the compressed section and swelling of the adjacent portion. In the wrist, the tendons most commonly affected are the abductor pollicis longus and extensor pollicis brevis tendons in the first dorsal compartment, as seen in de Quervain's disease.[13] The primary symptom of this condition is acute pain with any movement of the thumb; the pain often radiates into the forearm and is usually associated with weakness in pinch function. Tenderness of the radial border of the wrist is generally well localized, but in less advanced cases, it may be difficult to distinguish de Quervain's disease from conditions such as arthritis of the basal joint of the thumb. To make the distinction, Finkelstein's test[14] is performed by having the patient firmly enclose his thumb in his palm and then abruptly bend the wrist ulnad. This places maximum stress on the involved tendons and causes severe pain in those with de Quervain's disease. In contrast, Finkelstein's test is negative in patients with basal joint arthritis. Sonography demonstrates a hypoechoic halo surrounding the tendons in de Quervain's disease (Fig. 11-4).

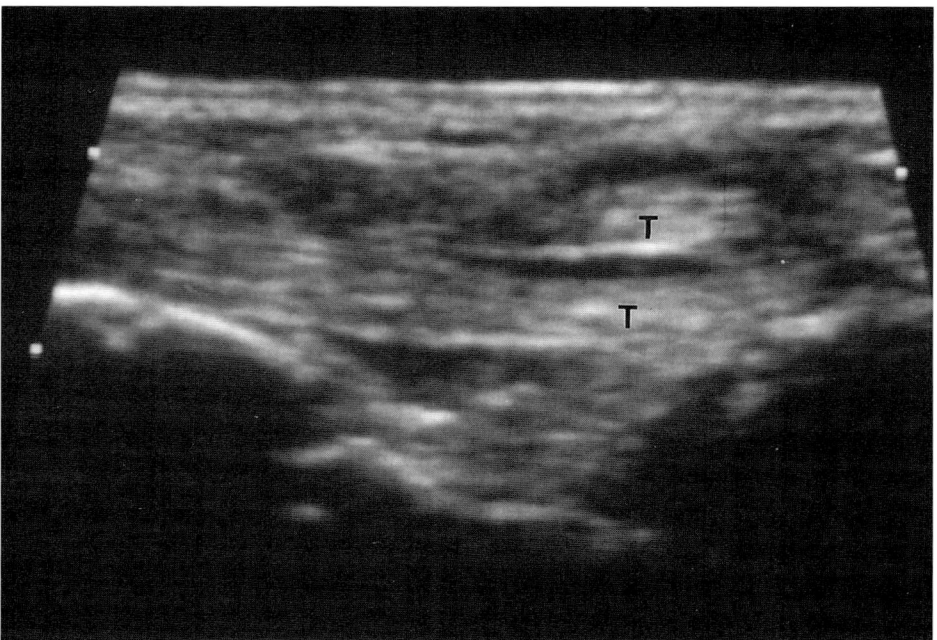

Fig. 11-4 Stenosing tenosynovitis of the abductor pollicis longus and the extensor pollicis brevis tendons (de Quervain's disease). Sonogram shows hypoechoic exudate around the tendons (T).

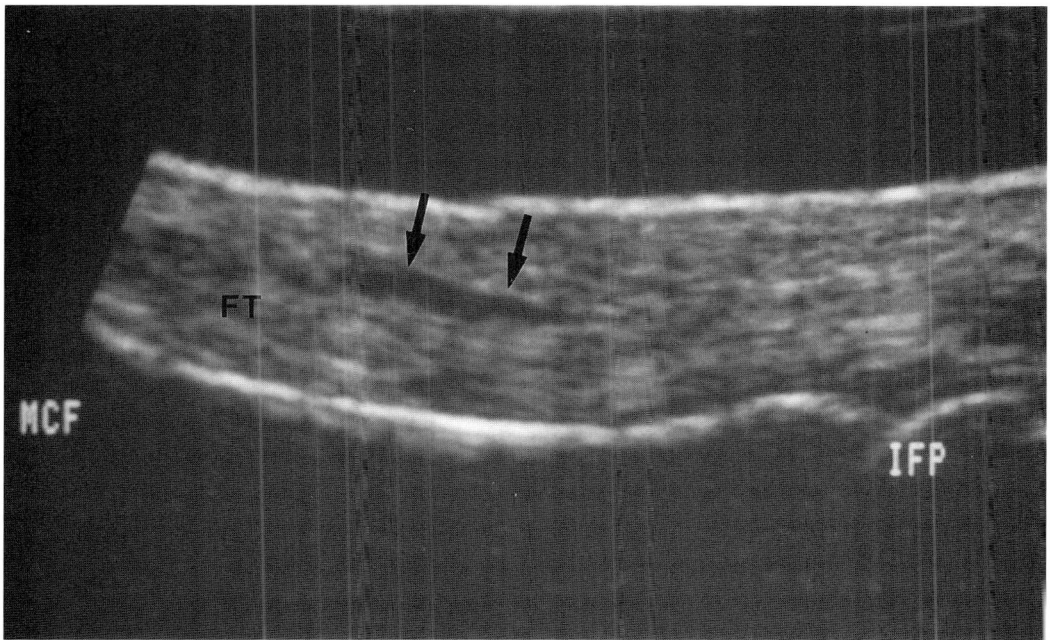

Fig. 11-5 Trigger finger. Longitudinal sonogram of the flexor tendons (FT) of the fourth finger at the level of the first phalanx. A slight nonpalpable swelling of the tendon sheath (*arrows*) is caused by early stenosing tenosynovitis. MCF, toward MP joint; IFP, PIP joint.

Less often involved tendons include the extensor carpi radialis longus and extensor carpi radialis brevis tendons, the extensor pollicis longus tendon at the radial tubercle, and the extensor carpi ulnaris tendon. Although tenosynovitis of the common extensor tendons often produces a swelling of the sheath, this condition is not usually associated with triggering, as observed in stenosing tenosynovitis of other tendons.

Triggering often involves flexor tendons, generally at the proximal digital pulley in the distal palm (trigger finger). Triggering results from the formation of a nodule at the surface of the flexor tendon that passes through a pulley. Even severe triggering may be associated with no pain, and locking of the digit in flexion is the major complaint. An increased incidence of triggering is recognized in rheumatic patients. On clinical examination, one may appreciate a tender nodule as one flexes and extends the involved finger. In these cases, sonography can visualize a hypoechoic, well-defined, oval nodule superficial to the tendon at the level of the pulley of the tendon sheath (Fig. 11-5).

Flexor Carpi Radialis Tendinitis

Flexor carpi radialis tendinitis is a distinct clinical entity but is not widely recognized. The major complaint is discomfort in the wrist, particularly after strenuous exercise. There is localized tenderness over the flexor carpi radialis tendon, just proximal to the scaphoid tubercle. Pain can be reproduced by direct palpation just next to the ridge of the trapezium bone. Longitudinal sonograms show an inhomogeneous, hypoechoic thickening of the tendon at the level of its distal attachment and in the vicinity of the

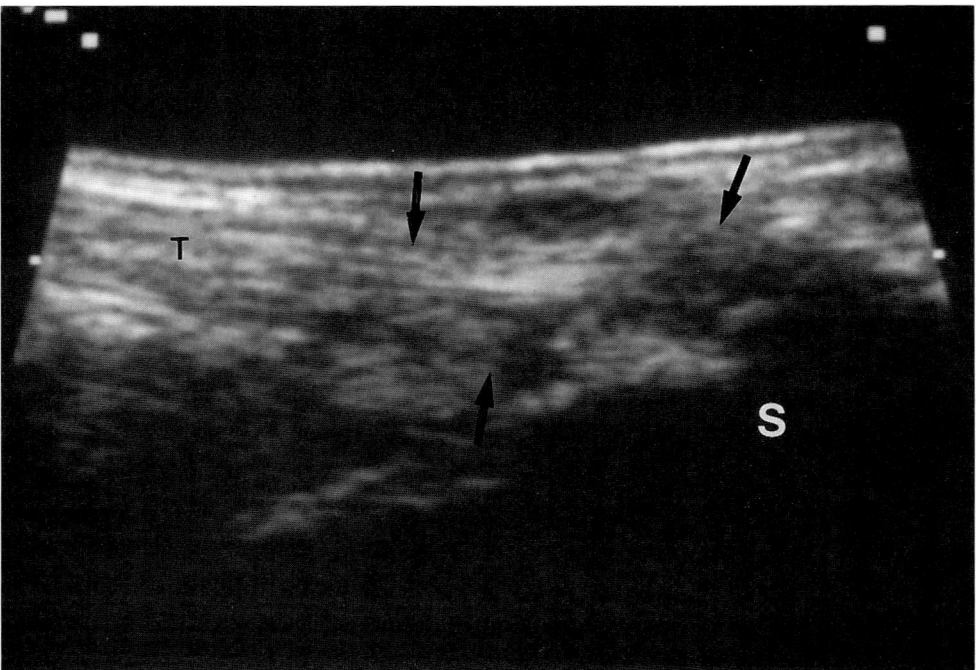

Fig. 11-6 Flexor carpi radialis tendinitis. Sonogram shows focal swelling (*arrows*) of the tendon (T) at the level of the scaphoid tubercle (S).

scaphoid tubercle (Fig. 11-6). Chronic tendinitis is characterized by the development of intratendinous calcifications.

COMPRESSIVE NEUROPATHIES

In certain sites of the wrist, nerves course through narrow passages formed by unyielding structures. Abnormal structures may overfill a tight compartment. The median nerve may be compressed in the carpal tunnel, as may the ulnar nerve in Guyon's canal. Fibrosis of the transverse carpal ligament and tenosynovitis of the flexor tendons are the most common causes of nerve compression in the carpal tunnel; the increased thickness of the synovial sheath cannot be accommodated because the transverse ligament does not expand, and compression of the median nerve results. Other sources of soft-tissue swelling, such as tumors or ganglia,[15] also can cause nerve compression by taking up already overcrowded space.

Carpal Tunnel Syndrome

In the wrist, the median nerve becomes superficial, lying between the palmaris longus and flexor carpi radialis tendons. The nerve passes under the transverse carpal ligament, through the carpal tunnel. The tendons in the carpal tunnel are surrounded by the radial and ulnar bursae. Any increase in the volume of these bursae can compress the median nerve (Fig. 11-7).[16] Less common causes of median nerve compression in the carpal tunnel include ganglia (Fig. 11-8), soft-tissue tumors (e.g., lipoma, Fig. 11-9), inflammatory diseases (e.g., rheumatoid arthritis [Fig. 11-10], gout, and amyloidosis), trauma (e.g., radial or carpal fractures and joint dislocations), supernumerary muscles, and preg-

HAND AND WRIST 159

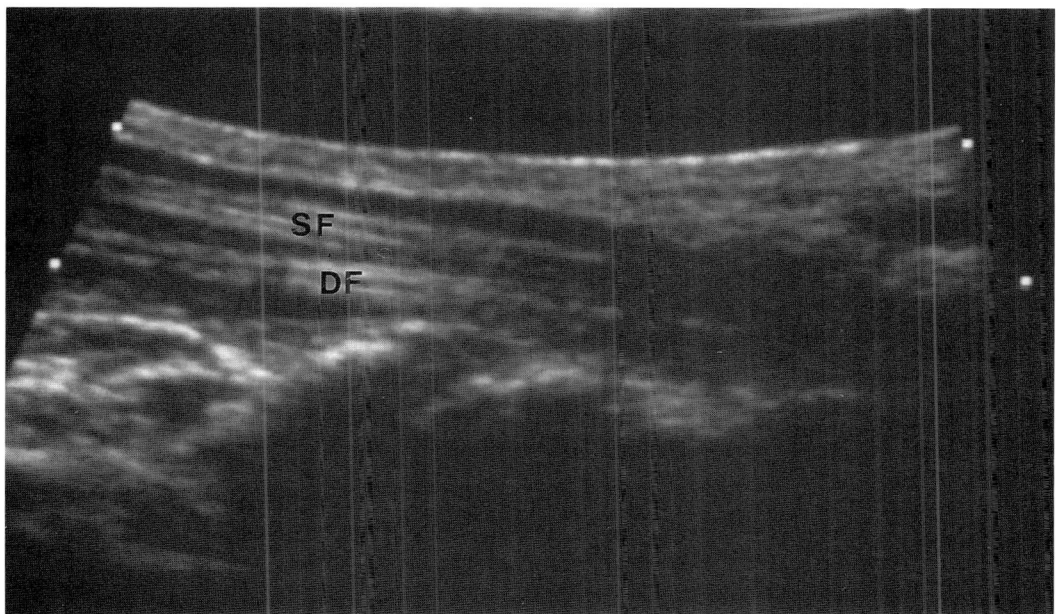

Fig. 11-7 Carpal tunnel syndrome. Longitudinal sonogram shows tenosynovitis of the superficial (SF) and deep flexor (DF) tendons in the wrist.

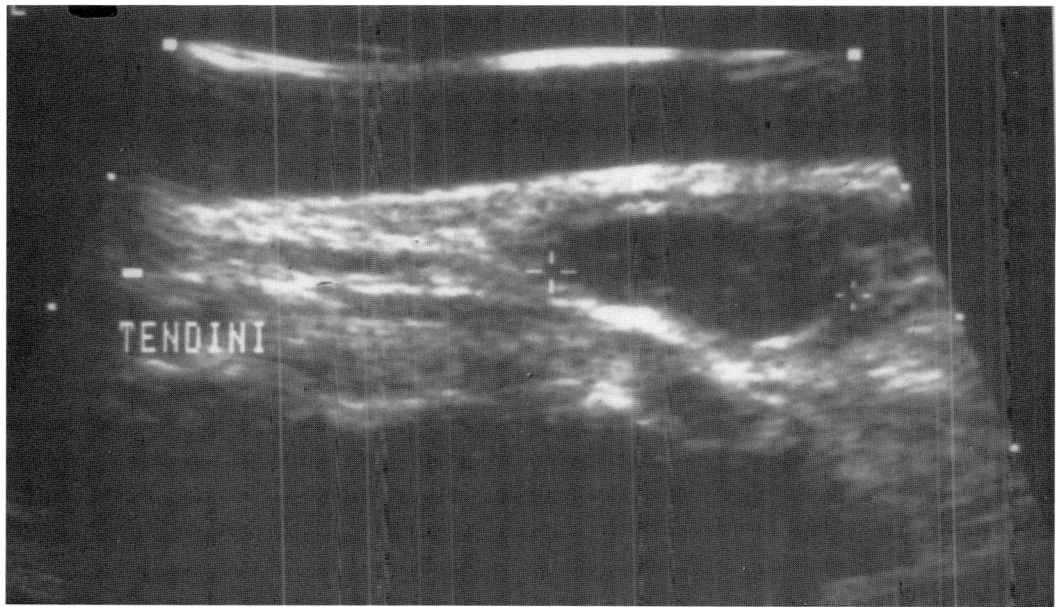

Fig. 11-8 Carpal tunnel syndrome. Longitudinal sonogram shows a hypoechoic ganglion with thick walls (*calipers*), compressing the flexor tendons (tendini).

nancy. In the palm, the median nerve divides into branches, innervating most of the thenar eminence muscles and supplying sensibility to the radial aspect of the palm and the volar aspects of the thumb, second, and third fingers and the radial half of the fourth finger.

A common symptom of carpal tunnel syndrome is numbness that most often involves the middle and index fingers. Patients afflicted with this syndrome complain of activity-aggravated weakness or clumsiness of the hand. Pain in the upper extremity is another common symptom. The major findings at physical examination include a positive response to Phalen's wrist flexion test and a positive Tinel's sign at the wrist. There also may be weakness or atrophy of the abductor pollicis brevis muscle in the thenar area.

Electromyography and nerve conduction studies can differentiate wrist syndromes from more proximal nerve compression, but sonography has the advantage of often being able to directly visualize the underlying cause of extrinsic compression of the nerve (Figs. 11-7 to 11-10). When the median nerve is subjected to long-term compression, it may decrease in diameter. This can be observed sonographically. If the compression occurs intermittently, constriction of the nerve itself is less frequent.

Ulnar Tunnel Syndrome

The ulnar nerve passes over the flexor retinaculum and under the palmar carpal ligament through Guyon's canal. Only the ulnar nerve, artery, and veins pass through this canal. For this reason, entrapment of the ulnar nerve is much less common than median nerve compression. Ganglia are a common cause of ulnar tunnel syndrome (Fig. 11-11). Ganglia appear as round fluid-filled cavities;

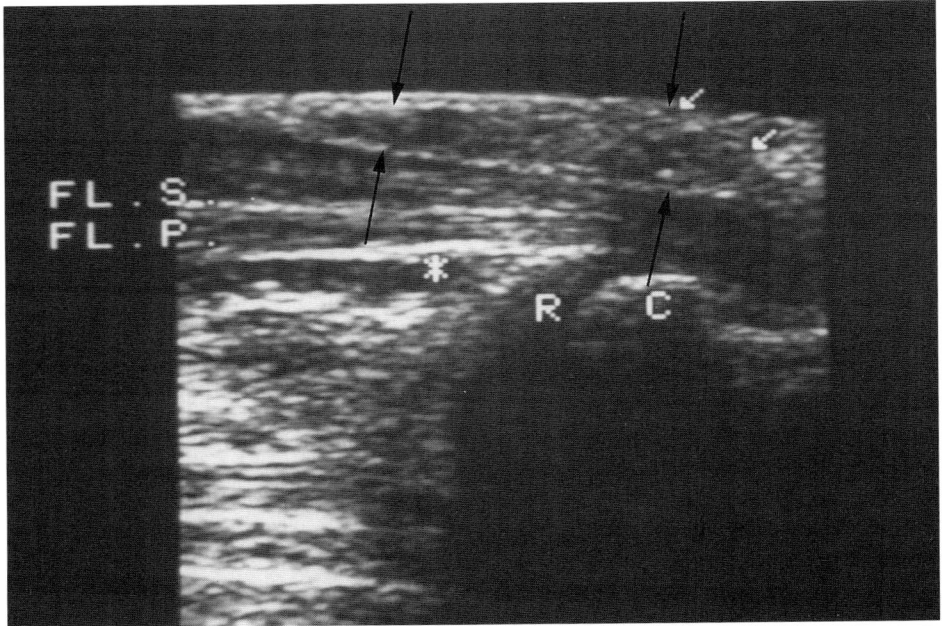

Fig. 11-9 Carpal tunnel syndrome caused by a lipoma. **(A)** Longitudinal sonogram shows an elongated hypoechoic mass (*arrows*) anterior to the superficial (FLS) and deep (FLP) flexor tendons in the wrist. There is also mild tenosynovitis (*asterisk*). RC, radiocarpal joint. (Figure continues.)

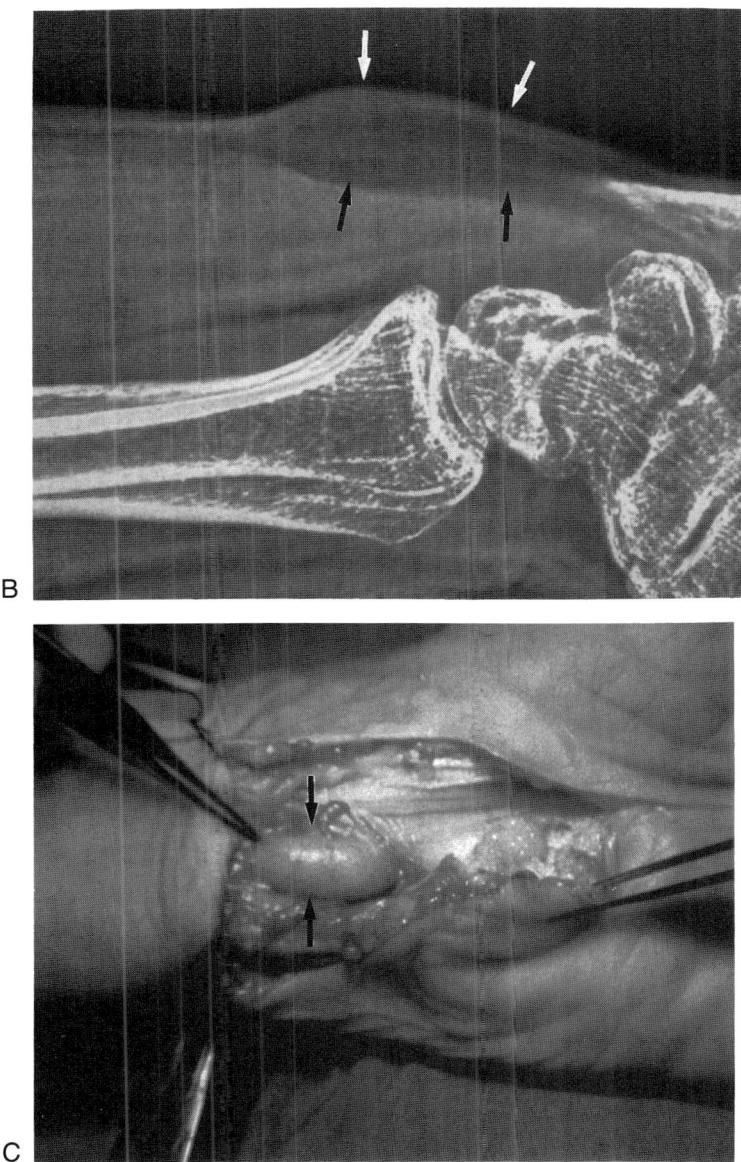

Fig. 11-9 (*Continued*). **(B)** Xeroradiograph confirms the radiolucent lipoma (*arrows*). **(C)** Intraoperative photograph of the lesion (*arrows*).

the presence of a communication with the adjacent joint space can be easily documented. Other causes of nerve compression, such as anomalous hypothenar muscles or tumors, are less common.

The major complaint may be pain, often of an ill-defined nature, in combination with weakness and dysesthesia of the fourth and fifth fingers. Guyon's canal syndrome more frequently results in motor deficit than does

162 MUSCULOSKELETAL ULTRASOUND

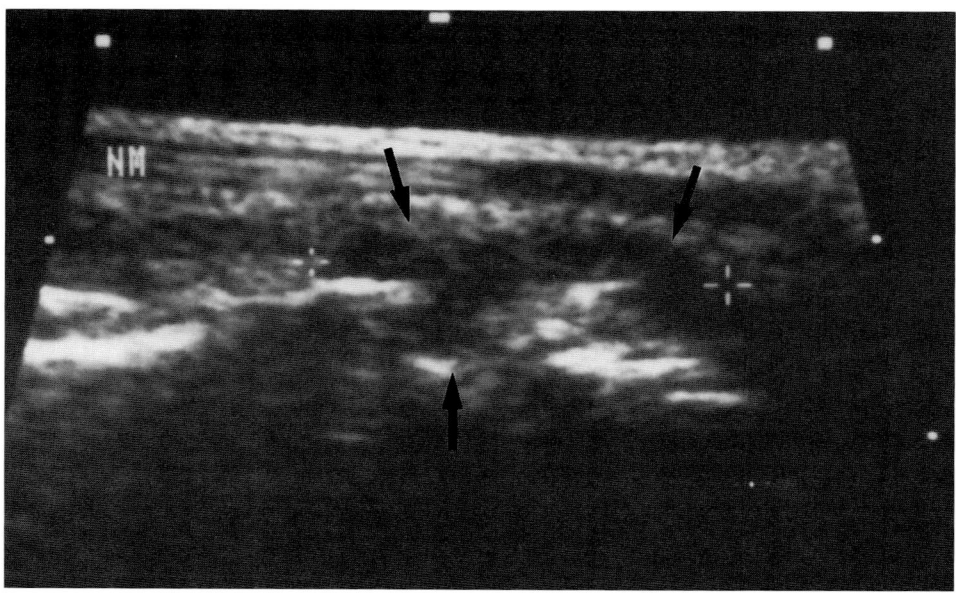

Fig. 11-10 Carpal tunnel syndrome in a patient with rheumatoid disease. Longitudinal sonogram of the wrist shows the hypoechoic pannus in the radiocarpal joint (*calipers and arrows*) pressing on the posterior aspect of the median nerve (NM).

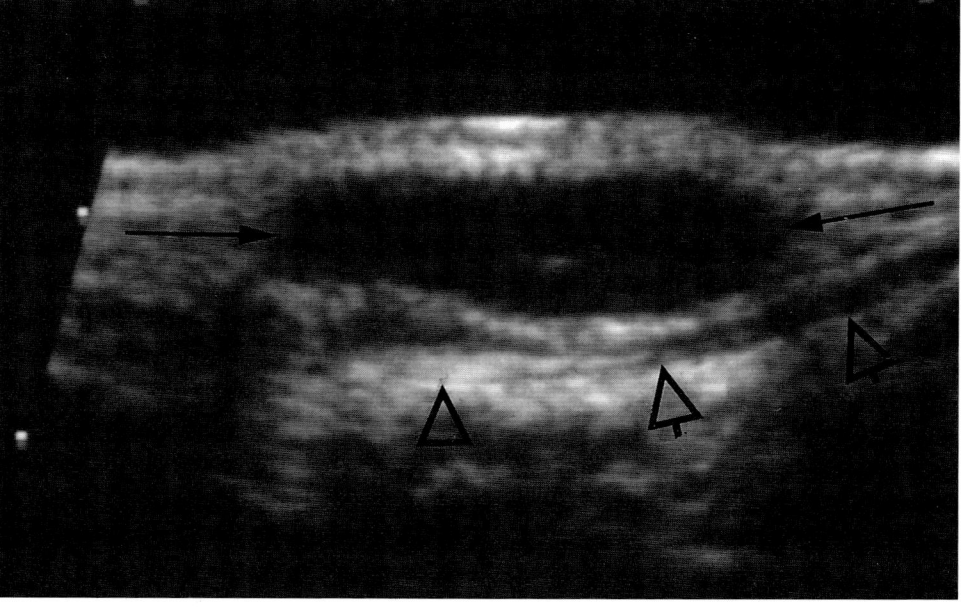

Fig. 11-11 Ulnar tunnel syndrome. Longitudinal sonogram of the ulnar aspect of the volar aspect of the wrist shows a ganglion (*arrows*) compressing the ulnar nerve (*arrowheads*) in Guyon's canal.

median nerve compression in carpal tunnel syndrome.

TRAUMA

Hand joint injuries often result in capsular-ligamentous lesions that produce painful swelling and long-lasting instability of the joints. Early diagnosis of a ligamentous rupture allows correct surgical and physical therapy.[17,18]

Traumatic Injuries to the Thumb

Traumatic lesions of the capsule and ligaments of the MP joint of the thumb occur frequently and are often misdiagnosed, even in professional athletes. The ligaments of this joint are particularly exposed to acute injuries and repetitive trauma ("gamekeeper's thumb").[19]

Stener's lesion is a rupture of the ulnar collateral ligament of the MP joint of the thumb in which the proximal stump of the ligament comes to lie superficial to the adductor pollicis aponeurosis, at the ulnar side of the MP joint, thereby preventing spontaneous healing.[20,21] Sometimes the ligament is avulsed with a fragment of bone, in which case the lesion will be detectable by standard radiographic examination. Stener's lesions lead to chronic instability, jeopardizing thumb function and requiring surgical repair. A Stener's lesion is suspected clinically on the basis of incompetence of the ulnar collateral ligament. Rupture of the radial collateral ligament of the MP joint of the thumb is less frequent; spontaneous healing is usually possible with splinting.

Because plain-film radiography cannot directly visualize injured capsular-ligamentous components of the MP joint of the thumb, the modality's basic value lies in its ability to exclude the possibility of fracture. Widening of the articular space is considered an indirect radiographic sign of ligamentous rupture. However, its demonstration requires conventional radiographs obtained during forced stress of the joint, and this test, usually very painful, cannot be used in cases of severe trauma.

Sonography yields very good images of the components of the MP joint, particularly the capsule, the collateral ligaments, and the joint space. The ulnar collateral ligament appears as a bandlike thickening of the capsular plane. The joint space appears as a hypoechoic fissure between the bones, the surfaces of which are sharply outlined by ultrasound; the width of the joint space can be measured at rest and under stress (Fig. 11-12). Capsular distension due to intra-articular hemorrhage is readily diagnosed by the presence of a fluid-filled hypoechoic cavity.

Following severe trauma, the width of the articular space increases during stress maneuvers; this increase can range from moderate widening to serious joint instability. Stress maneuvers can be performed by gently bending the finger under real-time sonographic monitoring. In the most severe trauma cases, rupture of the collateral ligament can be directly appreciated with ultrasound. In these cases, the capsular-ligamentous structures appear to be interrupted by an irregular hypoechoic tear. Its demonstration usually requires a careful time-consuming search by an experienced examiner. More often, some degree of inhomogeneity can be observed, without clear-cut evidence of rupture. In all cases, the distention of the joint capsules is well seen.

Other Traumatic Injuries

Injuries of the collateral ligaments of the other MP joints are less common and disabling. Usually they involve the radial collateral ligament of the third, fourth, or fifth finger.

164 MUSCULOSKELETAL ULTRASOUND

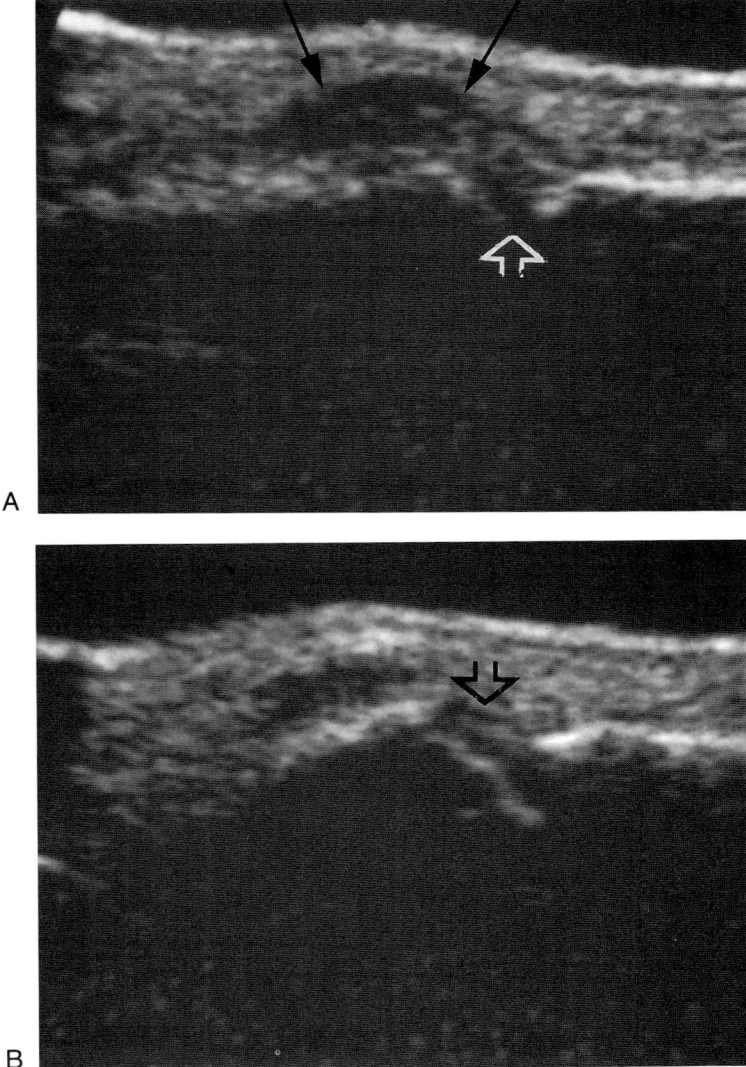

Fig. 11-12 Injured thumb. **(A)** Coronal sonogram taken at rest shows marked capsular distention by an intra-articular hemorrhage (*arrows*). *Open arrow* points to the joint space. **(B)** Under forced stress, the joint space is widened (*open arrow*).

Joint competence is preserved by intrinsic muscles, so instability is rarely a problem.

Ligament injuries to the proximal interphalangeal (PIP) joint are particularly common in the second finger; they are frequently associated with rupture of the volar plate.

Clinical findings consist of swelling and tenderness over the site of trauma. Complete ruptures can be diagnosed clinically during stress maneuvers.

Buttonhole (or boutonnière) deformity results from the rupture of the central slip of the extensor digitorum communis tendon over the dorsum of the PIP joint; the rupture causes a typical hyperextension of the distal interphalangeal (DIP) joint and flexion of the PIP joint.

The volar plate is formed by fibrocartilaginous tissue firmly attached to the base of the second phalanx. Volar plate injuries are commonly due to hyperextension of the PIP joint, which causes dorsal dislocation of the joint.

Mallet deformity ("baseball finger") of the DIP joint is one of the most common hand injuries noted in clinical practice. The lesion is caused by sudden forceful flexion of the finger when it is held in rigid extension. Avulsion of the dorsal lip of the distal phalanx may accompany subcutaneous rupture of the tendon. The patient complains of acute pain, with typical drop of the DIP joint. Sonography of the injured DIP joint is effective in the first days after trauma, before the associated hematoma resorbs. With the use of a high-frequency transducer, the tear of the attachment of the extensor tendon is clearly seen as a small fluid-filled collection around the distal portion of the tendon, adjacent to the bony phalanx. The hematoma often contains low-level internal echoes. A longitudinal sonogram provides a detailed view of the retracted upper tendon fragment. The tendon fragments are separated by a variable distance.

Avulsion of the flexor digitorum profundus tendon is rare. The condition is sometimes referred to as "jersey finger" because the lesion is often caused by entrapment of an actively flexing finger in a pullover shirt while dressing, resulting in forced extension. The avulsion may be associated with a chip fracture. This condition can be missed and the lack of DIP joint flexion misdiagnosed as being due to soft-tissue swelling.

FOREIGN BODIES

The presence of foreign bodies within the soft tissues is a common consequence of hand trauma (see also Ch. 3). Foreign bodies such as wood, glass, or plastic are very difficult to detect with conventional radiography owing to the usual small size of the fragments and to their low radiographic contrast with the surrounding tissues. Sonography allows a good representation of nearly all kinds of foreign bodies, irrespective of their radiographic density.[22,23] Because of the peculiar acoustic character of some materials, sonography sometimes allows determination of the nature of a foreign body.[24] Materials of strong acoustic impedance, such as pebbles, are markedly echogenic and cast an acoustic shadow like that associated with gallstones. Sonography provides a very sharp representation of glass and metallic bodies thanks to the associated comet-tail artifact.[25,26] Wood and other natural materials, such as sea urchin thorns, appear as echogenic structures with a moderate acoustic shadow. Foreign bodies that have been present for a long period may be surrounded by a hypoechoic halo due to the presence of a chronic inflammatory reaction (Fig. 11-13).

GANGLIA

Ganglia are cystic lesions arising from the synovial lining of a joint or tendon sheath; they are the most common swelling encountered in the hand.[9] Ganglia usually present as well-defined, cystic masses but on occasion may appear as more poorly defined, sessile, questionable masses. In the hand, ganglia develop most often in one of the following sites.

1. Dorsal aspect of the wrist: These ganglia generally arise from the scapholunate or midcarpal joints, are intimately related to the dorsal capsule, and are often associated with complaints of pain and weakness of the wrist.

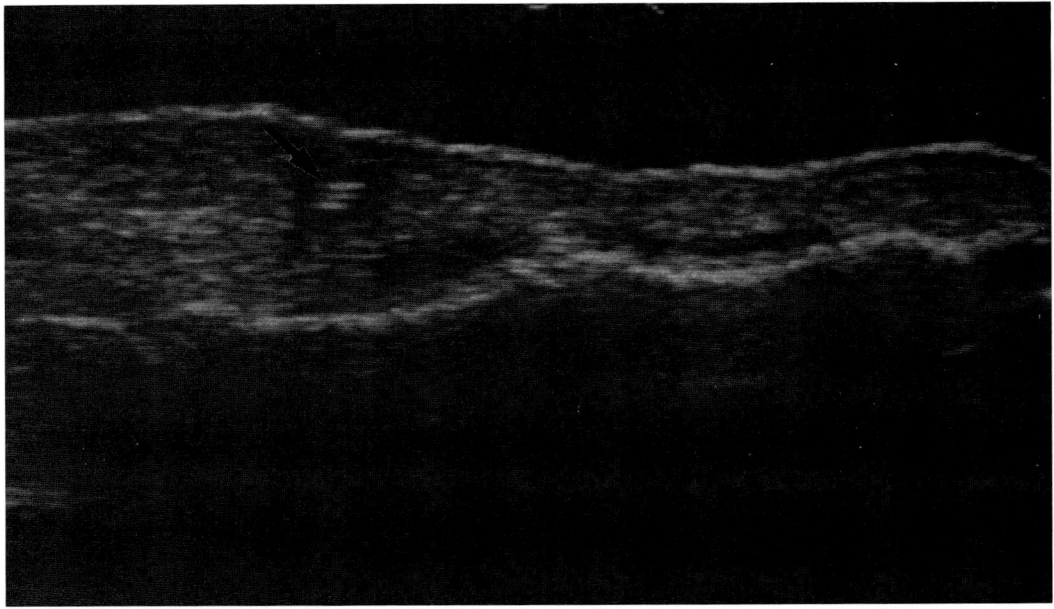

Fig. 11-13 Foreign body. Longitudinal sonogram of the fourth finger shows a small, echogenic, wooden foreign body (*arrow*) surrounded by a hypoechoic granuloma.

2. Anterior aspect of the wrist, lateral to the flexor carpi radialis tendon, close to the radial artery: These ganglia arise from the radiocarpal or inferior radioulnar joint.
3. Distal phalanx, overlying the nail bed: These ganglia originate in the distal interphalangeal joint.

Plain-film radiography should always be performed because an excess of bone, such as a carpometacarpal boss, may be misdiagnosed clinically as a ganglion. In addition, radiographs may detect bone erosion caused by a ganglion. Contrast arthrography can demonstrate a ganglion by its opacification through the communicating duct between the ganglion and the joint cavity. On the other hand, direct injection of contrast medium into the ganglion has generally failed to show such a communication, probably because of a valve mechanism.

High-resolution sonography has proved to be very effective in assessing ganglia.[10] Sonographically, ganglia appear as round or oval anechoic lesions with thin, regular margins (Fig. 11-14). Ganglia are sometimes grouped in clusters. A recurrent ganglion after surgery or conservative treatment (aspiration, squeezing, or local injection of drugs) may have thick blurred margins and internal echoes as a result of inflammation and changes in the cystic fluid. A diagnostic feature of ganglia is the presence of a small, anechoic communicating duct, originating from the deep aspect of the ganglion and extending to the articular space. In our experience, this duct, frequently longer than the width of the ganglion itself, can be demonstrated sonographically in 70 percent of cases.[10] This demonstration is very important for surgical planning because the duct should be carefully pursued at the time of surgery to reduce the risk of recurrence. Failure to demonstrate the duct sonographically is usually due to fibrosis of surrounding soft tissues and cyst margins, a common finding in long-standing ganglia. After surgery,

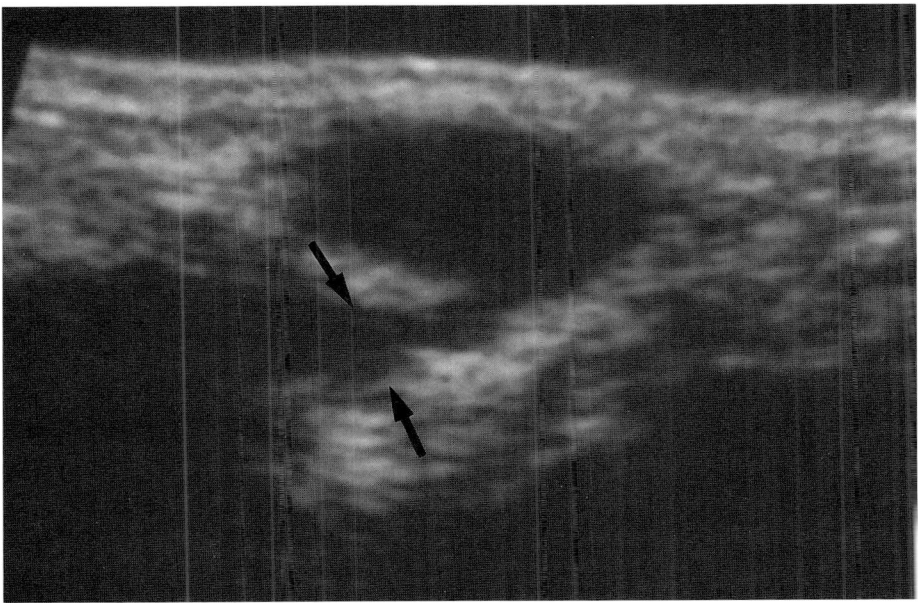

Fig. 11-14 Ganglion in the dorsal aspect of the wrist. Sonogram shows the cystic cavity and a thin duct (*arrows*) communicating with the articular space of the radiocarpal joint.

sonographic monitoring allows early detection of recurrent ganglia.

TUMORS

The hand is often affected by benign soft-tissue tumors. Neoplasms frequently interfere with hand function, requiring surgical excision. Sonography is often used, together with plain-film radiography, as the first-line diagnostic modality in patients with suspected soft-tissue tumors.[2] Although it does not offer the global imaging of CT or MRI, sonography is an effective and simple method for assessing soft-tissue neoplasms in the absence of bone abnormalities on plain radiographs. Sonography readily differentiates fluid-filled from solid masses, and it can delineate the extent of the lesion. It is also ideal for guiding percutaneous needle biopsy. Sonography usually cannot differentiate the histopathologic type of tumors.

Giant Cell Tumors

Giant cell tumor (the localized form of pigmented villonodular synovitis) is the second most frequent cause of swelling in the hand, after ganglia. These tumors most frequently arise at the palmar aspect of the fingers from the synovium of the flexor tendon sheath or the interphalangeal joints. They are generally asymptomatic and slow-growing, so the patient presents late, with a palpable mass that has the consistency of rubber. Sonographically, giant cell tumors appear hypoechoic with ill-defined contours (Fig. 11-15); there are no specific sonographic findings.[6]

Glomus Tumors

Glomus tumors arise from the neuromyoarterial glomus bodies, generally in the palm and the fingers.[27] The most common location is the fingertip, including the subungual space. Most patients present with spasmodic

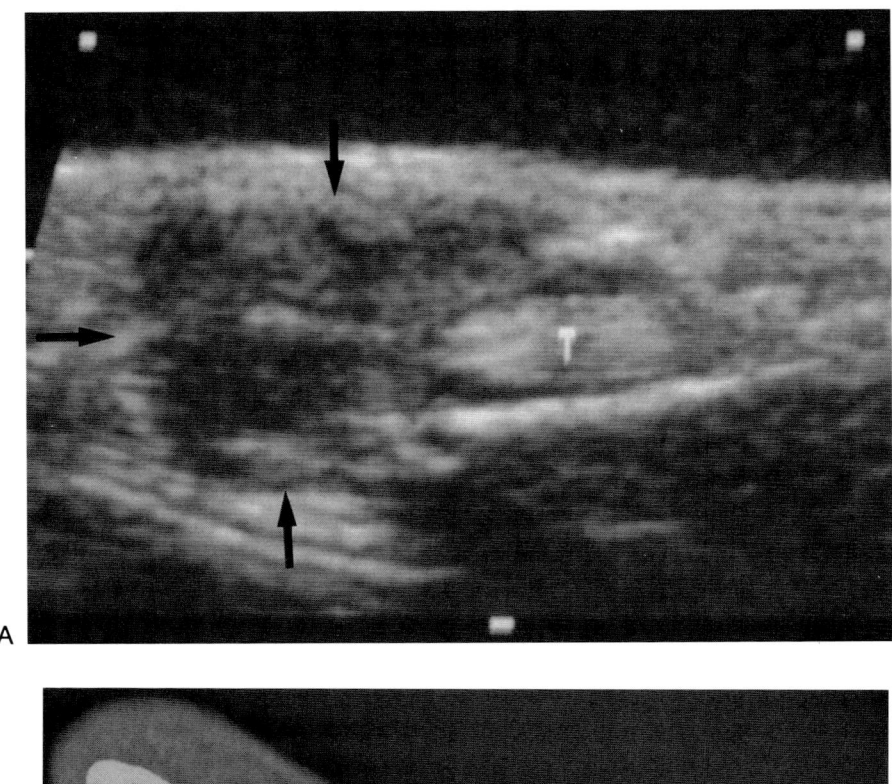

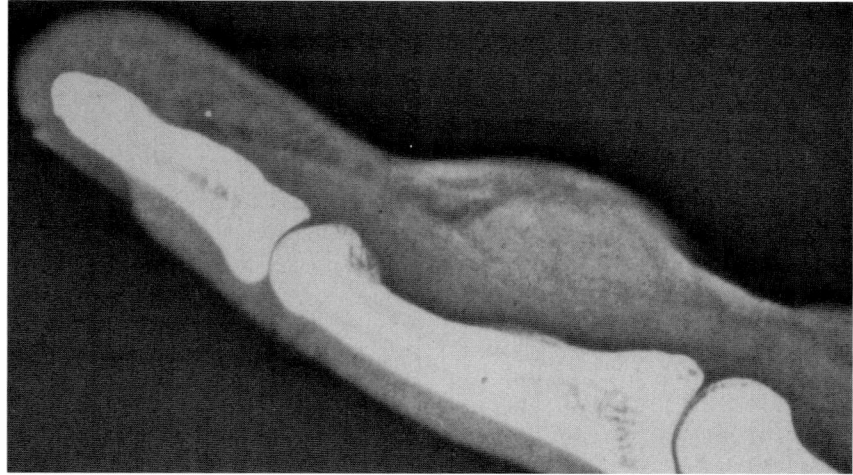

Fig. 11-15 Giant cell tumor. **(A)** Longitudinal sonogram shows a hypoechoic mass (*arrows*) arising from the synovium of the flexor tendons (T) of the third finger. **(B)** Xeroradiograph shows a nonspecific soft-tissue mass.

pain, temperature sensitivity, and focal tenderness. A soft-tissue mass is rarely noted because of the tumor's small size, seldom more than a few millimeters in diameter. High-frequency sonography can visualize a well-defined, round or oval, markedly hypoechoic mass (Fig. 11-16).[28] If the tumor is located in the subungual space, it is flattened and therefore more difficult to detect.

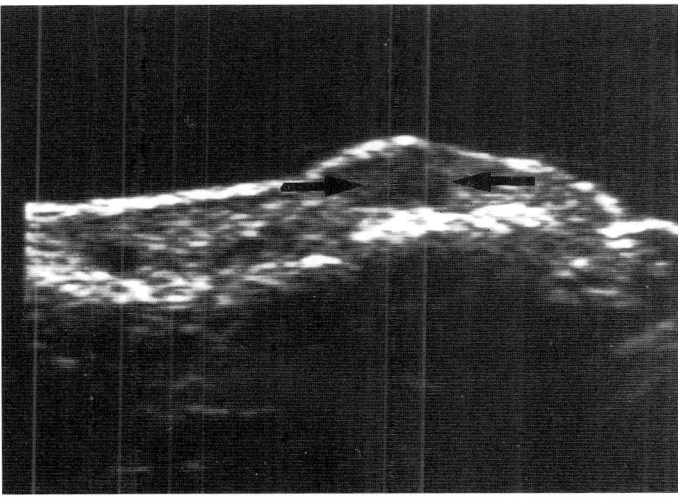

Fig. 11-16 Glomus tumor of the tip of the third finger. Longitudinal sonogram shows a rounded, hypoechoic, 6-mm mass (*arrows*).

Nerve Tumors

Neurofibromas and neurilemmomas are relatively rare in the hand.[29] There are no specific sonographic patterns, although such tumors are generally hypoechoic (see also Ch. 6). In the wrist, the median nerve also may be affected by a neuroma, particularly inside the carpal tunnel (Fig. 11-17). Although a nerve-related mass may be suspected when the lesion projects along the course of a major nerve trunk in a patient with typical neurologic symptoms,[30] when a nerve tumor arises from a small distal nerve branch, its neural origin cannot be suggested by sonography.

Lipomas

Lipomas are frequently found at the palmar surface, including the carpal tunnel[31,32]; here, they can reach considerable dimensions and cause compressive symptoms or interfere with the grasp mechanism.[33] White-yellow in color, lipomas are often (but not always) encapsulated. Although lipomas may be hypoechoic, isoechoic, or hyperechoic on sonograms,[34] they generally appear as oval masses, with the long axis oriented parallel to the muscle bundles and with well-defined margins. Typically, lipomas are hyperechoic, with thin, echogenic internal stripes parallel to the long axis of the tumor (Fig. 11-18). Low-kilovoltage radiography, xeroradiography, or CT is used to confirm the fatty nature of the mass.

Epidermoid Cysts

Epidermoid cysts arise following trauma that causes seeding of epidermal cells into the dermis, subcutaneous fat, or even bone. These cysts are often found in the fingertips. Epidermoid cysts appear as rounded subcutaneous masses, attached to the overlying skin. They are seldom painful. When the bone is involved, erosion is noted on radiographs. Sonographically, epidermoid inclusion cysts are round and hypoechoic, with rather regular margins. Sometimes, minute hyperechoic internal dots are seen, representing clusters of keratin.

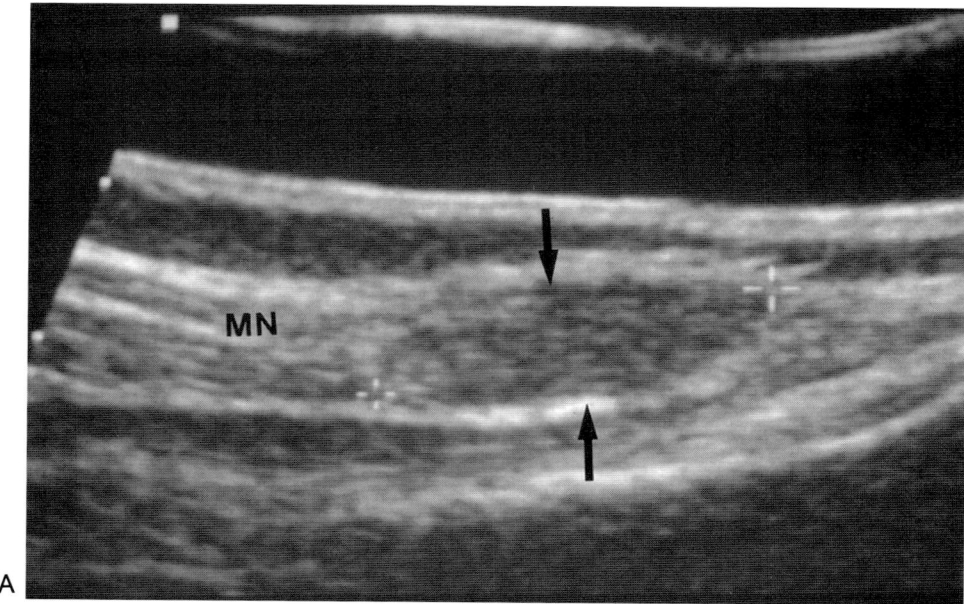

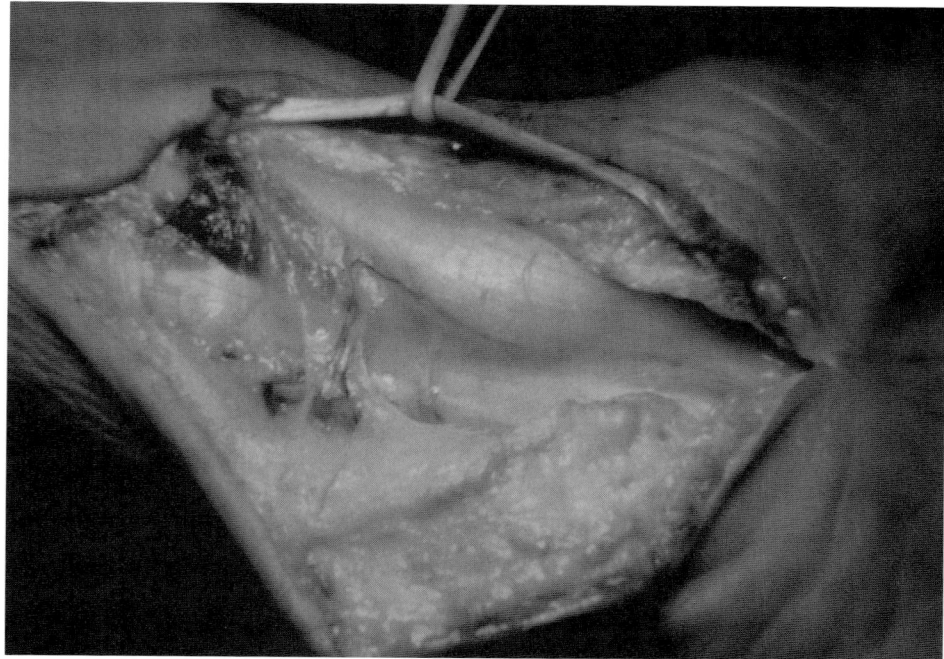

Fig. 11-17 Neuroma of the median nerve. **(A)** Longitudinal sonogram of the wrist shows an oval, hypoechoic mass *(calipers and arrows)* in continuity with the normal median nerve (MN) proximally. **(B)** Intraoperative photograph of the lesion.

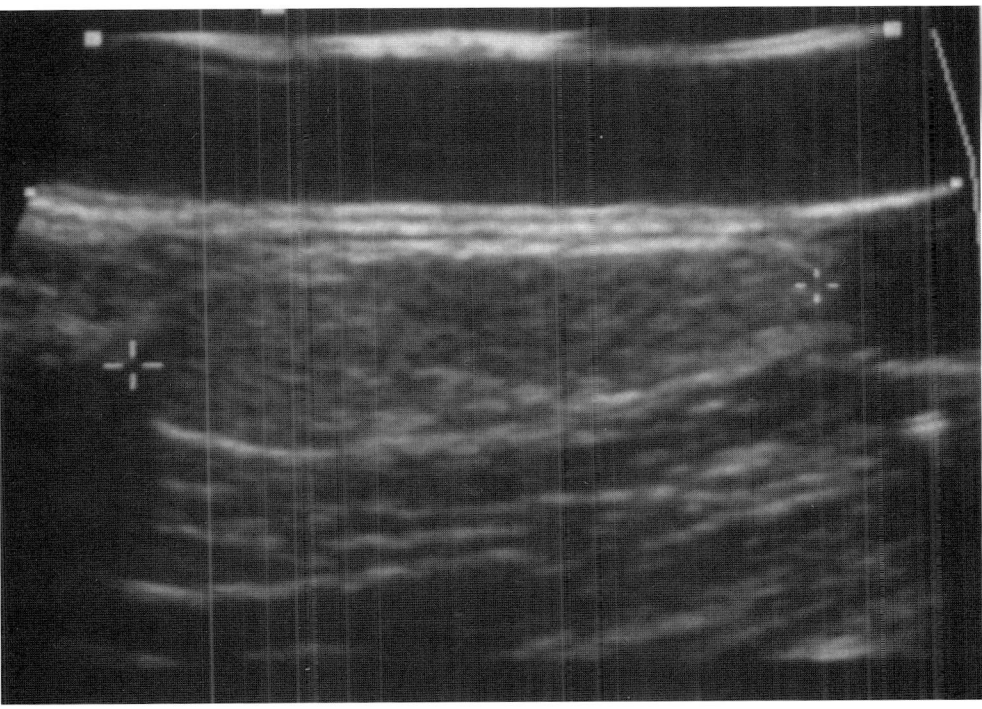

Fig. 11-18 Lipoma of the thenar area. Longitudinal sonogram shows an encapsulated, moderately echogenic, solid mass (*calipers*).

Hemangiomas

Hemangiomas are frequently observed in the hand. Some authors consider hemangiomas congenital vascular hamartomas.[35] Superficial angiomas appear as purple or sometimes brownish-red swellings, usually well-defined and with the consistency of soft rubber. These tumors are extremely variable in size. In some cases, the mass involves a large part of the hand.[36,37] Large intramuscular hemangiomas can infiltrate extensively into the muscle and may involve multiple muscles, thus creating surgical problems.[38,39] Plain radiographs may demonstrate phleboliths. Angiography is usually diagnostic. MRI has been reported as the first-choice imaging modality when an intramuscular hemangioma is suspected.[40] Sonography is of limited value because hemangiomas can be either hypoechoic or hyperechoic and are often poorly defined. In our experience, hemangiomas are most often hyperechoic.[41] Calcifications appear as bright foci with acoustic shadowing (Fig. 11-19). Color Doppler sonography generally cannot show detectable signals inside the mass because flow within the vascular spaces of these tumors is too slow.

Malignant Soft-Tissue Tumors

The hand is rarely involved by soft-tissue sarcomas, which are sometimes misdiagnosed as cysts or hematomas.[42] Because of the considerable amount of synovial tissue in the hand (in tendon sheaths and joints), synovial sarcoma is the most common soft-tissue sarcoma in the hand. Other, rarer

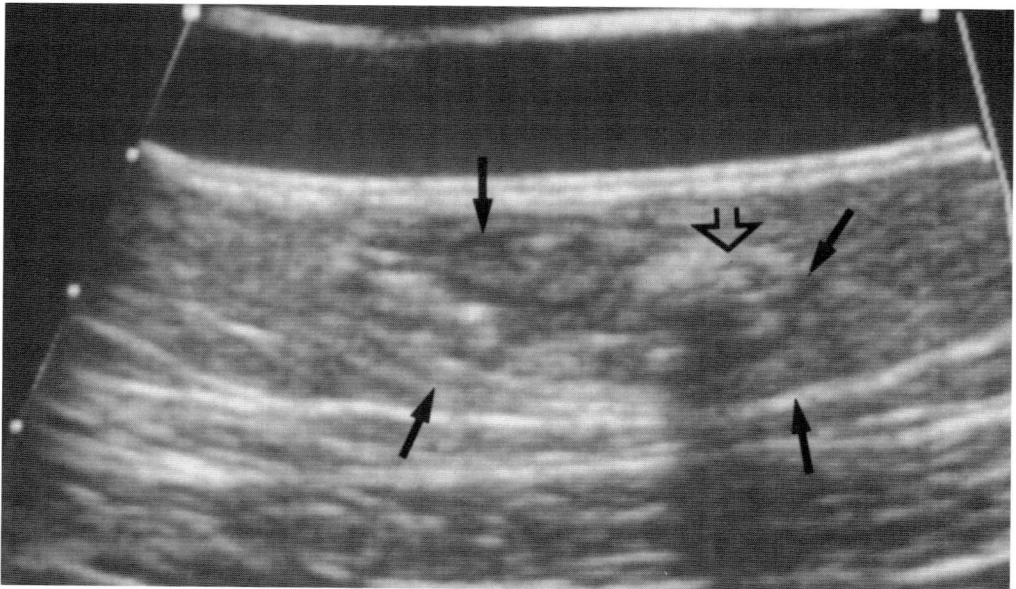

Fig. 11-19 Hemangioma in the subcutaneous tissues. Longitudinal sonogram demonstrates an inhomogeneous encapsulated mass (*arrows*). Internal calcification with distal acoustic shadowing can be seen (*open arrow*).

types of sarcoma are malignant fibrous histiocytoma, fibrosarcoma, liposarcoma, neurosarcoma, and angiosarcoma.[43]

Malignant soft-tissue neoplasms have been reported to display a variety of sonographic appearances, ranging from hypoechoic to highly echogenic (Fig. 11-20).[44,45] Tumors that do not contain fat or calcifications tend to be hypoechoic. Central necrosis, which commonly occurs with rhabdomyosarcomas, can simulate a complex cystic mass; sonography can guide the placement of the biopsy needle into the solid component.

Bone Tumors

Because bones markedly attenuate the ultrasound beam, CT rather than sonography is used to assess bone tumors. However, sonography can visualize soft-tissue involvement arising from malignant bone tumors (Fig. 11-21).

DUPUYTREN'S CONTRACTURE

Dupuytren's contracture is the retraction of the palmar aponeurosis due to fibromatosis, which gradually induces a flexion deformity of the MP and PIP joints. In the vast majority of cases, the patient (usually male) recalls repeated palmar microtrauma; in only a few cases is there evidence of familial occurrence of the contracture. The disease affects the fourth and fifth fingers most frequently.

Dupuytren's contracture often develops in alcoholics and in epileptics receiving long-term treatment with phenobarbital. The disease usually starts with the development of single or multiple palmar nodules. The disease can stop at this stage or, in a variable period of time, proceed to the formation of fibrous bands running through the palm and reaching the first phalanx, the PIP joint, and

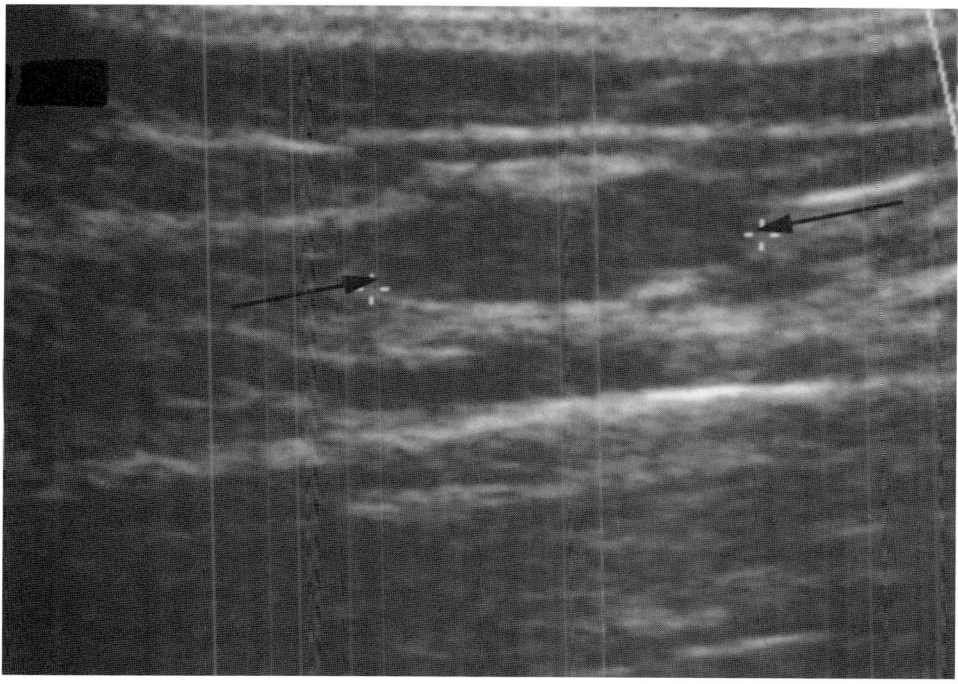

Fig. 11-20 Fibrosarcoma. Longitudinal sonogram shows a hypoechoic, lobulated mass in the distal forearm (*arrows*), involving the extensor muscles.

the second phalanx. Nodules and bands are usually asymptomatic. In the later stages of the disease, the palmar surface frequently shows dimples due to adhesion between the palmar aponeurosis and the skin. An early age at onset, a large number of digits involved, and the presence of knuckle pads are poor prognostic factors. Selective fasciectomy is the accepted treatment of established contracture.

Other fibromatoses, such as Peyronie's disease and Ledderhose's plantar fibromatosis, are related to Dupuytren's contracture.

Sonographically, fibromatous nodules appear as focal, fusiform, hypoechoic thickenings of the subcutaneous tissues, generally with well-defined margins (Fig. 11-22). In advanced cases, deeper planes may be involved, with retraction of the subcutaneous tissues.[46]

RHEUMATOID DISEASE

Although characteristic radiographic signs of rheumatoid arthritis are seen in advanced cases, the early diagnosis of rheumatoid disease is based mainly on soft-tissue changes involving tendons and joints.[47] These subtle changes may be difficult to appreciate with radiographs, even with xeroradiographs.[48] Although plain radiography remains the first diagnostic step in all patients, high-resolution sonography is an effective frontline diagnostic approach and a useful method for monitoring extra-articular rheumatoid disease.

174 MUSCULOSKELETAL ULTRASOUND

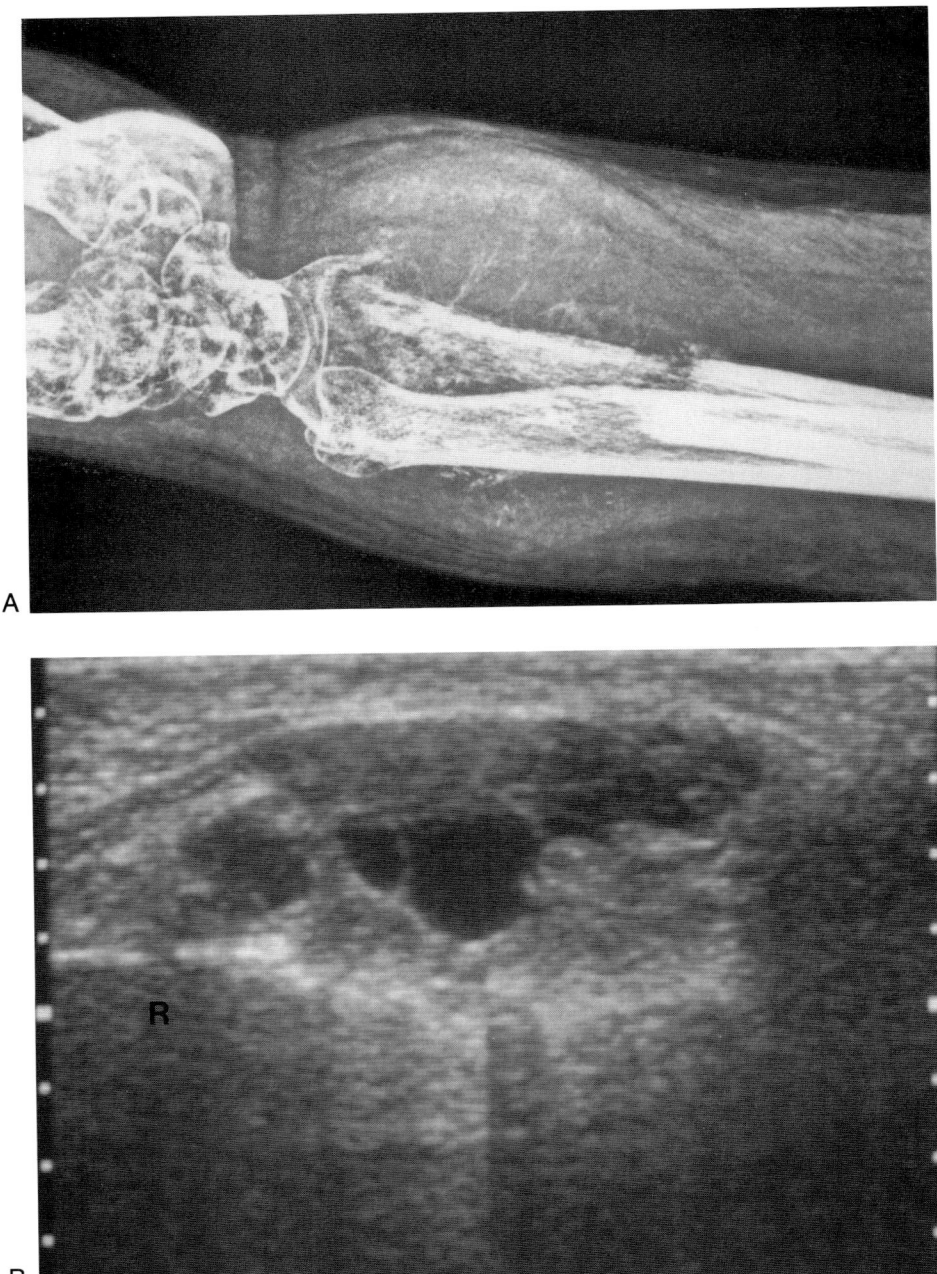

Fig. 11-21 Malignant Ewing sarcoma of the distal third of the radius. **(A)** Xeroradiograph shows an extensive destructive lesion with pathologic fracture and swelling of the surrounding soft tissues. **(B)** Sonogram shows a complex soft-tissue mass with liquefaction necrosis. Note the destruction of the cortex of the radius (R).

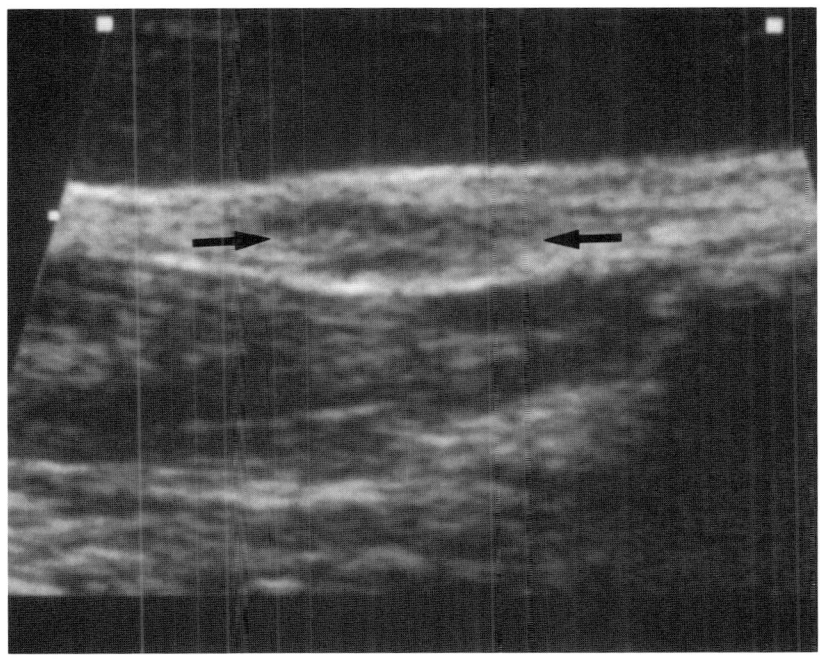

Fig. 11-22 Dupuytren's contracture. Longitudinal sonogram of the palm shows an inhomogeneous, spindle-shaped, hypoechoic mass (*arrows*) in the subcutaneous tissues.

In cases of joint swelling, sonography shows a widening of the articular space, which contains a hypoechoic or anechoic exudate (Fig. 11-23). Synovitis is best detected at the dorsal aspect of the joints. In addition, small subchondral bony erosions can be recognized.

In the wrist, a synovial swelling near the ulnar styloid process is a hallmark of rheumatoid disease (*caput ulnae syndrome*); this appears on sonograms as a hypoechoic mass surrounding the styloid process. Rheumatoid nodules in the periarticular soft tissues of the fingers appear as well-defined, hypoechoic, oval areas. The presence of such nodules has been shown to be a poor prognostic factor.[8]

Tenosynovitis of flexor or extensor tendons is one of the most important features of early rheumatoid disease and is observed in most patients with established disease. This condition may lead to rupture of the tendon itself. Although tenosynovitis and the resulting interference with the tendon blood supply predispose tendons to rupture, attrition due to friction against a rough or prominent bone is another significant factor. Tendon rupture in rheumatoid disease occurs only where tendons have a synovial sheath and the most frequent sites of rupture are close to bony prominences.

On longitudinal sonograms, enlarged tendon sheaths appear as well-defined anechoic or inhomogeneously hypoechoic, oval or spindle-shaped areas (Fig. 11-24).[11,12] The tendon is appreciated as a central hyperechoic ribbon, better studied during flexion or extension movements on real-time examination. Abnormalities of the tendon itself

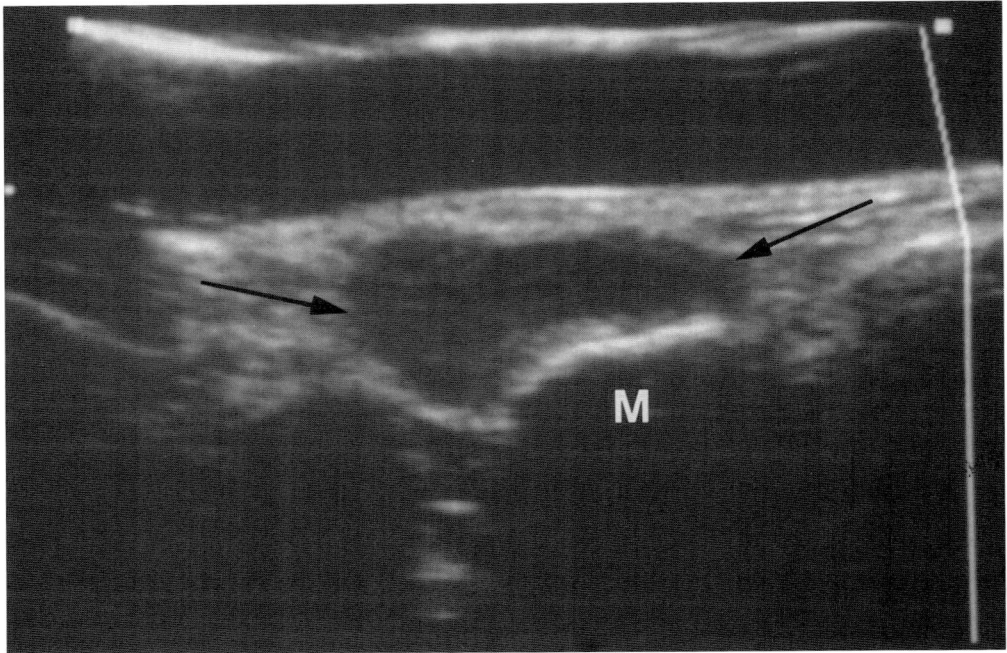

Fig. 11-23 Rheumatoid synovitis. Transverse sonogram of the dorsum of the hand in a patient with rheumatoid arthritis. The head of the third metacarpal (M) is in contact with a hypoechoic cavity (*arrows*) corresponding to synovitis and effusion of the MP joint.

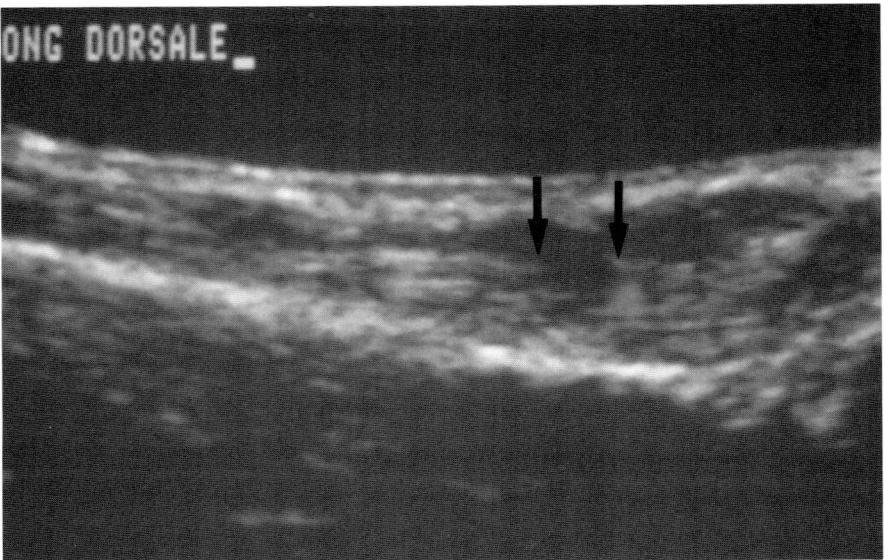

Fig. 11-24 Rheumatoid tenosynovitis with tendon rupture. Longitudinal sonogram of the dorsum of the wrist shows discontinuity of an extensor tendon of a finger (*arrows*).

can be recognized on longitudinal sonograms as focal areas of swelling or thinning or areas of inhomogeneous echotexture. In cases of complete rupture, the sonographic diagnosis is confirmed by the loss of tendon movement within the sheath on real-time examination. After surgical repair, sonography may provide useful information about the functioning of the tendon.

REFERENCES

1. Brunelli G, Saffar P: Wrist Imaging. Springer-Verlag, Paris, 1992
2. Braunstein EM, Silver TM, Martel W et al: Ultrasonographic diagnosis of extremity masses. Skeletal Radiol 6:157, 1981
3. Fornage BD, Rifkin MD: Ultrasound examination of tendons. Radiol Clin North Am 26:87, 1988
4. Vincent LM: Ultrasound of soft tissue abnormalities of the extremities. Radiol Clin North Am 26:131, 1988
5. Fornage BD: L'échographie de la main: technique et anatomie ultrasonore normale. Journal d'Echographie et de Médecine Ultrasonore 7:193, 1986
6. Fornage BD, Rifkin MD: Ultrasound examination of hand and foot. Radiol Clin North Am 26:109, 1988
7. Fornage BD: The hypoechoic normal tendon: a pitfall. J Ultrasound Med 6:19, 1987
8. Lister G: The Hand: Diagnosis and Indications. Churchill Livingstone, Edinburgh, 1984
9. Nelson CL, Sawmiller S, Phalen GS: Ganglions of the wrist and hand. J Bone Joint Surg 54A:1459, 1972
10. De Flaviis L, Nessi R, Del Bò P et al: High-resolution ultrasonography of wrist ganglia. J Clin Ultrasound 15:17, 1987
11. Fornage BD: Soft tissue changes in the hand in rheumatoid arthritis: evaluation with US. Radiology 173:735, 1989
12. De Flaviis L, Scaglione P, Nessi R et al: Ultrasonography of the hand in rheumatoid arthritis. Acta Radiol 29:457, 1988
13. De Quervain F: Uber eine Form von chronischer Tendovaginitis. Correspondenz-Blatt F Schweizer Aerzte 25:389, 1895
14. Finkelstein H: Stenosing tendovaginitis at the radial styloid process. J Bone Joint Surg 12:509, 1930
15. Harvey FJ, Bosanquet JS: Carpal tunnel syndrome caused by simple ganglion. Hand 13:164, 1981
16. Phalen GS: The carpal tunnel syndrome. J Bone Joint Surg 48A:211, 1966
17. Malone TR: Hand and Wrist Injuries and Treatment. Williams & Wilkins, Baltimore, 1989
18. Strickland JW, Retting AC: Hand Injuries in Athletes. WB Saunders, Philadelphia, 1992
19. Campbell CS: Gamekeeper's thumb. J Bone Joint Surg 37B:148, 1955
20. Moberg E, Stener B: Injuries to the ligaments of the thumb and fingers: diagnosis treatment and prognosis. Acta Chir Scand 106:166, 1953
21. Stener B: Displacement of the ruptured ulnar collateral ligament of the metacarpophalangeal joint of the thumb: a clinical and anatomical study. J Bone Joint Surg 44B: 869, 1962
22. Little CM, Parker MG, Callowich MC et al: The ultrasonic detection of soft tissue foreign bodies. Invest Radiol 121:275, 1986
23. Fornage BD: Sonographic diagnosis of foreign bodies of the distal extremities. AJR 147:567, 1986
24. De Flaviis L, Scaglione P, Del Bò P et al: Detection of foreign bodies in soft tissues: experimental comparison of ultrasonography and xeroradiography. J Trauma 28:400, 1988
25. Alfer'ev VI, Karpel'son AE: Echo-shadow method for visualizing foreign bodies. Med Tekh 5:52, 1978
26. Ziskin MC, Thickman DI, Goldberg NJ: The comet tail artifact. J Ultrasound Med 1:1, 1982
27. Varian JP, Cleack DK: Glomus tumours in the hand. Hand 12:293, 1980
28. Fornage BD, Schernberg FL, Rifkin MD et al: Sonographic diagnosis of glomus tumor of the finger. J Ultrasound Med 3:523, 1984
29. Strickland JW, Steichen JB: Nerve tumors of the hand and forearm. J Hand Surg (Am) 2:285, 1977
30. Fornage BD: Peripheral nerves of the extremities: imaging with US. Radiology 167:179, 1988
31. Morley GH: Intraneural lipoma of the me-

dian nerve in the carpal tunnel. Report of a case. J Bone Joint Surg 46B:734, 1964
32. Patel ME, Silver JW, Lipton DE et al: Lipofibroma of the median nerve in the palm and digits of the hand. J Bone Joint Surg 61A:393, 1979
33. Phalen GS, Kendrik JI, Rodriguez JM: Lipomas of the upper extremity, a series of 15 tumors in the hand and wrist and 6 tumors causing nerve compression. Am J Surg 121:298, 1971
34. Fornage BD, Tassin GB: Sonographic appearances of superficial soft-tissue lipomas. J Clin Ultrasound 19:215, 1991
35. Enzinger FM, Weiss SW: Soft-Tissue Tumors. Mosby, St. Louis, 1983
36. Bate TH: Hemangioma of the tendon sheath. J Bone Joint Surg 36A:104, 1954
37. Charache H: Tumors of tendon sheaths. Arch Surg 44:1038, 1942
38. Chauhan ND, Baird DS: Skeletal muscle hemangioma. An unusual case and a short review of the literature. J Ir Med Assoc 66:291, 1973
39. Cohen AJ, Youkey JR, Clagett GP: Intramuscular hemangioma. JAMA 249:2680, 1983
40. Yuh WTC, Kathol MH, Sein MA et al: Hemangiomas of skeletal muscle: MR findings in five patients. AJR 149:765, 1987
41. Derchi LE, Balconi G, De Flaviis L et al: Sonographic appearances of hemangiomas of skeletal muscle. J Ultrasound Med 8:263, 1989
42. Enneking WE: Staging of musculoskeletal neoplasms. Musculoskeletal Tumour Society. Skeletal Radiol 13:183, 1985
43. Russell WO, Cohen J, Cutler S et al: Staging System for Soft Tissue Sarcoma. American Joint Committee for Cancer Staging and End Results Reporting Task Force on Soft Tissue Sarcoma. American College of Surgeons, Chicago, 1980
44. Bruneton JN, Caramella E: Tumeurs des parties molles. In: Echographie en Pathologie Tumorale de l'Adulte. Masson, Paris, 1984
45. Peetrons P, Stienon M, Carlier L et al: Ultrasonographie des sarcomes des tissus mous. Journal d'Echographie et de Médecine Ultrasonore 5:305, 1984
46. Nessi R, Betti R, Bencini PL et al: Ultrasonography of nodular and infiltrative lesions of the skin and subcutaneous tissues. J Clin Ultrasound 18:103, 1990
47. Weston WJ, Palmer DG: Soft Tissues of the Extremities. A Radiologic Study of Rheumatic Disease. Springer-Verlag, Berlin, 1978
48. Verow PW, Dippy J: Soft tissue changes in early rheumatoid arthritis as seen on xeroradiography and non-screen radiographs. Clin Radiol 29:585, 1978

12
Hip in Infants and Children

H. Theodore Harcke

Real-time sonography is presently used in the diagnosis and management of developmental dislocation and/or dysplasia of the hip (DDH) in infants. Sonography offers several advantages over other imaging techniques, particularly in the first 6 months of life, when the femoral head and acetabulum are composed mainly of cartilage. On sonograms, one is able to distinguish the cartilaginous components of the acetabulum and the femoral head from the other soft-tissue structures. This is not possible with conventional radiographs, which show only the bony parts of the pelvis and the femur. Real-time sonography also permits a multiplanar evaluation, which clearly determines femoral head position with respect to the acetabulum. Hence, another limitation of the routine radiograph, the compression of three dimensions into two, is eliminated. The ability to observe changes in hip position during movement is a further advantage of sonography. Finally, by its ability to replace most radiographic studies, ultrasound evaluation can reduce the radiation exposure of the young infant. It is therefore easy to understand the wide acceptance of hip sonography throughout the world.

In older children, hip sonography is performed for a different purpose, the detection of joint effusion. Hip pain is a common presenting symptom in pediatric patients and can reflect a number of conditions, including those in which radiographic findings are absent or subtle early in their course. The presence of fluid in the hip joint is an important finding that may lead to diagnostic aspiration.

This chapter describes the use of sonography in DDH and in the detection of joint effusions. Techniques of examination are discussed, and normal and pathologic anatomy is illustrated.

DEVELOPMENTAL DYSPLASIA OF THE HIP

As ultrasound methods were developed, two basic philosophies evolved regarding the technique for studying the infant hip. Graf, who began the in-depth use of hip sonography, based evaluation on a coronal image of the hip obtained laterally with the femur in neutral position. His method emphasized angular measurements of acetabular landmarks in addition to assessment of hip position.[1-6] In contrast to a single-view approach, the Harcke technique emphasized a dynamic approach that assessed the hip in positions that mirror the Ortolani and Barlow maneuvers used by clinicians in the physical examination of the infant hip. Although the dynamic approach also considers acetabular development, its greatest emphasis is on position and stability of the femoral head.[7-11]

Despite differences in the approach to the examination, the two methods have common features, and in 1993 Harcke, Graf, and Clarke proposed a dynamic minimum standard examination that incorporates aspects of both techniques. A proper examination must identify the critical landmarks of the femur and the acetabulum. The examiner must ascertain the positional relationships in three dimensions and observe the changes that take place when the hip is stressed. The four-step examination presented here contains all the elements that are described in the proposed minimum standard hip examination.

Anatomy and Examination Technique

Knowledge of the anatomy of the infant hip is essential. At the time of birth, the femoral head and neck are composed of cartilage, which is hypoechoic compared with surrounding soft tissues and thus easily distinguished. On technically correct sonograms obtained with higher-frequency transducers, some scattered specular echoes can be seen within the cartilage.[11]

Between the second and eighth months of life, the ossification center develops within the femoral head. It usually appears earlier in females than in males. There is a wide range in time of appearance, and some asymmetry between the left and right hips in size and time of appearance of the ossification center may be seen. Sonography will show the ossification center several weeks before it is visible radiographically. This is due to two factors: (1) the initial confluence of blood vessels that precedes ossification produces increased echoes, and (2) when ossification begins, there is insufficient calcium to produce a visible radiographic density, yet sound waves are readily reflected. It has been shown that ultrasound can be used to assess the size of the ossification center.[12]

The acetabulum is composed of both bone and cartilage. Ossification centers develop in the ilium, ischium, and pubis, and by birth these centers create the acetabular cup. The individual components are separated by the three rays of the triradiate, or Y, cartilage. The cartilaginous acetabular labrum extends beyond the bony cup and is similar to the bony head in echogenicity, except at the lateral margin. The lateral labrum margin is composed of fibrocartilage and shows increased echogenicity. The hip capsule, composed of fibrous tissue, is anchored to the labrum and pelvis. It covers the femoral head and attaches to the femur. On sonograms, the capsule produces an echogenic band. The gluteus minimus and gluteus medius muscles overlie the hip joint.

Acoustic interfaces between the acetabulum and the femoral head are created by the articular surfaces, the ligamentum teres femoris, and connective tissue. These produce echo patterns that permit differentiation of the femoral head cartilage from the acetabular cartilage. Moving the hip can create echoes within the joint space, probably because of the formation of microbubbles.

Real-time linear-array transducers are preferred for hip examinations. Although it is possible to use curved linear and sector probes with success, the flat linear configuration produces less distortion of acetabular and pelvic anatomic landmarks. The highest possible frequency that permits penetration of the soft tissues should be used. For infants up to 3 months of age, a 7.5-MHz transducer is most successful. It usually offers sufficient penetration and the best resolution. A 5.0-MHz transducer is generally required between 3 and 7 months of age. (With improvements in focusing capabilities, a 5.0-MHz frequency can be used on some models of equipment for younger infants as well.) After 7 months of age, many infants are large enough to require a 3.0-MHz transducer for adequate penetration. Use of sonography to assess hip position becomes unreliable after 1 year of age in most children. By this time, the ossification center is large enough to inhibit visualization of acetabular landmarks. However, some infants with congenital hip dislocation, dysplasia, or both can be studied after 1 year of age because of the ossification center's delayed growth.

Four-Step Examination Technique

The complete examination is based upon a four-step (four-view) procedure. When it is performed in the following sequence, there is a natural progression that incorporates both dynamic and morphologic aspects of hip sonography.

Step 1. A coronal view with the hip in neutral position

Step 2. A coronal view with the hip flexed

Step 3. A transverse view with the hip flexed

Step 4. A transverse view with the hip in neutral position

The first view corresponds to the projection advocated by Graf for assessing acetabular development.[1] In the two views performed with the hip flexed, movement and application of stress to the femur are important parts of the evaluation. One maneuver is abduction-adduction, and the other maneuver is a push-pull or pistoning motion in the anteroposterior direction. Hip position in abnormal cases will vary depending on position and stress. A hip that is subluxated or dislocated in the neutral position may be reduced with flexion and abduction. Conversely, a hip that appears normal in neutral position and on passive flexion-abduction may subluxate or dislocate with the push-pull maneuver. The fourth view allows comparison of femoral head position at rest in a plane orthogonal to the plane used in the first view.

The four-step assessment is designed to determine the following information:

1. The position of the femoral head at rest in neutral and flexed positions (normal, subluxated, dislocated);
2. The relative stability of the hip with motion and stress (normal, lax, dislocatable, reducible, not reducible); and
3. The anatomic development of the hip components (the depth and configuration of the bony and cartilaginous portions of the acetabulum and the presence and size of the proximal femoral ossification center).

The four-step examination constitutes a complete assessment in which findings can be confirmed in more than one plane. When the examiner is learning to perform hip sonography, this confirmation adds to the accuracy of the examination. The examiner, particularly when experienced, has the option of omitting some steps and performing only the two views specified in the dynamic minimum standard examination (described later). A sufficient experience with both normal and abnormal hips should be acquired before the sonographer gives official reports. This will require at least 100 examinations of infants in the age range of birth to 6 months, with a representative spectrum of cases. Proper experience ensures that the accuracy of sonography remains high and that all can have confidence in the method.

CORONAL-NEUTRAL VIEW (STEP 1)

The hip is maintained in physiologic neutral position; in the infant, this represents about 20 degrees of flexion. The examiner should not force the femur into the horizontal position (hip extension). The transducer is maintained in the coronal plane with respect to the acetabulum (Fig. 12-1A). It is possible to obtain this view with the infant supine or in a lateral decubitus position. If the supine position is used, the transducer is held in the left hand when the right hip is examined and in the right hand for the left hip examination. If the lateral decubitus position is used, infants are placed on their left sides for the right hip examination and on their right sides for the left hip examination.

The standard plane is defined by a straight iliac line and the point where the iliac bone and triradiate cartilage join in the medial part of the acetabulum (Figs. 12-1B and C). The cartilaginous labrum extends from the acetabulum laterally over the femoral head. Most of the labrum is hypoechoic like the femoral head; however, at the tip of the labrum, the cartilage becomes fibrous and has increased echogenicity. The echogenic tip of the labrum must be visible in the standard plane for the image to be considered technically adequate.

The coronal-neutral view is used principally to assess development of the acetabulum. Considerable experience exists with landmarks in this view; these have been the basis for the alpha and beta angular measurements advocated by Graf (Fig. 12-1D).[2] Measurement is considered optional in the standard examination, and a complete discussion of measurement is not included in this chapter because we have not relied upon it in classifying DDH. Although acetabular morphology can be determined according to alpha and beta angles, it is also possible to use a verbal description that includes the appearance of the bony acetabular slope, the configuration of the lateral margin, and the appearance of the labrum. The issue of measuring acetabular angles has been controversial. Although Graf has stated that precise measurements can be obtained if his techniques are applied correctly, he also recognizes that classification can be determined visually. Measurement variations of from 4 to 6.5 degrees are reported for alpha angles, while beta angle measurement is more variable.[4] I experienced difficulty with interobserver and intraobserver variation and subscribe to the principle that hip position and stability are the major classification criteria. However, acetabular morphology is not ignored; I employ a visual assessment that notes the contour and depth of the acetabulum, the configuration of the labrum, and the presence or absence of an ossification center.

Normally, only the tip of the labrum, where the hyaline cartilage has changed to fibrocartilage, is echogenic. In the presence of acetabular dysplasia, instability, and femoral head displacement, changes in the appearance of the labrum can occur. The labrum may become deformed and fibrous throughout, and the extent of the echogenic component will increase. When the hip is dislocated, the labrum may be interposed between the femoral head and the acetabulum, preventing reduction (Fig. 12-1E).

In correlating the sonographic appearance of the acetabulum with the radiographic appearance, cases have been noted in which sonography showed the acetabulum to be better developed than it appeared radiographically. Less commonly, an acetabulum that appeared better developed radiographically than sonographically has been seen.

CORONAL-FLEXION VIEW (STEP 2)

The hip is flexed to 90 degrees; the transducer is maintained in the coronal orientation used in the coronal-neutral view (Fig. 12-2A). Again, with respect to the acetabu-

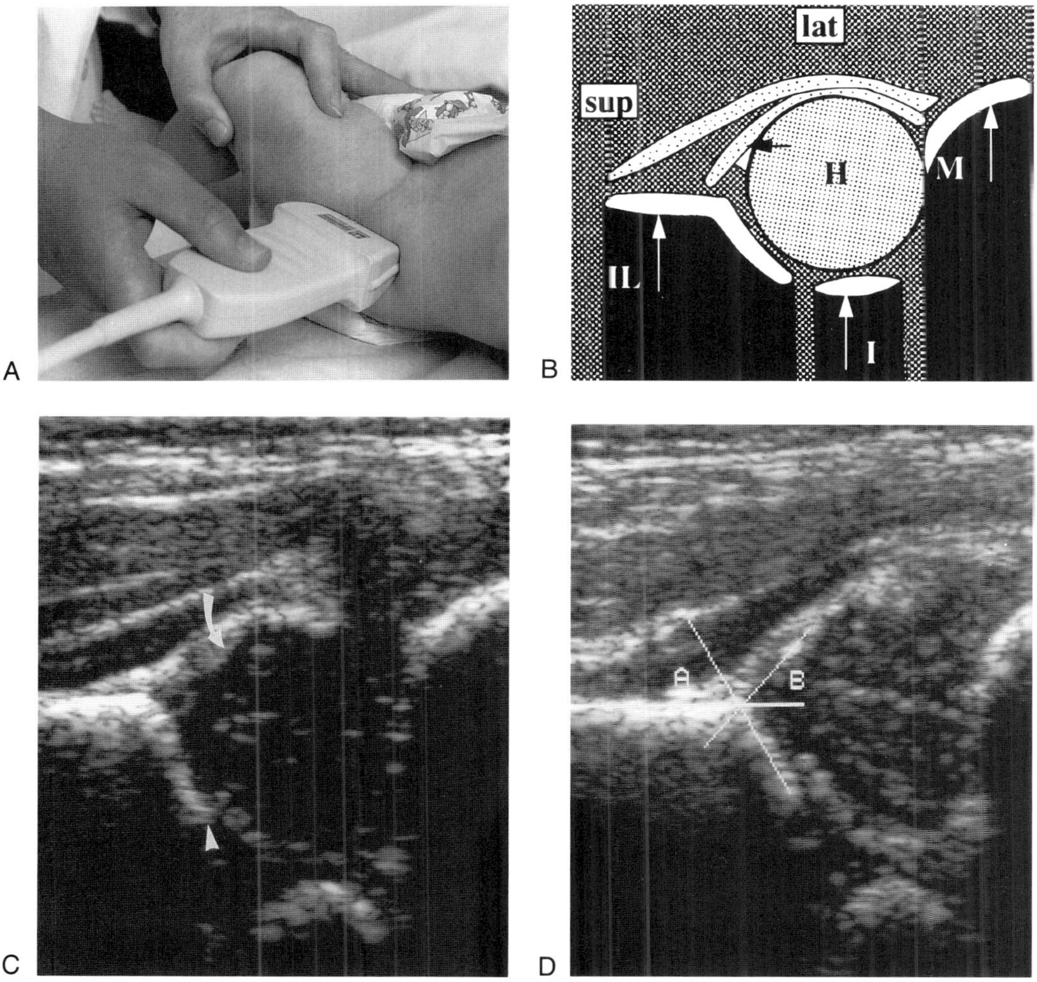

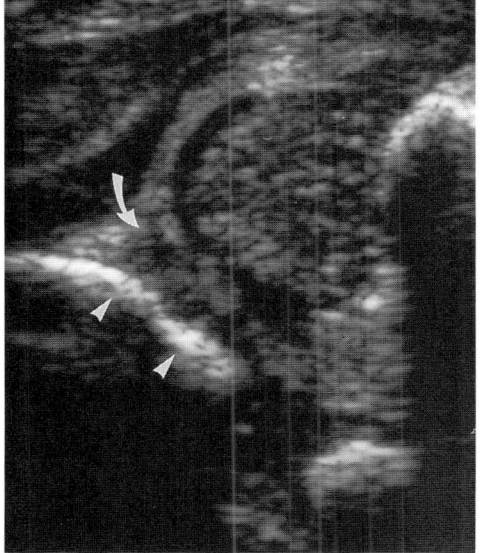

Fig. 12-1 Coronal-neutral view. **(A)** Transducer position over lateral aspect of the hip, which is in physiologic neutral position, slightly flexed. **(B)** Schematic of midacetabular (standard plane) sonogram. H, femoral head; I, ischium; IL, iliac bone; lat, lateral; M, femoral metaphysis; sup, superior; *black arrow,* tip of labrum. **(C)** Normal sonogram. Note tip of the labrum (*curved arrow*) and junction of the iliac bone and triradiate cartilage (*arrowhead*). **(D)** Sonogram showing alpha (A) and beta (B) angles used for optional acetabular measurement. **(E)** Sonogram showing hip dislocation. Note bony acetabular dysplasia (*arrowheads*) with deformity and increased echogenicity of the labrum (*curved arrow*).

183

184 MUSCULOSKELETAL ULTRASOUND

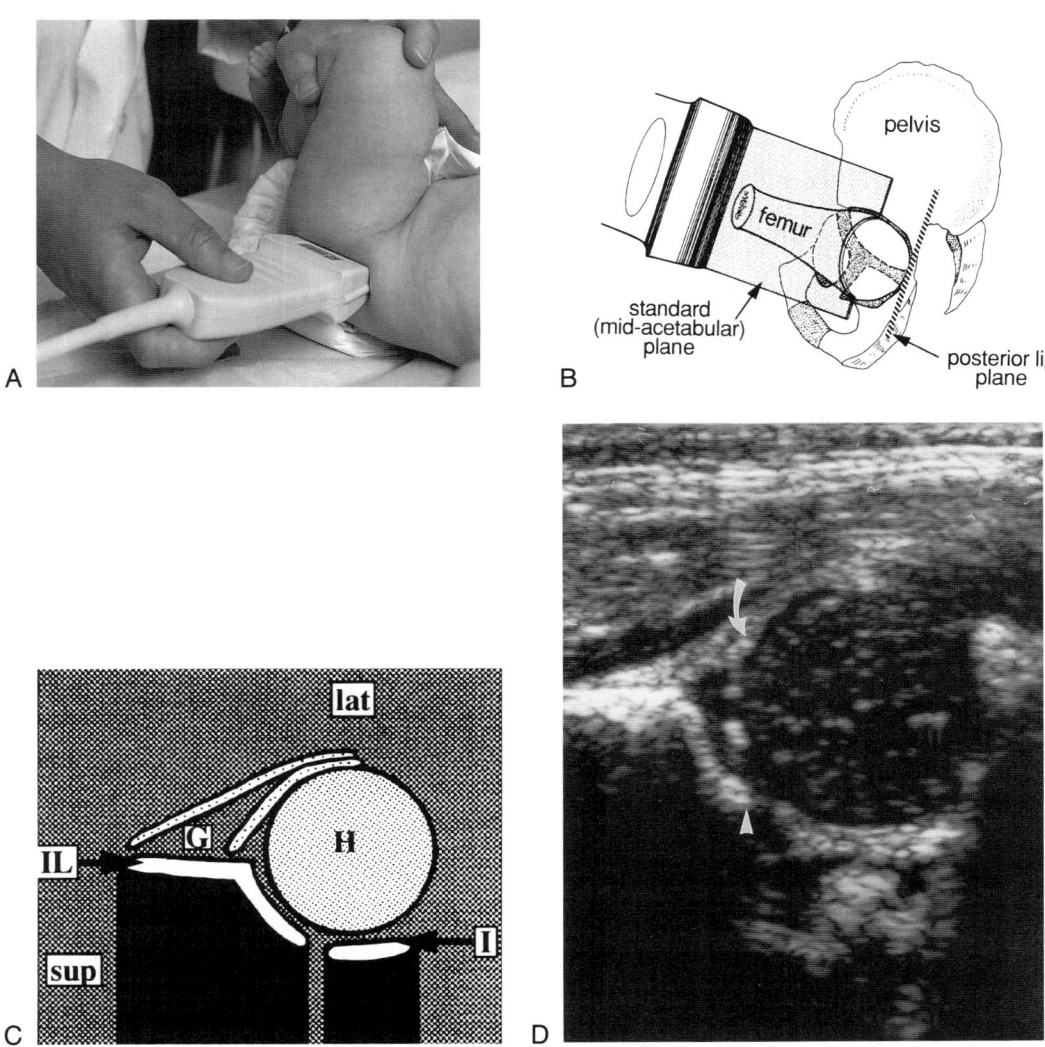

Fig. 12-2 Coronal-flexion view. **(A)** Transducer position over lateral aspect of the hip, which is flexed to 90 degrees. **(B)** Schematic showing transducer orientation for the standard plane (midacetabular) sonogram. For the posterior lip plane, transducer should be positioned as indicated by the dotted line. **(C)** Schematic of midacetabular (standard plane) sonogram. Landmarks are the same as in coronal-neutral view (see Fig. 12-1B) except for loss of femoral metaphyseal echoes. G, gluteal muscle; H, femoral head; I, ischium; IL, iliac bone; lat, lateral; sup, superior. **(D)** Normal sonogram in midacetabular (standard) plane. Note tip of the labrum (*curved arrow*) and junction of iliac bone and triradiate cartilage (*arrowhead*). (*Figure continues.*)

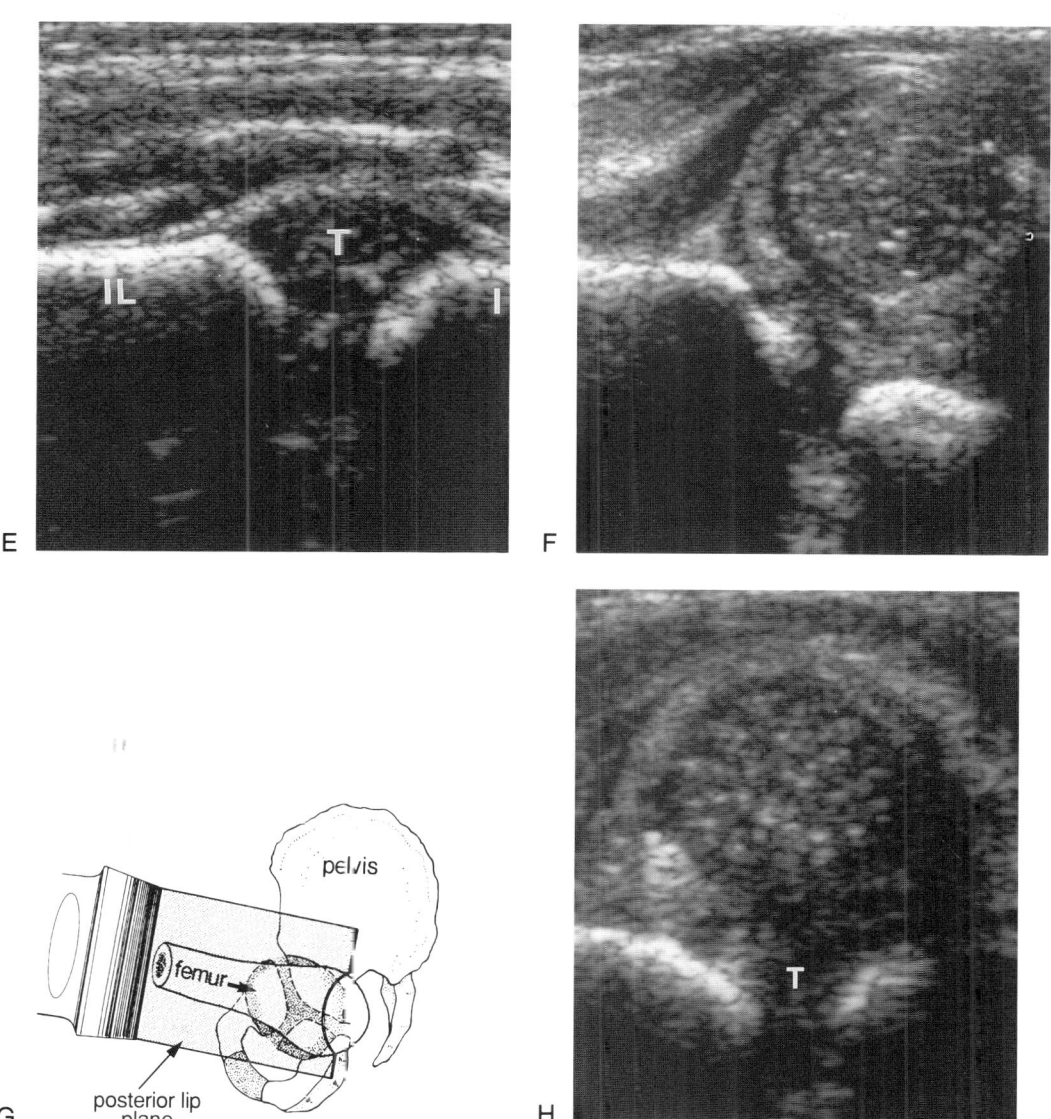

Fig. 12-2 (*Continued*). **(E)** Normal sonogram in posterior lip plane. IL, ilium; I, ischium; T, triradiate cartilage. **(F)** Sonogram showing hip dislocation. Note deformity and increased echogenicity in the labrum. **(G)** Schematic showing transducer orientation in posterior lip plane. An unstable hip will displace posteriorly with stress (*arrow*), causing the femoral head to appear in the plane. **(H)** Sonogram in posterior lip plane showing posterior dislocation of the femoral head. The head appears over the posterior lip of the triradiate cartilage (T).

lum, the standard plane is desired (Fig. 12-2B). During the examination, the sonographer will move the transducer in an anteroposterior direction so that the entire hip is evaluated. Anterior to the femoral head, the curvilinear margin of the bony shaft of the femur is encountered. Posteriorly, the back edge of the acetabulum and the posterior lip of the triradiate cartilage become a critical landmark (see Fig. 12-2B, dotted line). When the femoral head is positioned normally and the scan plane is directed into the midportion of the acetabulum, the medial part of the femoral head is surrounded by echoes from the bony acetabulum. Superiorly, the lateral margin of the iliac bone is seen (Fig. 12-2C and D). The appearance is likened to that of a ball on a spoon. The femoral head represents the ball, the acetabulum forms the bowl portion of the spoon, and the iliac line is the handle of the spoon. The configuration of the iliac bone changes as the transducer moves from anterior to posterior. Anteriorly, the bone is inclined laterally; in the midportion of the acetabulum, it has a straight configuration; and posteriorly, it exhibits some concavity. Knowledge of these changes in configuration helps the sonographer identify the correct coronal orientation for this view and the coronal-neutral view. Over the posterior lip of the triradiate cartilage, the femoral head will not be seen when the hip is in normal position and has no instability (Fig. 12-2E).

In subluxation, the femoral head is displaced laterally or posteriorly—or both—with respect to the acetabulum. Soft-tissue echoes are seen between the medial aspect of the femoral head and the bony reflection from the medial acetabulum. In dislocation, the femoral head is positioned posteriorly and/or superiorly (Fig. 12-2F). In posterior dislocation, the femoral head is seen lateral to the posterior lip of the triradiate cartilage, and the bony shaft of the femur will block echoes from reaching the acetabulum. In superior dislocations, the femoral head rests against the iliac bone.

The stress maneuver uses a push-pull or piston movement with the hip flexed and the transducer maintained in the coronal plane. The purpose of this maneuver is to demonstrate instability of the femoral head. Instability can vary from mild capsular laxity to dislocation. It is important that the infant be relaxed when this maneuver is performed. The transducer is maintained over the posterior lip of the acetabulum, and a firm but gentle push in the posterior direction is made against the knee. This will establish the presence of instability, which is shown by the appearance of a portion of the femoral head over the posterior lip of the triradiate cartilage (Fig. 12-2G and H).

In the coronal-flexion view, it is also possible to assess the configuration of the acetabulum. It is important to maintain the standard plane that best represents the acetabulum. The criteria are the same as with the coronal-neutral view. The iliac line must be straight and horizontal on the display monitor. The junction of the ilium and the triradiate cartilage and the echogenic tip of the labrum at the lateral termination of the acetabular cartilage must be visible. The principal difference between the standard plane images in the coronal-neutral and coronal-flexion views is the presence of the femoral shaft-metaphyseal echo in the coronal-neutral view. Optional measurement of alpha and beta angles can be done in the coronal-flexion view. Sonographic measurement of the coverage of the femoral head by the bony acetabulum has been studied in this view and correlated with radiographic measurements of acetabular angle.[13] The data show that femoral head coverage exceeding 58 percent correlates with normal radiographic measurements and that a coverage of less than 33 percent correlates with clearly abnormal radiographic measurements. There is a significant zone of indeterminate values for which sonographic and radiographic measurements do not correlate well.

TRANSVERSE-FLEXION VIEW (STEP 3)

The flexed hip is maintained at 90 degrees (without change from step 2), and the transducer is rotated to an axial or transverse plane. The transducer is situated posterolaterally over the hip joint; to facilitate this, the supine infant is rotated into an anterior oblique position, which is maintained by a rolled towel placed under the back (Fig. 12-3A). Anteriorly, the bony shaft and the metaphysis of the femur give bright reflected echoes adjacent to the sonolucent femoral head. Bony echoes from the acetabulum are seen surrounding the femoral head posteriorly. In the normal hip, this gives a U configuration (Fig. 12-3B to D). When the hip is subluxated, the femoral head is laterally displaced in relation to the acetabulum but remains in contact with the lateral part of the ischium. Subluxation becomes dislocation when the head moves laterally or posteriorly to the acetabulum (Fig. 12-3E).

In frank dislocation, the femoral head is positioned laterally, posteriorly, or superiorly. The U pattern cannot be obtained. A dislocated hip is tested for reducibility by pulling forward and abducting the flexed femur (the Ortolani test). Moving the flexed hip from maximum adduction to wide abduction demonstrates the relative stability of the femoral head. Wide abduction gives the most stable configuration. Adduction stresses the hip and, with the addition of a posterior push, can demonstrate subluxation in an otherwise normal-appearing hip. This maneuver (analogous to the Barlow test) can also show dislocation of a subluxated hip. With unstable hips, the process of dislocation and reduction is well visualized in the transverse-flexion view as the degrees of abduction and adduction are varied (Fig. 12-3E and F).

TRANSVERSE-NEUTRAL VIEW (STEP 4)

With the transducer remaining in the transverse plane, the flexed hip is returned to the physiologic neutral position. The transducer is now directed horizontally into the acetabulum. The plane of interest is one that passes through the femoral head into the acetabulum at the center of the triradiate cartilage. This plane can be found by starting the examination distally over the bony shaft of the femur and moving the transducer proximally until the cartilaginous structures become apparent (Fig. 12-4A). In the normal hip, the sonolucent femoral head is positioned against the bony acetabulum; the midpoint is approximately centered over the back edge of the triradiate cartilage. The echoes seen on the sonogram create the form of a flower: the femoral head is the blossom, the acetabular echoes are the leaves, and the stem is formed by the echoes that pass through the triradiate cartilage (Fig. 12-4B to D). If an ossified nucleus is present, one must angle the plane of the transducer above or below the nucleus to identify the triradiate cartilage. Acoustic shadowing by the ossification center must not be mistaken for the triradiate cartilage gap.

When the hip is subluxated, the transverse-neutral view will reveal soft-tissue echoes between the femoral head and the acetabulum (Fig. 12-4E). The width and configuration of the gap depend on the degree of displacement. Cartilage over the pubic (anterior) component of the acetabulum (Fig. 12-4D) is normally thick (and hypoechoic). This should not be mistaken for displacement. With instability, the femoral head commonly subluxates posteriorly and, in mild cases, remains in contact with the posterior aspect of the acetabulum. With more

188 MUSCULOSKELETAL ULTRASOUND

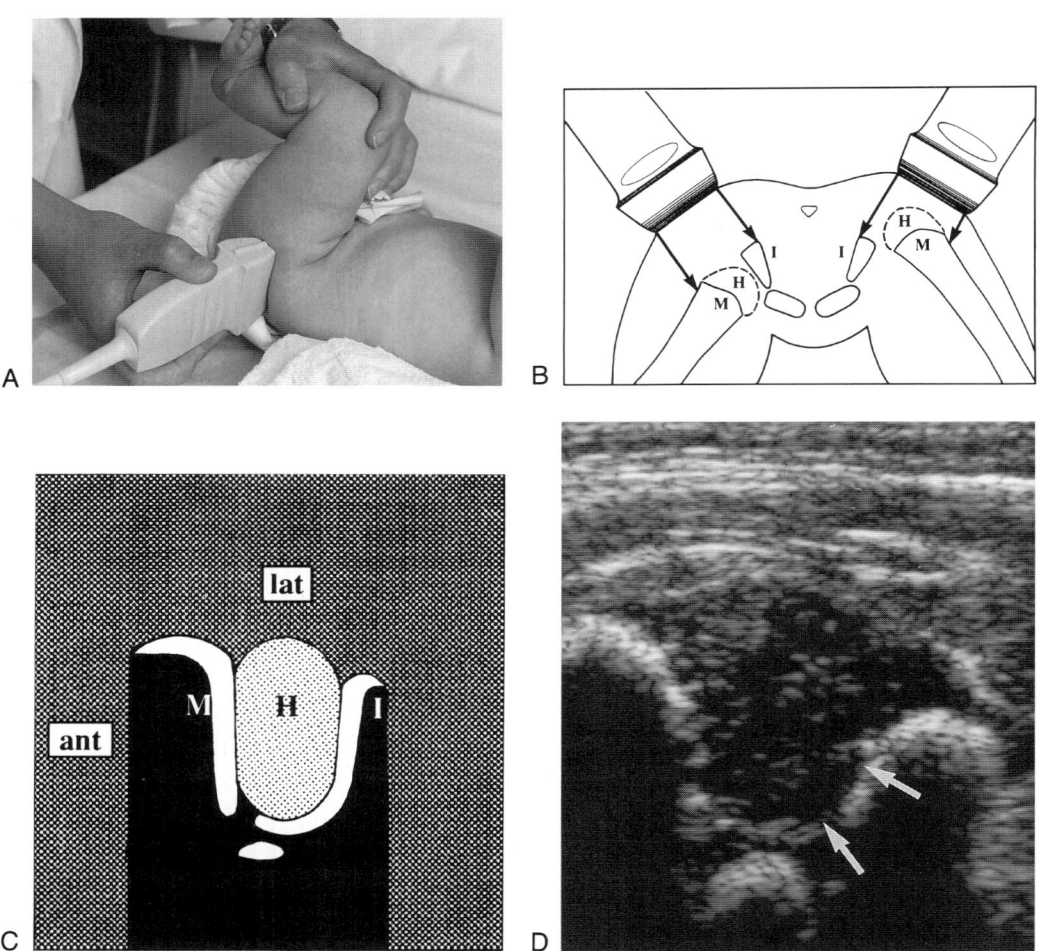

Fig. 12-3 Transverse-flexion view. **(A)** Transducer position over posterolateral aspect of the hip. Hip is flexed to 90 degrees. **(B)** Schematic showing transducer orientation and anatomic relationships in a normal hip (left) and dislocated hip (right). H, femoral head; I, ischium (acetabular component); M, femoral metaphysis. Arrows delineate the ultrasound field of view. **(C)** Schematic of sonogram. ant, anterior; H, femoral head; I, ischium (acetabular component); lat, lateral; M, femoral metaphysis. **(D)** Normal sonogram. Note that the femoral head is in contact with the ischium (*arrows*). (*Figure continues.*)

HIP IN INFANTS AND CHILDREN **189**

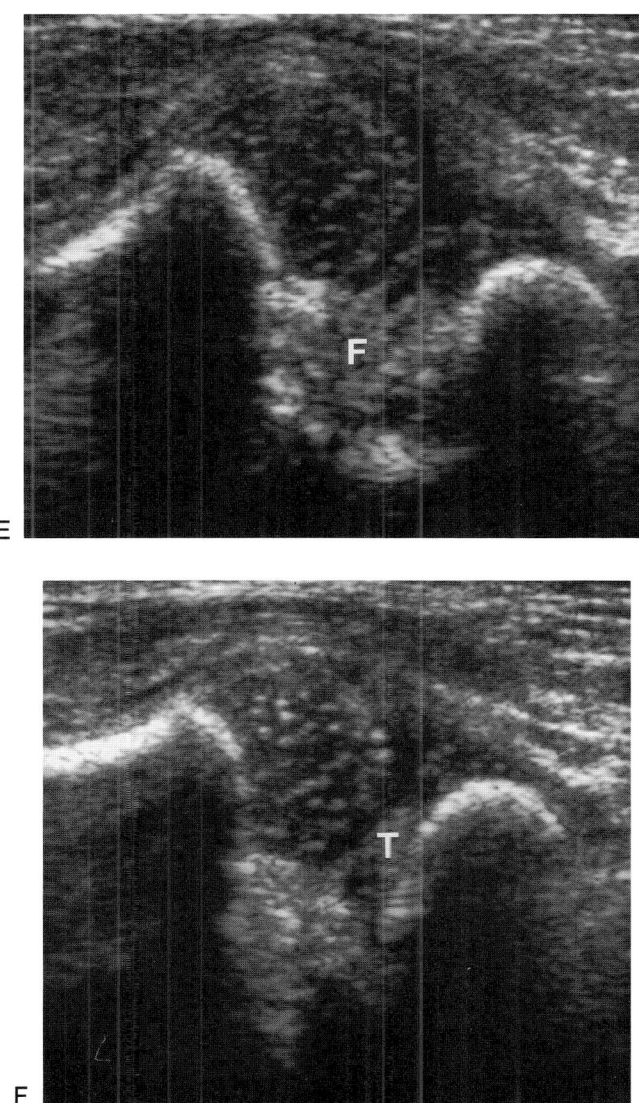

Fig. 12-3 (*Continued*). **(E)** Sonogram showing dislocation. The femoral head is displaced from the acetabulum, which is filled with fibroadipose (echogenic) tissue (F). **(F)** Partial reduction of the dislocated hip in Fig. 12-3E accomplished by abduction and pull maneuver. Some tissue (T) remains in the acetabulum and is interposed between the head and the ischium.

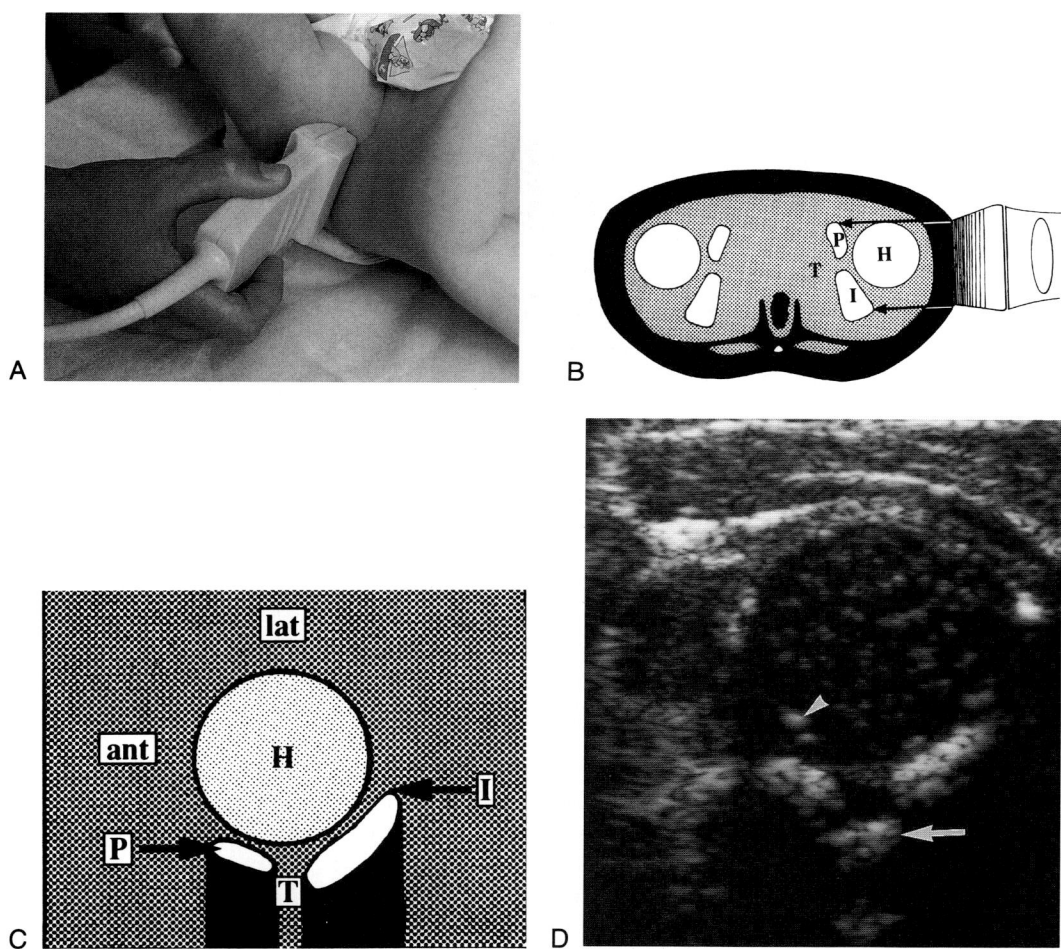

Fig. 12-4 Transverse-neutral view. **(A)** Transducer position over the lateral aspect of the hip. The hip is in physiologic neutral position, slightly flexed. **(B)** Schematic showing transducer orientation and anatomic relationships in a normal hip. H, femoral head; I, ischium; P, pubis; T, triradiate cartilage. Arrows delineate the ultrasound field of view. **(C)** Schematic of sonogram. ant, anterior; H, femoral head; I, ischium; lat, lateral; P, pubis; T, triradiate cartilage. **(D)** Normal sonogram. Note that the sound beam is able to pass through the triradiate cartilage (*arrow*) into an area where acoustic shadowing is produced by the bony pubis and ischium. Some physiologic echoes in the joint (*arrowhead*) allow differentiation of the femoral head cartilage from the articular cartilage of the pubis. (*Figure continues.*)

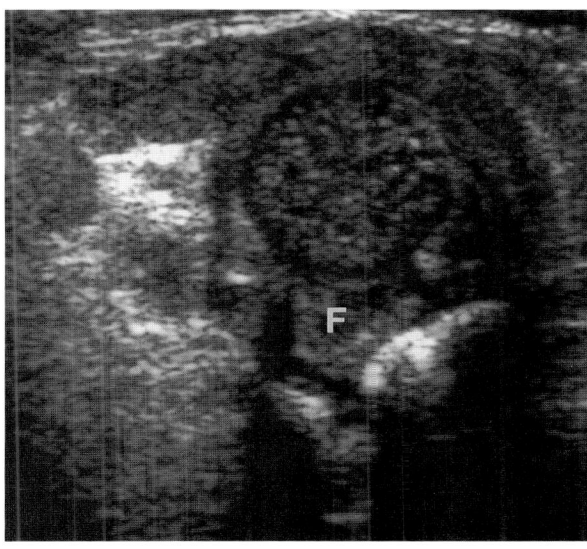

Fig. 12-4 (*Continued*). **(E)** Sonogram showing dislocation. The femoral head is displaced from the acetabulum, which is filled with echogenic fibroadipose tissue (F).

severe subluxation, lateral displacement accompanies the posterior migration. Most dislocations are posterior and superior. Because the dislocated femoral head will lie against some portion of the bony ilium, reflected echoes from the bone will be apparent medial to the head. The presence of dislocation, however, is distinguished from the normal hip by the inability to find the hypoechoic gap of the triradiate cartilage. Some dislocations exhibit significant lateral displacement, and since the femoral head does not rest against the bone, it will be completely surrounded by soft-tissue echoes (Fig. 12-4E). The size of the ossified nucleus can be evaluated in this view. In DDH, delayed appearance of the ossification center and retarded development compared with the normal contralateral hip are expected.

Dynamic Minimum Standard Examination

The complete four-step examination described in this chapter includes the essential elements proposed as the minimum standard for hip sonography by Harcke, Graf, and Clarke in 1993 (Table 12-1).[14] Adoption of a standard that includes both a morphologic evaluation and an assessment of stability was felt to provide all the information needed for adequate diagnosis and treatment. Shortening the examination to two views permits a faster study, particularly for experienced sonographers. For those just learning hip sonography there is value in doing a complete examination, particularly when a hip is abnormal or the examiner is uncertain of the findings. Sound technical elements of sonography are followed: a real-time linear-array transducer should be used, and examination in orthogonal planes is specified.

The hip is examined both at rest and when stressed. Acetabular morphology is assessed in a coronal-midacetabular plane (the standard plane). This can be done with the infant supine or on his or her side (lateral decubitus) with the hip either in neutral position or flexed to 90 degrees. One must therefore perform either the coronal-neutral view (step 1) or the coronal-flexion view (step 2). Note

Table 12-1. Dynamic Sonography 4-Step Method

	Hip Classification			
View and Maneuver	Normal	Laxity w/ Stress	Subluxated	Dislocatable/ Dislocated
Coronal-neutral[a,b] (standard plane)	N	N	A	A
Coronal-flexion[a,b] (standard plane)	N	N	A	A
Coronal-flexion (posterior lip)	N	N→A	A	A
No stress→piston stress				
Transverse-flexion[c]	N	N→A	N→A	A
Abduction→adduction				
Transverse-neutral	N	N	A	A

Abbreviations:
    N, normal;
    A, abnormal.

[a] Measurement of acetabular landmarks (angles/coverage) is optional. If performed, either the coronal-neutral or coronal-flexion view in the standard plane can be used.

[b] The dynamic minimum standard exam must contain either a coronal-neutral or a coronal-flexion view.

[c] The dynamic minimum standard exam must contain this view and include a stress maneuver.

that stress and angle measurements are optional in this part of the minimum standard examination.

The transverse-flexion view with stress (step 3) is a mandatory part of the standard examination. It can be performed with the infant in the supine (oblique) or lateral decubitus position. Stressing the flexed hip is a requirement. Adduction of the femur together with a posterior push constitutes the condition most likely to displace the femoral head from the acetabulum. This test of stability can be reversed with abduction of the femur and an anterior pull.

In classifying hip dysplasia, both the stability of the femoral head and the characteristics of the acetabulum are considered. Although these observations can be determined from two views as outlined, an examination should be supplemented with other views as required. The goal of this examination is to be accurate, not just to perform the minimum.

Other Views

Other approaches to imaging of the infant hip have been proposed. Although these will not be illustrated in this chapter, they are mentioned for the sake of completeness.[15-18]

Several anterior views have been described and utilized. In one of the first articles on real-time sonography, Novick et al described an anterior view performed with the hip flexed and abducted.[15] Gomes et al modified this approach with an anterior view that also is performed with flexion and abduction.[16] In addition, they have incorporated a dynamic stress test to demonstrate the presence of instability. Dahlstrom and associates' anterior view is done in an axial or transverse plane with the femoral head flexed and abducted.[17]

Suzuki et al proposed a method in which both hips are imaged simultaneously from an anterior approach.[18] A physically larger probe (3.5-MHz frequency) enables the hips to be seen on the same image with the femurs extended or flexed and abducted.

I have tried anterior views but use them only in unusual cases, always in addition to the lateral views presented above. I have found that anterior views can be helpful in patients with teratologic dislocations or congenital anomalies.

Clinical Experience

To evaluate the accuracy of hip sonography, sonographic findings have been compared with clinical and radiographic findings.[2,5,8,9,19,20] Sonography has been found to be highly specific and sensitive, with false-positive and false-negative rates in the 1 to 2 percent range. Dislocations have not been missed, and sonography has detected dislocations and other abnormalities that were not detected clinically or radiographically. The few false-positive and false-negative sonographic results have been in cases of subluxation. Subluxation encompasses a wide range of capsular laxity and displacement of the hip, and the assessment of the degree of subluxation is to some extent subjective. To be consistent, assessment requires that the patient be relaxed and that the examiner be experienced in the amount of stress applied to the hips. I have attributed my false-negative studies for subluxation to a lack of patient relaxation or to other technical difficulties in obtaining a satisfactory scan, e.g., the size of the patient. I feel that our false-positive cases with sonographic evidence of subluxation and normal clinical and radiographic findings reflect the greater sensitivity of sonography. It has been well documented that neither clinical nor radiographic examinations are highly accurate in the evaluation of DDH.[8]

In examining infants for DDH, one subset of particular interest is infants younger than 30 days. In this group, I have observed a high frequency of capsular laxity, which usually resolves without treatment in the first month of life. I believe this transient laxity represents one part of the spectrum of normal. This has influenced my protocols for screening and recommendations for treatment. When capsular laxity is seen in infants younger than 30 days, I recommend observation without treatment; repeat sonographic evaluation at 4 to 6 weeks of age can confirm normalcy, preventing unnecessary treatment.

Current applications of sonography for DDH can be divided into two categories, initial assessment and management. In most cases, referral for initial assessment is based upon a questionable or abnormal physical examination or an indicator of increased risk, such as breech delivery.[21,22] Routine mass screening for DDH has been done in small populations[6,19,20,23] and is now advocated in some areas of Europe. This is a controversial issue because of the resources that are required and the indications that newborn screening programs lead to overtreatment.[20,24]

An alternative to mass screening of all newborns is to screen a smaller segment of the population known to be at increased risk for DDH. In Clarke and colleagues' study of risk-factor-based screening,[21] not all cases of DDH in the population were detected. However, risk-factor-based screening seems to be a reasonable way to incorporate sonography into present protocols for infant screening that are based on clinical examination in the newborn period. My algorithm for hip sonography (Fig. 12-5) is based on the clinical examination and presence or absence of risk factors.[14] This scheme will require about 10 percent of all newborns to undergo an ultrasound study. It should be noted, however, that few will require treatment or follow-up examination because the ultrasound study is usually not obtained until 4 to 6 weeks of age, by which time minor and questionable abnormalities have resolved.

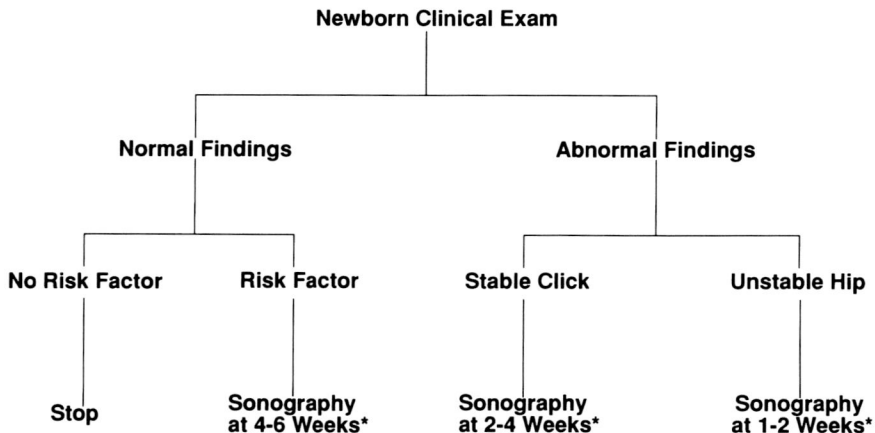

Fig. 12-5 The recommended protocol for integration of sonography with clinical screening programs depends on the results of the screening examination and the presence or absence of DDH risk factors. *Observe or treat based on results and clinical findings.

In infants with diagnosed DDH, sonography can be used to monitor hip position and acetabular development. This helps the orthopaedic surgeon determine if, when, and how to institute treatment. When the decision to treat is made, most of the devices used permit evaluation while the hip is held in flexion and abduction. A dynamic splint (the Pavlik harness) is commonly employed in the treatment of DDH, and sonography has been very helpful in monitoring harness treatment. Hip sonography can be repeated frequently without concern for radiation exposure, which can be significant when radiographs are used to monitor treatment of DDH.[25] Because of the dynamic aspects of the examination and because it provides three-dimensional information, continuing instability and posterior displacement are easily detected.[26] Conversely, the examination can determine when stability has been achieved and use of the harness can be discontinued.[22,25] A dislocated hip should seat itself within 3 weeks or the harness should be abandoned in favor of traction and closed reduction.[27] It should be noted that at the conclusion of full-time harness treatment, I obtain a radiograph of the pelvis to document bony acetabular development.

This sonographic technique has been adapted for infants treated with rigid casting; the modified examination requires either alternative anterior-groin views or cutting a window in the cast over the lateral aspect of the hip. Although this technique was used in the past, limited computed tomography with one or two axial sections through the hip is used at this time. This is fast and accurate, and it eliminates disturbing the molded cast.[22]

Summary

Many sonographers have had positive experiences with sonography in the diagnosis and management of DDH, and use of the technique is accepted throughout the world. Variations in technique exist: the four-step method presented incorporates a dynamic examination and a morphologic assessment of the acetabulum that allows for measurement if desired. It includes the elements of

HIP JOINT EFFUSION

Although sonography becomes unreliable for evaluation of DDH typically at about 1 year of age, it can still be used to assess the painful hip. Although radiography is the initial imaging technique of choice and is often diagnostic, there are frequent instances in which radiographs are normal in the presence of a joint effusion. Detection of such an effusion is important because a wide variety of conditions that cause hip pain in pediatric patients (e.g., transient synovitis, osteomyelitis, Perthes disease, slipped capital femoral epiphysis, fracture, and arthritis) have joint effusion as an early sign. Sonography can be used to determine whether an effusion is present and to guide a diagnostic or therapeutic arthrocentesis.

Anatomy and Examination Technique

The patient is examined in the supine position with the hips in neutral position without flexion (if possible). Both hips are examined because it is helpful to compare the symptomatic hip with the asymptomatic contralateral one. A high-frequency linear-array transducer is used to scan the hip in a ventral, oblique plane aligned with the long axis of the femoral neck (Figs. 12-6A and B). Bright echoes are produced by the anterior bony cortex of the femoral head and neck, with the intervening cartilage of the physis appearing sonolucent; the anterior margin of the bony acetabulum is visualized superiorly. Over the anterior recess of the joint, the capsule parallels the femoral neck. The outer margin forms an echogenic line anterior to the cortex of the femoral neck and extending over the femoral head. The iliopsoas muscle is superficial to the capsule (Fig. 12-6C).

In the normal hip, the joint capsule has a concave contour, and the thickness of the capsule from the outer margin to the cortex of the femoral neck measures from 2 to 5 mm. The hip capsules should be symmetric within 2 mm. When there is an effusion, the anterior recess of the capsule becomes distended, and a convex outer margin is produced. In the abnormal joint, the width of the joint space is more than 2 mm thicker than in the normal contralateral capsule (Fig. 12-6D).[28-30] Visual comparison is usually adequate for diagnosis, with measurement confirming the observations. The use of measurement alone to make a diagnosis without fulfillment of other criteria is not recommended but becomes necessary when both hips are abnormal. This is an infrequent occurrence.

Fluid of varying echogenicity can be seen within the capsule. The echoes are created by inflammatory debris or hemorrhage.[29] Some studies have indicated specificity with regard to the appearance of the fluid. Zieger et al reported the fluid to be anechoic or hypoechoic in transient synovitis and more echogenic in septic arthritis.[31] They concluded that, if the fluid is anechoic, the diagnosis of septic arthritis can be excluded. Other investigators disagree and have found the character of the fluid to be nonspecific.[28,32] These studies describe echoes (probably representing hemorrhage) in the fluid in transient synovitis and, conversely, report cases of septic arthritis in which the infected fluid was anechoic.

When fluid is detected, arthrocentesis can be performed under ultrasound guidance, if needed; a saline lavage should be used if fluid cannot be withdrawn. Although the procedure requires the patient's cooperation, it is relatively easy to execute, and it avoids the ionizing radiation required in fluoroscopi-

196 MUSCULOSKELETAL ULTRASOUND

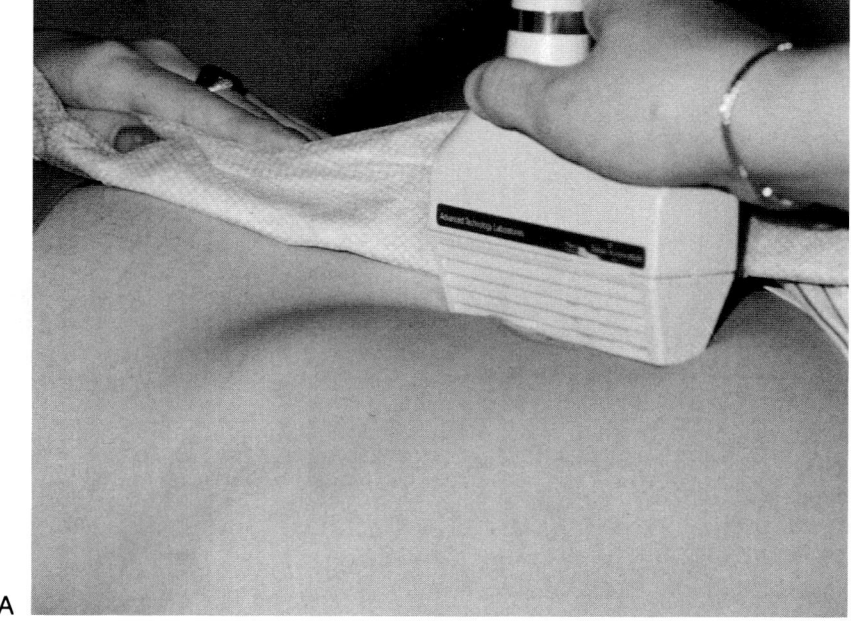

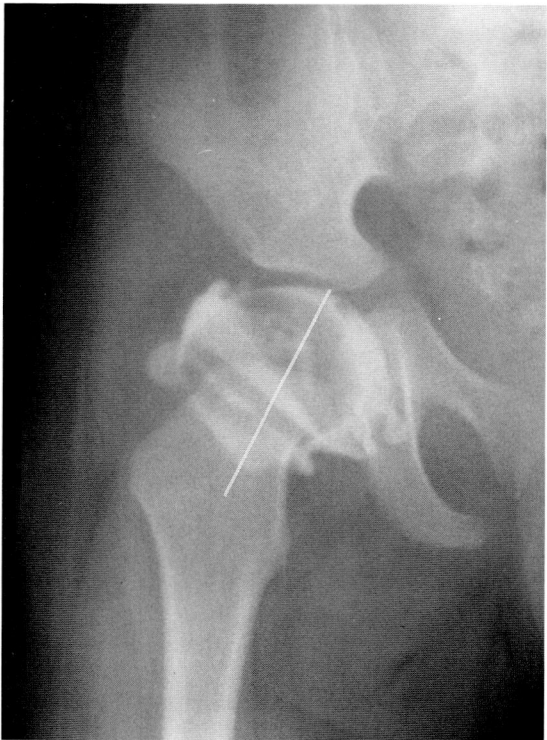

Fig. 12-6 Assessment of joint effusion. **(A)** Transducer position over the anterior aspect of the hip. **(B)** Scan plane is oblique to the true sagittal plane over the femoral neck (*white line*) as shown on arthrogram. (*Figure continues.*)

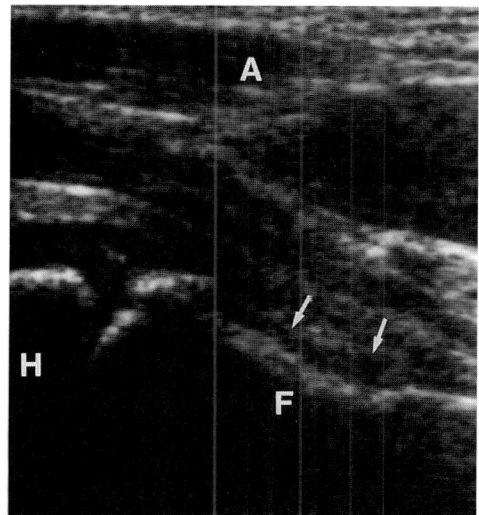 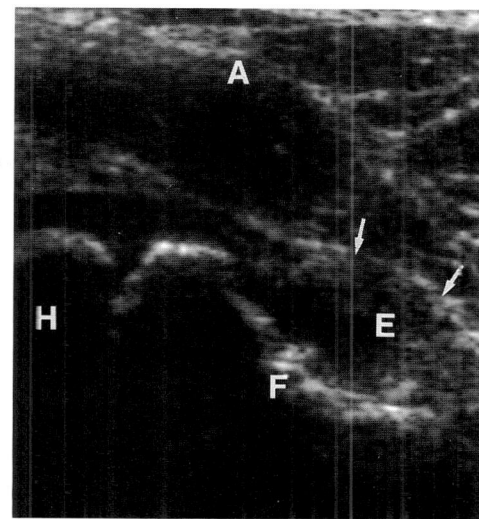

Fig. 12-6 (*Continued*). **(C)** Normal sonogram. Soft-tissue echoes follow contour of the femur (*arrows*). A, anterior; F, femoral neck; H, femoral head. **(D)** Abnormal sonogram. The joint effusion (E) causes convex bulging of capsule (*arrows*). A, anterior; F, femoral neck; H, femoral head.

cally guided arthrocentesis. Some clinicians use arthrocentesis therapeutically in Perthes disease because it reduces pain and allows a more normal range of motion.[29]

During the course of ultrasound examination for a joint effusion, other anatomic changes in the femur may be observed. These observations include fragmentation of the femoral head in Perthes disease, slippage of the head in slipped capital femoral epiphysis, and cortical disruption in fracture or osteomyelitis. These findings are usually better evaluated radiographically, and the use of sonography without a prior radiograph is not advocated. Soft-tissue swelling and other soft-tissue abnormalities outside the joint capsule have also been diagnosed.

Clinical Experience

Studies of hip joint effusion detection by ultrasound report the technique to be easy to master and rapid to perform. The results show a high sensitivity for the detection of effusion, with as little as 1 ml of fluid recognized experimentally.[28] At my institution, if a joint effusion is diagnosed by sonography and arthrocentesis yields no fluid, we insist on a repeat procedure under ultrasound guidance (to be sure the needle is in the proper place) and ask that a large needle be used and lavage attempted. Very thick material in the joint can be difficult to aspirate, prompting one to initially call the sonographic findings false-positive. False-negative examination results have been reported in infants who are younger than 1 year,[31] probably because it is difficult to distinguish hypoechoic cartilage from adjacent hypoechoic joint fluid.

Although hip sonography is sensitive in the detection of effusion, its place in the workup of the painful hip is not clearly defined. In one large series, although sonography facilitated early diagnosis or prompted further investigation in some patients, it altered the therapy or outcome in only 1 percent of the patients.[32] One group recommends the use of a protocol for evaluation of the painful hip that employs radiography, hip sonography, and scintigraphy.³³ Radiography is performed initially; when the findings are negative, sonography is performed. This is followed by aspiration if there is an effusion or by bone scan if there is no effusion.

At our institution, the workup of the painful hip is individualized. We have found sonography to be helpful in confirming joint pathology, but it has not led to a decrease in the use of other modalities, such as bone scan. For example, in the patient with clinical and laboratory signs of synovitis, hip sonography may be used to demonstrate an effusion; however, it cannot detect early Legg-Calvé-Perthes disease. In a patient with signs of sepsis and an effusion, a bone scan is necessary regardless of the results of aspiration to exclude the possibility of underlying osteomyelitis. A negative sonographic study of the hip in a septic patient rapidly and accurately excludes the possibility of septic arthritis but not osteomyelitis, so the bone scan is still required.

Summary

When the clinical picture is unclear, ultrasound determination of the presence or absence of an effusion can guide the clinician in the diagnosis and need for further evaluation. Sonography can be used to guide arthrocentesis, thus avoiding the ionizing radiation accompanying fluoroscopically guided aspiration.

REFERENCES

1. Graf R: The diagnosis of congenital hip-joint dislocation by the ultrasonic compound treatment. Arch Orthop Trauma Surg 97:117, 1980
2. Graf R: Classification of hip joint dysplasia by means of sonography. Arch Orthop Trauma Surg 102:248, 1984
3. Zieger M, Hilpert S, Schulz RD: Ultrasound of the infant hip. I. Basic principles. Pediatr Radiol 16:483, 1986
4. Zieger M: Ultrasound of the infant hip. II. Validity of the method. Pediatr Radiol 16:488, 1986
5. Zieger M, Schulz RD: Ultrasonography of the infant hip. III. Clinical application. Pediatr Radiol 17:226, 1987
6. Graf R: Ultrasonography of the infantile hip. In: Sanders RC, Hill MC (eds): Ultrasound Annual 1985. Raven, New York, 1985
7. Harcke HT, Clarke NMP, Lee MS et al: Examination of the infant hip with real-time ultrasonography. J Ultrasound Med 3:131, 1984
8. Clarke NMP, Harcke HT, McHugh P et al: Real-time ultrasound in the diagnosis of congenital dislocation and dysplasia of the hip. J Bone Joint Surg 67B:406, 1985
9. Boal DKB, Schwenkter EP: The infant hip: assessment with real-time US. Radiology 157:667, 1985
10. Keller MS, Chawla HS, Weiss AA: Real-time sonography of infant hip dislocation. Radiographics 6:447, 1986
11. Harcke HT, Grissom LE: Performing dynamic sonography of the infant hip. AJR 155:837, 1990
12. Harcke HT, Lee MS, Sinning L et al: Ossification center of the infant hip: sonographic and radiographic correlation. AJR 147:317, 1986
13. Morin C, Harcke HT, MacEwen GD: The infant hip: real-time US assessment of acetabular development. Radiology 157:673, 1985
14. Harcke HT: Screening newborns for developmental dysplasia of the hip: the role of sonography. AJR 162:395, 1994
15. Novick G, Ghelman B, Schneider M: Sonography of the neonatal and infant hip. AJR 141:639, 1983
16. Gomes H, Menanteau B, Motte J, Robiliard P: Sonography of the neonatal hip: a dynamic approach. Ann Radiol (Paris) 30:503, 1987
17. Dahlstrom H, Oberg L, Friberg S: Sonography in congenital dislocation of the hip. Acta Orthop Scand 57:402, 1986
18. Suzuki S, Kasahara Y, Futami T et al: Ultrasonography in congenital dislocation of the hip. Simultaneous imaging of both hips from in front. J Bone Joint Surg 73B:879, 1991
19. Langer R: Ultrasonic investigation of the hip

in newborns in the diagnosis of congenital hip dislocation: classification and results of a screening program. Skeletal Radiol 16:275, 1987
20. Tonnis D, Storch K, Ulbrich H: Results of newborn screening for CDH with and without sonography and correlation of risk factors. J Pediatr Orthop 10:145, 1990
21. Clarke NMP, Clegg J, Al-Chalabi AN: Ultrasound screening of hips at risk for CDH. J Bone Joint Surg 71B:9, 1989
22. Harcke HT, Kumar SJ: The role of ultrasound in the diagnosis and management of congenital dislocation and dysplasia of the hip. J Bone Joint Surg 73A:622, 1991
23. Berman L, Klenerman L: Ultrasound screening for hip abnormalities: preliminary findings in 1001 neonates. BMJ 293:719, 1986
24. Szoke N, Kuhl L, Heinricks J: Ultrasound examination in the diagnosis of congenital hip dysplasia of newborns. J Pediatr Orthop 8:12, 1988
25. Polanuer PA, Harcke HT, Bowen JR: Effective use of ultrasound in the management of congenital dislocation and/or dysplasia of the hip. Clin Orthop 252:176, 1990
26. Grissom LE, Harcke HT, Kumar SJ et al: Ultrasound evaluation of hip position in the Pavlik harness. J Ultrasound Med 7:1, 1988
27. Harding MGB, Harcke HT, Bowen JR et al: Ultrasound monitoring in the management of the congenitally dislocated hip treated with the Pavlik harness. (Submitted for publication)
28. Marchal GJ, van Holsbeeck MT, Raes M et al: Transient synovitis of the hip in children: role of US. Radiology 162:825, 1987
29. Alexander JE, Seibert JJ, Glasier CM et al: High-resolution hip ultrasound in the limping child. J Clin Ultrasound 17:19, 1989
30. Kallio P, Ryoppy S, Jappinen S et al: Ultrasonography in hip disease in children. Acta Orthop Scand 56:367, 1985
31. Zieger MM, Dorr U, Schulz RD: Ultrasonography of hip joint effusions. Skeletal Radiol 16:607, 1987
32. Mirales M, Gonzalez G, Pulpeiro JR et al: Sonography of the painful hip in children: 500 consecutive cases. AJR 152:579, 1989
33. Alexander JE, Seibert JJ, Aronson J et al: A protocol of plain radiographs, hip ultrasound, and triple phase bone scans in the evaluation of the painful pediatric hip. Clin Pediatr 27:175, 1988

13
Knee

Glenn M. Strome
J. Antonio Bouffard
Marnix van Holsbeeck

The advantages of sonographic examination of the knee parallel the broader advantages of musculoskeletal sonography: the modality's dynamic and interactive nature and its ability to be performed in a rapid and cost-effective manner. In addition, the noninvasive nature of sonography avoids complications associated with arthroscopy. Furthermore, the periarticular soft tissues cannot be evaluated by arthroscopy but are easily examined with sonography. In general, in the evaluation of musculoskeletal pathology, radiographs of the symptomatic area are obtained first. For imaging of the knee, additional modalities include magnetic resonance imaging (MRI) and sonography. MRI is useful for imaging pathology referable to menisci, bone marrow, and tumors of soft tissue or bone. Sonography is the imaging modality of choice for musculotendinous lesions, ligamentous tears (especially of the collateral ligaments), popliteal masses, swelling, localized knee pain, and evaluation of the synovium and bursae.[1,2]

JOINT EFFUSIONS

Increased intra-articular fluid is a nonspecific indicator of joint pathology associated with inflammatory processes, osteonecrosis, osteoarthritis, and trauma.[3] Sonography has been shown to detect effusions as small as 1 ml in the hip.[4] In the knee, fluid is readily demonstrated in the suprapatellar bursa (Fig. 13-1). The presence of an effusion provides an acoustic window that allows easier evaluation of the synovium and detection of loose bodies. Sonographically guided joint aspiration is more accurate than blind aspiration, thus reducing the occurrence of a "dry tap."[5] In addition, sonography can provide information about the etiology of the effusion. Debris can commonly be identified floating within an effusion. This debris may represent purulent material, blood clots, fat lobules, or loose bodies.[6] In rheumatoid disease, synovial villi are covered with precipitated fibrous material that can become dislodged, eventually settling into the sy-

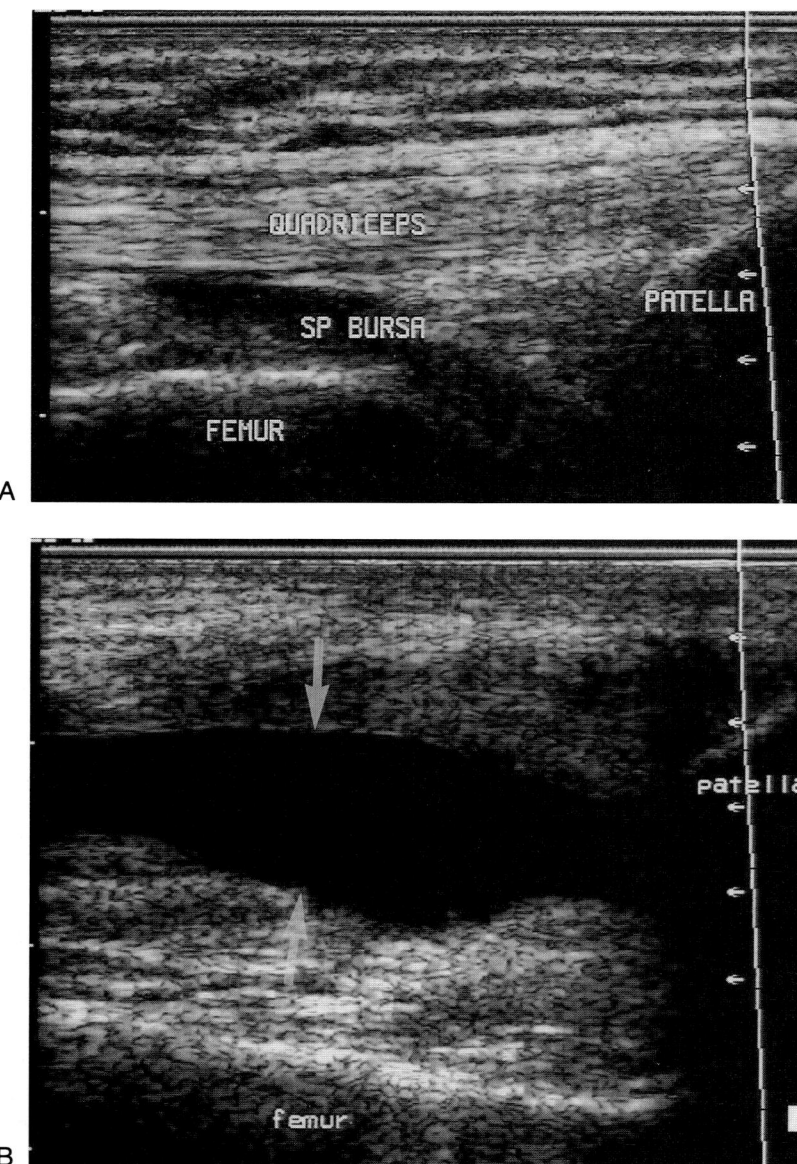

Fig. 13-1 Longitudinal sonograms of the suprapatellar bursa. **(A)** Normal knee. Sonogram shows the hypoechoic bursa (SP BURSA) between the quadriceps tendon anteriorly (QUADRICEPS) and the femur posteriorly. Low-level echoes within the bursa represent normal synovium and should not be mistaken for fluid. **(B)** Knee effusion. Sonogram shows a large suprapatellar effusion (*arrows*). The anechoic nature of the fluid argues against a chronic inflammatory etiology.

novial recesses of the joint.[7] The debris may appear as hyperechoic foci floating within the effusion. Loose bodies representing cartilage fragments may be seen following intra-articular trauma or meniscal tears. As with other loose bodies, these fragments, which often have a polygonal shape,[6] tend to migrate into the synovial recesses. In the setting of intra-articular steroid injections for rheumatoid disease, punctate hyperreflective foci can be demonstrated that correspond to resulting calcifications. These calcifications can be easily seen with sonography but are rarely identified on conventional radiographs.[6,8]

In addition, sonography can be used to monitor response to treatment of the inflammatory arthritides.[9] The first sign of a positive response to therapy has been shown to be a decrease in the quantity of intra-articular fluid, most often seen within 48 hours of the initiation of treatment.[10]

SYNOVIAL ABNORMALITIES

Synovial hypertrophy (pannus) can be seen in many inflammatory arthropathies, such as rheumatoid arthritis, hemophilic arthropathy, and amyloidosis with joint involvement. Synovial thickening also may be seen in the setting of tumoral involvement of the synovium.[3] Although pannus is most commonly seen in the setting of the inflammatory arthritides, it also may be seen with chronic infections such as tuberculosis or fungal infections[3]; in these cases, fibrous adhesions and synechiae are not rare.

Sonographic evaluation of synovial changes is most easily performed at the suprapatellar bursa (Fig. 13-2). Pannus is hypoechoic relative to surrounding soft tissues. As mentioned above, fine particulate debris representing fibrinous clots may be seen in rheumatoid arthritis. Reliable and reproducible sonographic measurements of synovial thickness necessitate maximal compression to expel all free fluid from the region.[6] Synovial thickness is defined as half the sum of the thicknesses of the anterior and posterior walls of the suprapatellar bursa. Serial measurements of synovial thickness have been shown to be useful for comparing the efficacy of different therapies for inflammatory arthritis.[10]

Sonography may be useful for the evaluation of synovial tumors. Pigmented villonodular synovitis and synovial chondromatosis appear as irregular thickenings of the synovium. With synovial osteochondromatosis, hyperechoic shadowing foci may be seen. Sonography often is able to detect these tumors at an early stage of development.

ARTICULAR CARTILAGE

On sonography, normal articular cartilage appears as a thin hypoechoic layer (Fig. 13-3) that is juxtaposed with the subchondral cortical bone (Fig. 13-4).[11] It is believed that the hypoechoic appearance is due to the homogeneous hydrophilic structure of hyaline cartilage. Because of the sharp margins at both the articular and deep surfaces, measurement of cartilage thickness is usually easily performed.

The earliest sign of pathologic change of the cartilage is edema.[9] Sonographically, cartilage margins become poorly defined, and the cartilage substance appears inhomogeneous. Later, there will be roughening of the surface of the involved cartilage, and often there will be a measurable loss in thickness.[11] These changes can be seen in the setting of both inflammatory arthritis and osteoarthritis. Serial sonographic examinations of cartilage thickness and surface changes may prove useful in evaluation of therapeutic response.

Focal traumatic lesions may appear as hyperechoic foci (Fig. 13-5). However, the ability of sonography to evaluate the patellar

204 MUSCULOSKELETAL ULTRASOUND

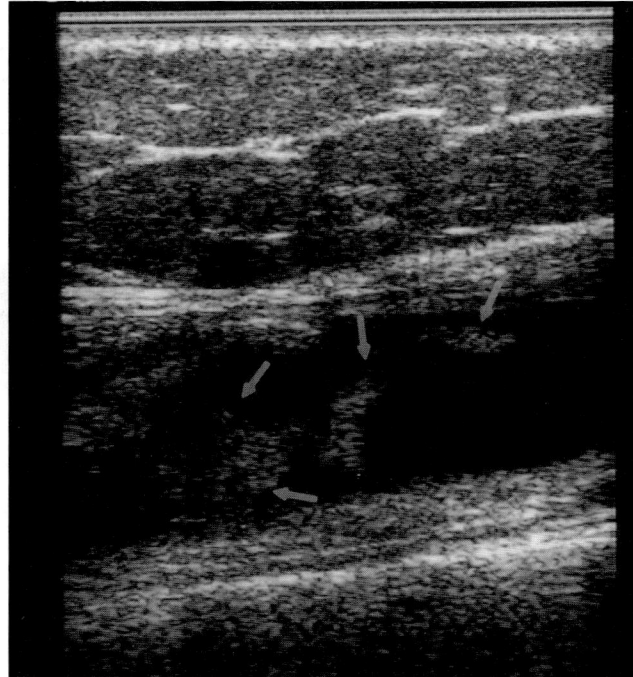

Fig. 13-2 Knee effusion and synovial thickening in a patient with rheumatoid arthritis. Longitudinal sonogram shows a moderate effusion in the suprapatellar bursa associated with lobulated thickening of the synovium (*arrows*). This synovial thickening clearly arises from the superior aspect of the bursa and is unlikely to be confused with mobile debris.

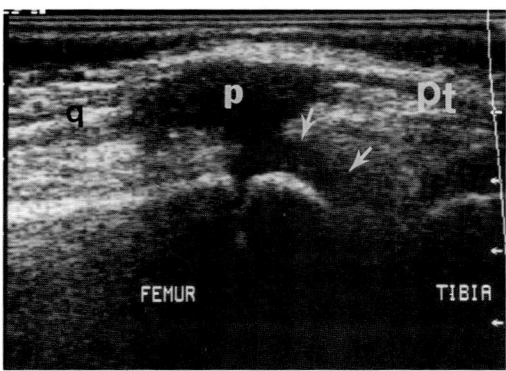

Fig. 13-3 Normal cartilage. Longitudinal sonogram shows the hypoechoic, nonossified patella (p) in a young child. Hypoechoic cartilage is also seen in the region of the distal femoral epiphysis (*arrows*). q, quadriceps tendon; pt, patellar tendon.

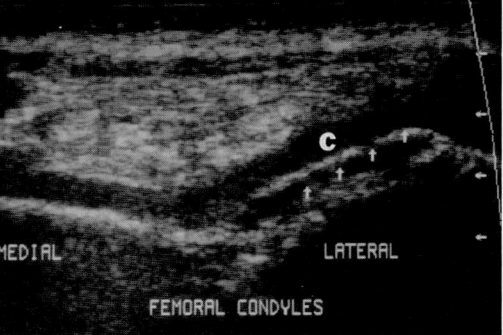

Fig. 13-4 Subchondral fracture of the lateral femoral condyle. Transverse sonogram shows the cortex of the lateral femoral condyle (*arrows*). The underlying hypoechoic cleft represents a subchondral fracture. Note the intact overlying articular cartilage (c).

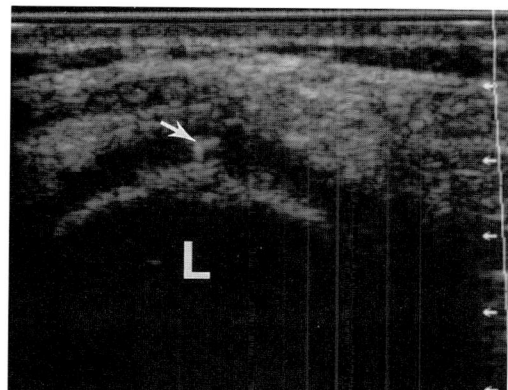

Fig. 13-5 Cartilaginous defect in the lateral femoral condyle. Longitudinal sonogram shows a hyperechoic focus (*arrow*) in the hyaline articular cartilage of the lateral femoral condyle (L). This lesion corresponded to an area of focal tenderness that developed after arthroscopy.

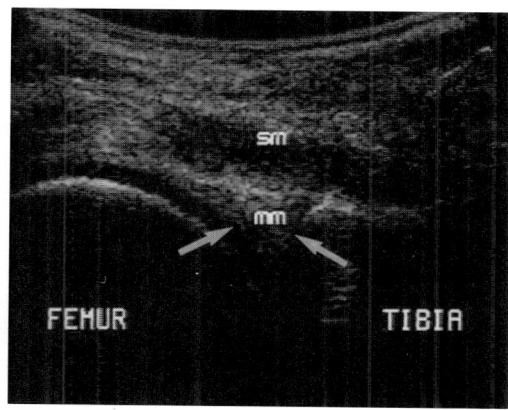

Fig. 13-6 Posterior horn of normal medial meniscus. Longitudinal sonogram obtained with the patient prone shows the normal hyperechoic appearance of meniscal fibrocartilage (mm and *arrows*). sm, semimembranous tendon.

articular surface and cartilage of the tibial plateau is limited because there is no acoustic window. Similarly, in patients who are unable to optimally flex their knees, portions of the articular cartilage will not be accessible to sonographic examination.[9]

KNEE MENISCI

The menisci are composed of semilunar cartilage that is attached on its periphery to the capsule of the knee joint. The menisci can be injured as a result of twisting, hyperflexion, and rotational injury resulting from ligament insufficiency.[12] Horizontal tears are often found in association with meniscal cysts.[13]

Sonographic examination of the menisci is performed with the patient in the lateral decubitus position, with the transducer oriented along the long axis of the leg. Subsequently, transverse images are obtained by rotating the transducer 90 degrees. Mild valgus and varus stress is applied when examining the medial and lateral menisci, respectively. The meniscus appears as a homogeneous hyperechoic triangle interposed between hypoechoic layers of articular cartilage (Fig. 13-6). Tears of the menisci are identified as hypoechoic clefts within the normally homogeneous hyperechoic fibrocartilage (Fig. 13-7).[6,14] Sonography reliably demonstrates posterior and peripheral meniscal tears, which are not well demonstrated with arthroscopy. Sonography is

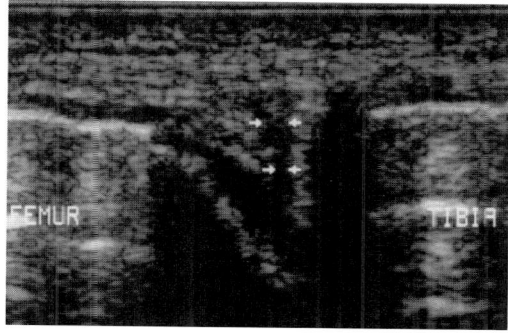

Fig. 13-7 Horizontal tear of the medial meniscus. Coronal sonogram shows a horizontally oriented hypoechoic cleft (*arrows*) in the body of the meniscus.

limited, however, in its ability to detect small tears along the inner margin of the menisci. MRI remains the gold standard for evaluation of meniscal tears.

The role of sonographic examination of the menisci includes the detection of meniscal tears in the setting of Baker's cysts; detection of meniscal capsular separation, a notoriously difficult diagnosis with MRI; and evaluation of the knee when other studies are negative but symptoms are still referable to the menisci. Sonography is also able to identify injuries of the collateral ligaments or periarticular bursae that may mimic meniscal lesions clinically. Baker's cysts are associated with tears of the medial and lateral menisci in 70 percent and 40 percent of cases, respectively.[15] The presence of a Baker's cyst creates a unique acoustic window that allows for excellent evaluation of the posterior horns of the menisci. Sonographic examination of the menisci should always be undertaken in the setting of a Baker's cyst.

MRI and sonography are both useful for diagnosing meniscal cysts.[16,17] Meniscal cysts are multiloculated cystic masses containing mucinous material that maintain a relationship with the meniscal margin. There are two main theories regarding the origin of meniscal cysts. One theory is that such a cyst originates from a horizontal cleavage in the meniscus that extends into the parameniscal region, causing cyst formation. The other theory contends that cyst formation is due to myxoid degeneration of the meniscus.[13] Support exists for both theories. Lateral meniscal cysts are frequently seen in association with horizontal tears of the lateral meniscus. In this setting, joint motion is thought to pump intra-articular fluid through the tear, causing cyst development. Medial meniscal cysts, on the other hand, are rarely associated with meniscal tears; the etiology of cyst formation in these cases is thought to be primary myxoid degeneration of the meniscus.[3]

On sonograms, meniscal cysts most frequently appear as hypoechoic or anechoic loculated cystic structures adjacent to the meniscus (Fig. 13-8). Frequently, displacement of fluid out of the cyst can be demonstrated by applying pressure on the transducer. In areas of myxoid degeneration, the meniscus appears swollen and shows decreased echogenicity. The meniscus may bulge outward, protruding from the joint space and causing knee swelling and pain. An enlarging meniscal cyst can mimic a soft-tissue tumor. It can also cause erosion of the lateral tibial plateau. Such erosions may simulate those of an inflammatory arthropathy. In these cases, sonography can provide valuable information to the orthopaedic surgeon regarding the full extent of the lesions.

BURSAE

On sonograms, normal bursae appear as hypoechoic clefts in the soft tissues and often are bounded by a hyperechoic line representing a tissue-fluid interface. The hypoechoic cleft is no larger than 2 mm. Right-left comparison is helpful to determine if the quantity of bursal fluid is increased.

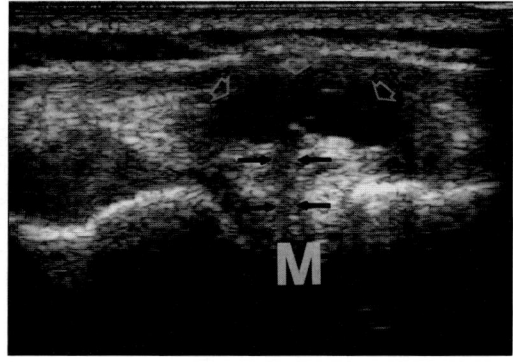

Fig. 13-8 Meniscal cyst. Coronal sonogram shows the lateral portion of the lateral meniscus (M) bulging into the cyst (*open arrows*). Note the horizontal tear (*arrows*).

Bursal inflammation can result from hemorrhage, chronic trauma, infection, or systemic processes such as rheumatoid disease, seronegative spondyloarthropathies, gout, sarcoidosis, and tuberculosis. Traumatic bursitis can be differentiated from bursitis due to systemic causes because the latter results in synovial proliferation, which can be documented sonographically. Nonspecific sonographic findings of bursitis include an enlarged fluid-filled bursa with blurred margins (Fig. 13-9). Synovial proliferation, synechiae, and loose bodies also may be demonstrated.

In athletes, the superficial and deep infrapatellar bursae may become inflamed as a result of acute traumatic bursitis or frictional bursitis. The latter is frequently associated with sports that require repetitive motions, such as rowing. Sonography has a role in differentiating between bursal and tendon pathologies in this setting. There is considerable overlap in their clinical presentations[18] but the therapies for these conditions will differ markedly. For example, bursitis may be treated with an injection of steroids. If a tendon injury is mistakenly diagnosed as bursitis and inadvertently injected with steroids, however, the tendon will be prone to subsequent rupture, with devastating consequences for the patient.

Communicating bursae are not found in young children but develop over time from noncommunicating bursae.[19] Communicating bursae can serve as potential reservoirs for joint effusions and loose bodies. In addition, these bursae provide a marker of intraarticular pathology. For example, the effects of an inflammatory arthritis and its response to treatment can be followed by measuring the quantity of fluid and the thickness of the synovium within the suprapatellar bursa.[10,20]

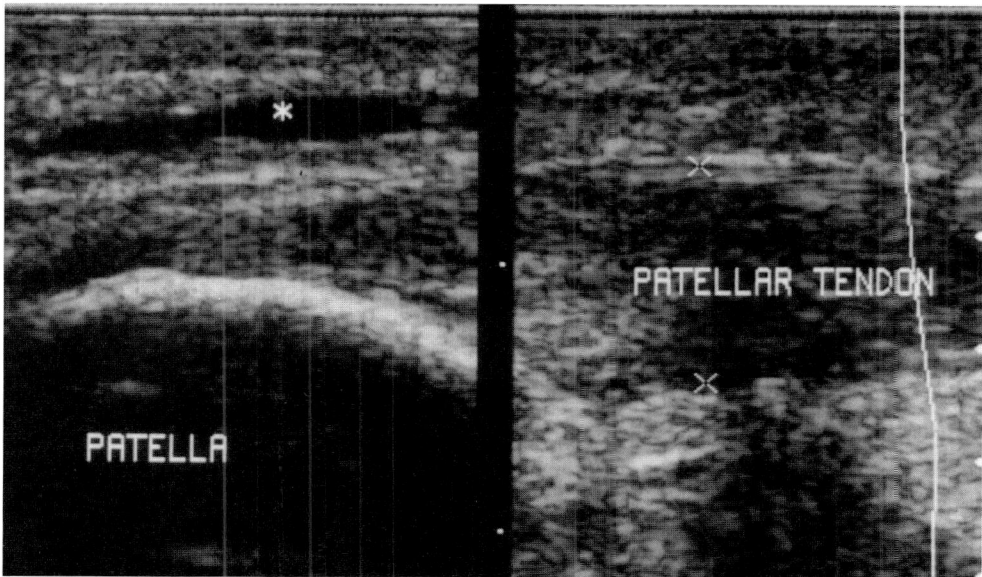

Fig. 13-9 Prepatellar bursitis and patellar tendinitis. Montage of two contiguous longitudinal sonograms shows anechoic fluid in the prepatellar bursa (*asterisk*) and the swollen and hypoechoic patellar tendon (*calipers*).

BAKER'S CYST

Because of its importance, Baker's cyst is discussed as a separate entity. A Baker's cyst is an enlarged gastrocnemiosemimembranous bursa. This bursa is found at the medial aspect of the popliteal fossa. Communication with the joint space is not present at birth and likely develops during adult life. Anatomically, the bursa can be divided into three portions: the base, superficial extent, and neck.[21] The base is located between the joint capsule and the gastrocnemius tendon; the neck lies between the gastrocnemius and semimembranous tendons; and the superficial portion is located beneath the fascia. The narrow channel of communication between the bursa and the joint space is closed during extension and open during flexion of the knee, allowing fluid to move into the joint by a ball-valve mechanism. Likewise, loose bodies also have a propensity to move into the gastrocnemiosemimembranous bursa.

Bursal distention is associated with two general categories of disease: pathology associated with increased articular fluid and the arthritides. Disorders that result in an increase in intra-articular fluid all result in irregularity of the joint surfaces; among these disorders are osteochondral fractures, osteonecrosis, cartilage defects, meniscal lesions, and loose bodies. Inflammatory arthropathies create increased fluid production and synovial proliferation in both the midjoint and the communicating bursa.

Sonographically, a Baker's cyst appears as a fluid collection in the medial aspect of the popliteal fossa (Fig. 13-10). When the cyst enlarges, it can cross the midline. On a transverse scan, the fluid collection surrounds the medial gastrocnemius tendon on its anterior, posterior, and medial surfaces like a horseshoe. Baker's cysts developing in the setting of rheumatoid arthropathy characteristically have marked irregularity of the synovium as well as internal echoes within the fluid.[10]

These cysts tend to be larger than those developing from other causes. They can dissect the soft tissue of the leg in two ways. More commonly, the Baker's cyst extends superficially between the gastrocnemius muscle and the superficial fascia. In rare cases, the Baker's cyst dissects between the medial gastrocnemius and soleus muscles and may not be palpable on physical examination.[6]

Baker's cysts associated with rheumatoid disease can become massively distended with fluid and synovial proliferation; they can extend from the knee as far distally as the ankle. As in other communicating bursae, sonography can effectively diagnose loose bodies, septations, hemorrhage, and inflammation complicating a Baker's cyst and giving rise to intracystic echoes (Fig. 13-11). Such cysts are also prone to rupture, resulting in a severe inflammatory response from the surrounding tissues. This presentation can mimic deep venous thrombosis of the lower extremity and has been called pseudothrombophlebitis.[22] Sonography can reliably establish the correct diagnosis. On sonograms, a Baker's cyst is usually rounded; a pointed appearance suggests rupture.[2] Occasionally, fluid can actually be expressed through the ruptured cyst wall by the application of gentle pressure.

LESIONS OF TENDONS AND LIGAMENTS

Patellar Tendon

The patellar tendon, which inserts distally on the tibial tuberosity, is an extension of the extensor mechanism distal to the apex of the patella. The proximal insertion has a conical shape and is slightly larger than the remainder of the tendon. The proximal insertion may be noticeably more prominent in people who are physically active; in sedentary individuals, it may be quite thin.[9] The normal

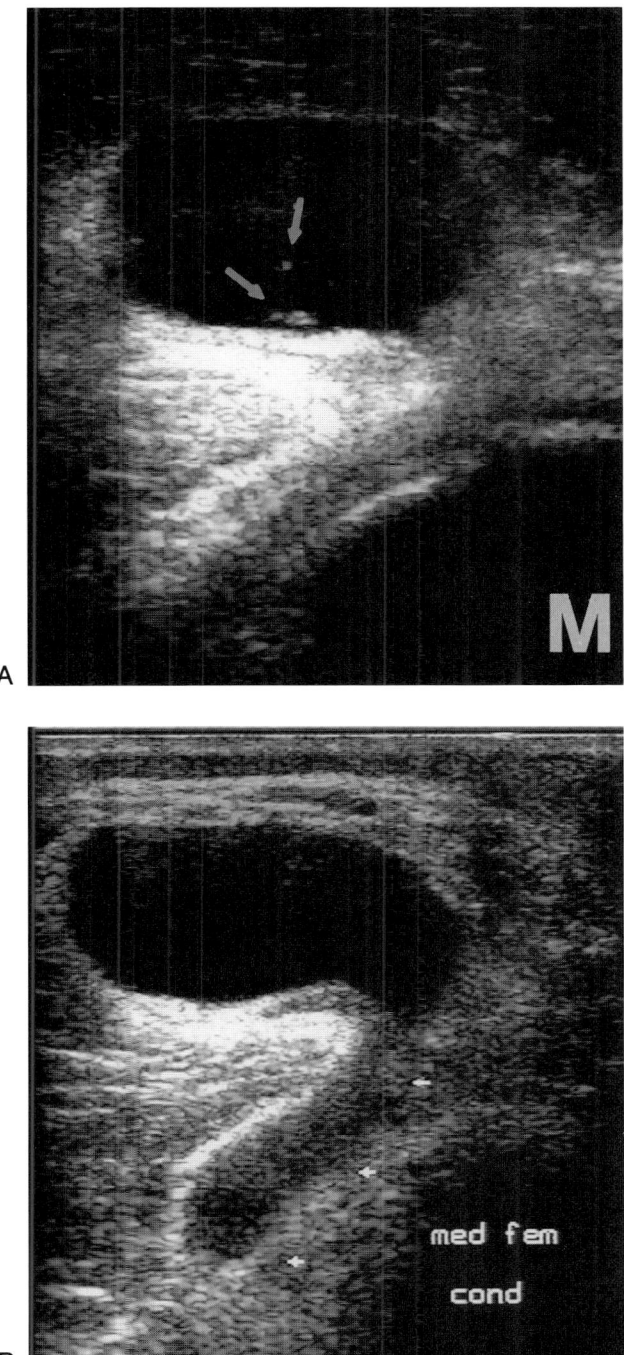

Fig. 13-10 Baker's cysts. **(A)** Transverse sonogram shows small echogenic debris (*arrows*) within the fluid-filled Baker's cyst. M, medial femoral condyle. **(B)** Transverse sonogram in another patient shows the fluid collection surrounding the medial gastrocnemius tendon posteriorly, medially, and anteriorly (*arrows*) in a horseshoe fashion. med fem cond, medial femoral condyle. (*Figure continues.*)

210 MUSCULOSKELETAL ULTRASOUND

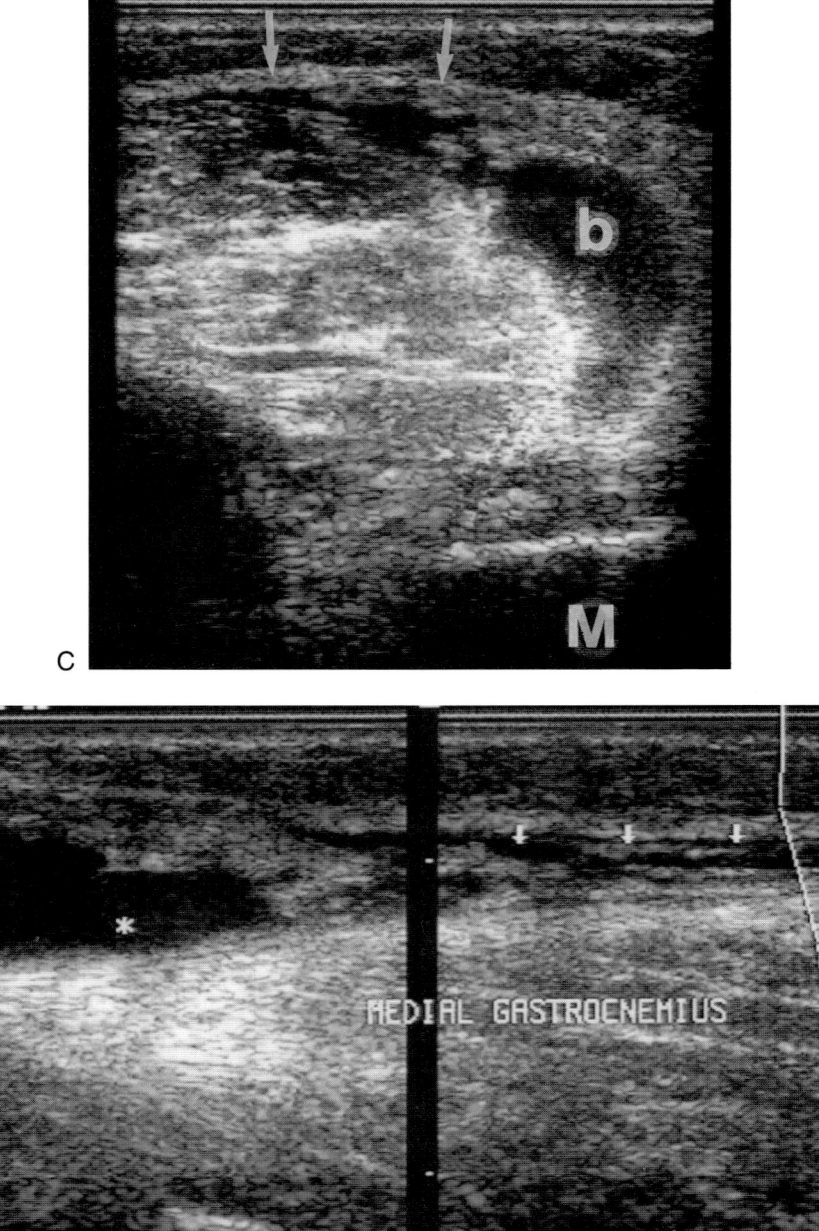

Fig. 13-10 (*Continued*). **(C)** Ruptured cyst. On transverse sonogram, the medial margin of the cyst (b) is no longer rounded but has become pointed, indicating rupture. The fluid has extravasated under the lower leg fascia (*arrows*). M, medial femoral condyle. **(D)** Ruptured cyst. Montage of two contiguous longitudinal sonograms shows the pointed inferior margin of the ruptured cyst (*asterisk*). *Arrows* point to the fluid extravasation along the lower leg fascia.

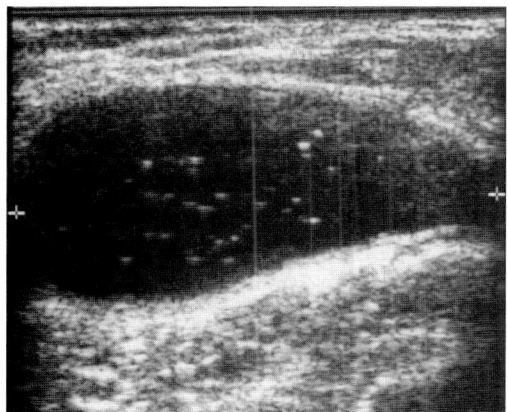

Fig. 13-11 Baker's cyst. Transverse sonogram shows minute echogenic foci in the cyst.

tendon appears echogenic and has a typical fibrillar echotexture.[23,24]

Patellar tendon tears occur most commonly in patients with total knee prostheses, athletes using injected steroids, and patients with systemic lupus erythematosus. On sonograms, a tear appears as a hypoechoic cleft of variable size in the course of the tendon.[24] The edges of the torn tendon may appear curled up. In addition, the free tendon edges frequently create a refractive artifact with shadowing. Complete evaluation of the extent of tendon rupture necessitates imaging in both the longitudinal and transverse planes.

Patellar tendinitis (jumper's knee), a common sports-related injury, is associated with sports that require repetitive action of the extensor mechanism of the knee, e.g., football, soccer, and basketball. Focal patellar tendinitis can also be seen in patients who have a history of blunt trauma to the patellar tendon. Sonographic examination of the asymptomatic side may be helpful in subtle cases (Fig. 13-12). The abnormal tendon will appear focally thickened just under the apex of the patella. This region will appear hypoechoic with increased sound-through transmission. Such hypoechoic foci have been shown to correspond to areas of fibromyxoid degeneration on histologic analysis[2,25]; this degeneration predisposes the tendon to rupture. Local steroid injection also increases the propensity of the tendon to rupture. If tendinitis is prolonged, such as in an athlete who continues activity despite injury, the

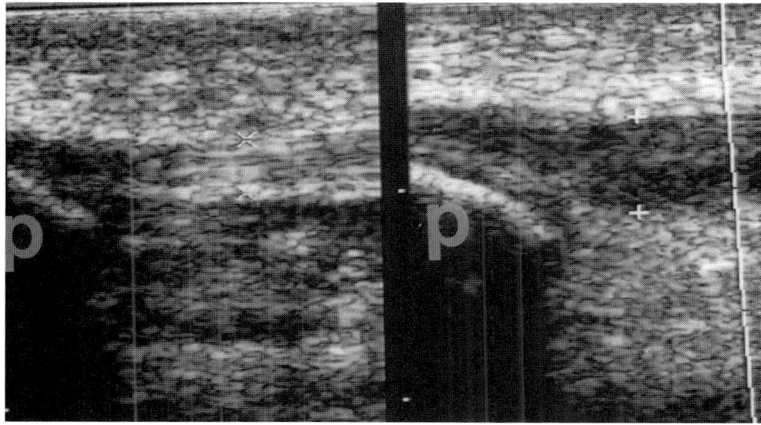

Fig. 13-12 Patellar tendinitis. Longitudinal sonograms of the insertion of the patellar tendon on the patella (p). The normal patellar tendon on the left has a normal thickness and echogenicity. The symptomatic tendon on the right is swollen and diffusely hypoechoic, with loss of the normal fibrillar echotexture.

entire tendon may be abnormally thickened and hypoechoic. Calcifications within the substance of the tendon can be seen in chronic cases.[24] Focal patellar tendinitis is also occasionally seen following arthroscopy (Fig. 13-13). Included in the differential diagnosis of patellar tendinitis are two other disease processes that can have similar clinical presentations but that have different prognoses and therapeutic managements. Localized tenderness over the tibial tuberosity can be due to bursitis involving a peritendinous bursa, without pathologic change of the tendon itself. This condition appears sonographically as an oval, anechoic, fluid-filled structure adjacent to the patellar tendon. Insertion tendinopathy also may mimic patellar tendinitis clinically.[26] This disease is found almost exclusively in athletes. Sonographically, there is enlargement of the fibrocartilaginous insertion of the tendon. Right-left comparison may be helpful in making this diagnosis.

Quadriceps Tendon

The quadriceps tendon represents the common tendon insertion of the rectus femoris, vastus lateralis, vastus medialis, and vastus intermedius muscles. The quadriceps tendon inserts onto the upper lateral and medial borders of the patella. On sonograms, the normal tendon appears hyperechoic. There may be an area of false hypoechogenicity near the patellar insertion; this should not be mistaken for pathology.[24]

Quadriceps tendon tears are most commonly seen in patients with knee prostheses and in patients with end-stage renal disease. Quadriceps tendon tears in these patients often occur after only minor trauma. On sonograms, a tendon tear appears as a hypoechoic cleft with bunched-up tendon on either side (Fig. 13-14). It is often possible to see fluid whirling through the defect in the tendon. As with patellar tendon rupture, there may

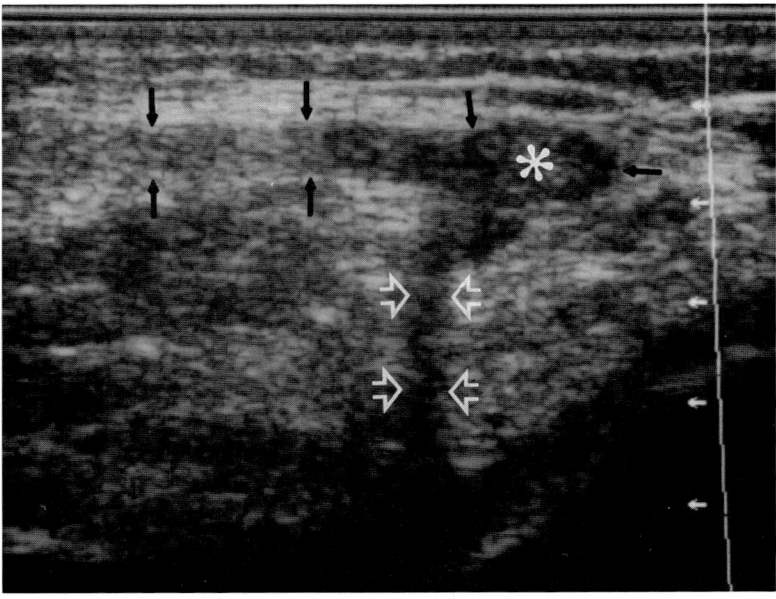

Fig. 13-13 Patellar tendinitis following arthroscopy. Transverse sonogram shows a hypoechoic area (*asterisk*) within the patellar tendon (*black arrows*). The arthroscopy tract through Hoffa's fat pad appears as a hypoechoic cleft (*open arrows*) deep to the area of inflammation.

KNEE 213

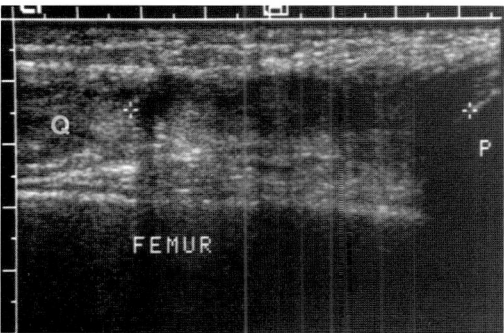

Fig. 13-14 Longitudinal sonogram of the quadriceps tendon in a patient who presented with inability to extend the knee. The quadriceps tendon rupture appears as a gap filled with hypoechoic hematoma (*calipers*). Note the retracted proximal portion of the ruptured tendon (Q). P, patella.

be a refractive artifact creating an acoustic shadow in the region of the free edges of the torn tendon. Early diagnosis of a quadriceps tendon rupture is essential because surgery is no longer possible after only 4 to 5 weeks following the injury. Sonography can be very useful in evaluating a patient with a total knee prosthesis who presents following an injury or with acute swelling; the modality allows for the differentiation of a quadriceps tendon rupture from an infected prosthesis.

The tendons of the biceps femoris muscle and pes anserinus also can be evaluated with sonography. Injuries to these tendons are most commonly seen in athletes. The biceps femoris tendon forms part of the lateral joint capsule and helps to stabilize the joint. Often, injuries of this tendon are misdiagnosed as lateral meniscal tears because of their similar clinical presentation. Injuries of the pes anserinus tendons are often seen in baseball players who present with swelling and tendon tears. We have also observed several cases of this type of musculotendinous tear in football wide receivers.

LIGAMENTS

Injuries to the collateral ligaments are important to document as they can mimic meniscal lesions clinically.

Medial Collateral Ligament

The medial collateral ligament, also known as the tibial collateral ligament, is a broad, flat structure extending from the medial femoral condyle to the medial aspect of the proximal tibia. The ligament has a deep layer, which is attached to the medial meniscus, and a superficial layer (Fig. 13-15). Sonographically, the medial collateral ligament is a trilaminar structure composed of two hyperechoic layers separated by a hypoechoic zone representing loose areolar tissue.[6] Occasionally, a bursa can be found within the hypoechoic zone.[27]

Injuries to the medial collateral ligament are quite common in athletes. Rupture of the anterior cruciate ligament and tear of the medial meniscus often accompany complete rupture of the medial collateral ligament.[28] The mechanism of injury is usually a blow to the lateral aspect of the knee, creating valgus stress. This type of injury is frequent in football and soccer players. Occasionally, an iso-

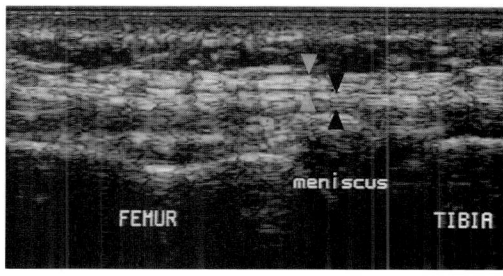

Fig. 13-15 Normal medial collateral ligament. Coronal sonogram shows the two echogenic layers of the ligament. The deep layer is attached to the medial meniscus. *White arrowheads,* superficial layer; *black arrowheads,* deep layer.

lated tear of the ligament is seen in the setting of arthroscopic surgery if the capsule is perforated. As with other ligament ruptures, acute rupture of the medial collateral ligament can be identified by an interruption in the hyperechoic bands that represent the superficial and deep components of the ligament. Fluid fills the gap, which may be either hypoechoic or anechoic. The most frequent site of rupture is the junction of the deep portion of the ligament with the medial meniscus. If only the deep portion of the ligament is ruptured, hematoma may be observed within the hypoechoic middle layer. In more severe injury, both the superficial and deep components may be ruptured (Fig. 13-16).[29]

Sonography can detect intrasubstance ruptures of the medial collateral ligament, which appear as thickening of the ligament. Right-left comparison with the contralateral uninjured ligament may be helpful in equivocal cases. Hypoechogenicity and thickening of either layer of the ligament without an identifiable gap are most compatible with a diagnosis of intrasubstance rupture. A meniscal cyst should be considered if only the central hypoechoic portion of the ligament is thickened. In this setting, the integrity of the deep layer of the ligament as well as the meniscus should be evaluated. If the deep ligament and meniscus appear to be intact, an enlarged bursa within the central region of the tendon should be considered.

Chronic ligament injury also may affect the medial collateral ligament. In many cases, this is a sequela of an untreated partial ligament rupture. Chronic injury may lead to a

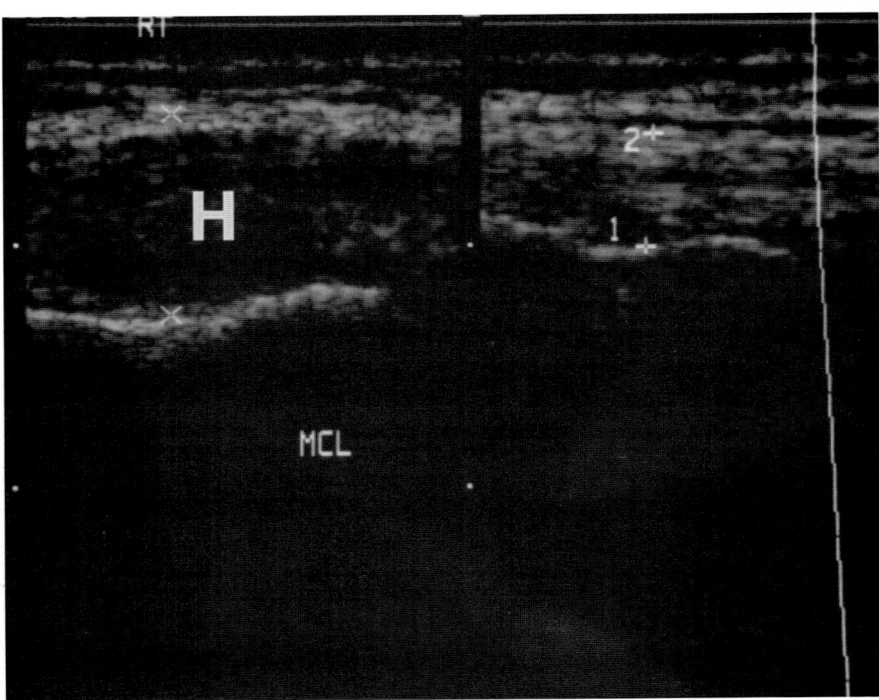

Fig. 13-16 Acute rupture of the medial collateral ligament. *Left:* Coronal sonogram shows the absence of identifiable ligament and the associated hematoma (H). *Right:* Normal contralateral ligament with the two echogenic layers for comparison.

hypertrophic response with excessive formation of granulation tissue and mucoid degeneration. The patient may complain of chronic pain and tenderness over the site of the lesion. This type of injury can be seen in football, basketball, and hockey players, but it is most characteristically found in soccer players, probably because the soccer-style kicking motion creates a tremendous amount of valgus stress on the inside of the knee. Soccer players typically describe increased pain immediately before contact with the ball when kicking. Meniscal pain has most often been characterized by an increase in pain immediately after contact with the ball.[29]

Sonographically, hypertrophic changes of chronic ligament injury produce a hypoechoic mass within the echogenic substance of the ligament.[6] A thickening and disruption of the normal trilaminar structure of the ligament can be demonstrated This thickening tends to be more prominent in the proximal portion of the ligament. Calcification associated with Pellegrini-Stieda disease can be found in a similar location, suggesting a common pathophysiology.[30]

Lateral Collateral Ligament

The lateral collateral ligament is a thin band that is attached superiorly to the lateral condyle of the femur and inferiorly to the head of the fibula. The tendon of the popliteal muscle is interposed between the ligament and the lateral meniscus. On sonograms, the normal lateral collateral ligament usually appears hypoechoic because of its oblique course with resultant anisotropy (Fig. 13-17). The biceps femoris tendon encircles the lateral collateral ligament from its lateral side, creating a horseshoe appearance on transverse sonograms. These two structures are difficult to distinguish at the fibular head attachment. A complete tear of the lateral collateral ligament may be identified as a discontinuity of the ligament associated with hematoma (Fig. 13-18). A partial tear may be demonstrated as an area of hypoechogenicity and swelling of the ligament.

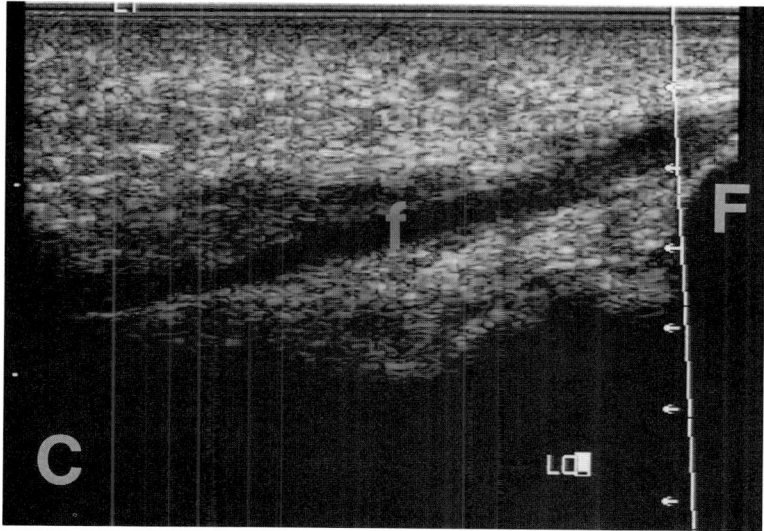

Fig. 13-17 Normal lateral collateral ligament. On longitudinal sonogram, the ligament (f) appears falsely hypoechoic because of its oblique course. C, lateral femoral condyle; F, head of the fibula.

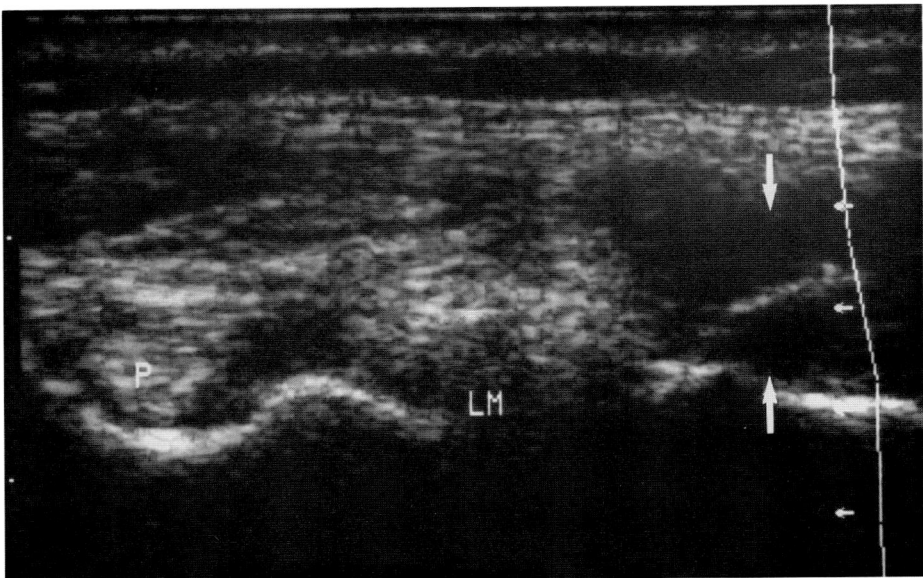

Fig. 13-18 Tear of the lateral collateral ligament. Coronal sonogram shows a hypoechoic hematoma (*arrows*) in the region of the distal ligament. LM, lateral meniscus; P, popliteal tendon.

Cruciate Ligaments

Examination of the anterior cruciate ligament is performed with the knee in full flexion and the transducer positioned along the length of the ligament.[31] However, because of pain, placing the knee in full flexion may be problematic in an acutely injured patient. The posterior cruciate ligament is examined while the patient is prone. Sonographically, the cruciate ligaments usually appear hypoechoic relative to surrounding fat, probably because of their oblique orientation to the ultrasound beam.

Following injury, the anterior cruciate ligament is nearly always avulsed from its proximal insertion. Adjacent hematoma is often seen in the intercondylar space. Injury to the posterior cruciate ligament is rarely isolated. The curled-up edges of an acutely ruptured posterior cruciate ligament can be seen on sonograms (Fig. 13-19).

One report from Europe[31] has claimed that sonography is accurate in the detection of anterior cruciate ligament injury; however, further corroborative studies are needed, and currently, MRI remains the imaging modality of choice for the diagnosis of cruciate ligament injury.

POPLITEAL FOSSA MASSES

The popliteal fossa is bounded laterally by the tendons of the biceps femoris muscle and medially by the tendons of the semimembranous and semitendinous muscles. Inferiorly, the fossa is bounded on each side by the heads of the gastrocnemius muscle. The popliteal fossa contains a neurovascular bundle in addition to the above-mentioned tendons and their associated bursae. The differential diagnosis of popliteal masses includes Baker's cyst,[32] aneurysm of the popliteal artery, venous thrombosis, hematoma, and

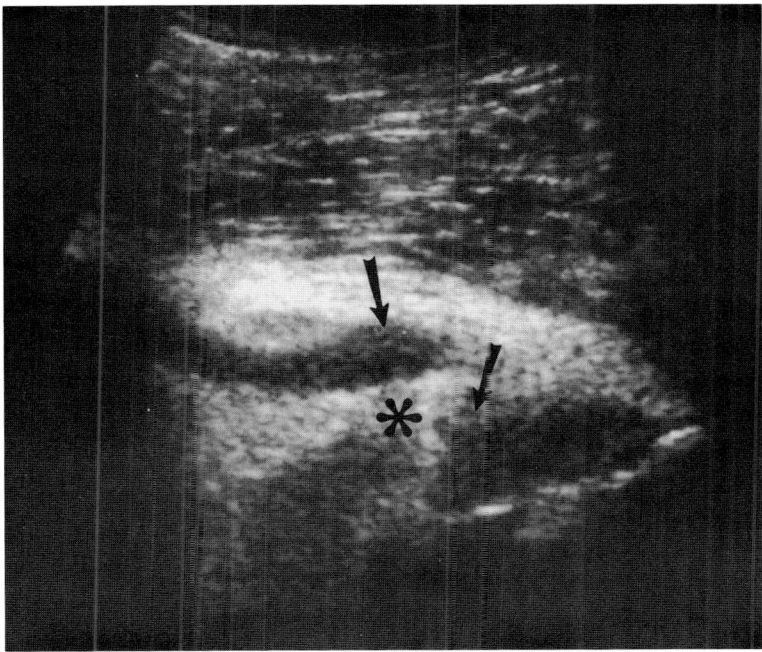

Fig. 13-19 Tear of the posterior cruciate ligament. Longitudinal sonogram with the transducer positioned over the popliteal fossa shows the torn ends of the ligament (*black arrows*). The ligament appears hypoechoic relative to the surrounding intercondylar fat (*asterisk*).

nerve tumors. Baker's cyst is by far the most common popliteal mass. Its sonographic appearance is described earlier in the chapter.

Aneurysms of the popliteal artery account for 70 percent of all peripheral aneurysms. These aneurysms are frequently associated with atherosclerotic aneurysms elsewhere, and approximately 50 percent are bilateral, making evaluation of the opposite popliteal artery, iliofemoral arteries, and aorta necessary once an aneurysm has been identified. The majority of popliteal aneurysms present as an asymptomatic, nonpulsatile mass, making them difficult to distinguish from a Baker's cyst by clinical examination alone.[33] One important differentiating point is that the location of these aneurysms differs from that of Baker's cysts: popliteal aneurysms are located over the midline or lateral aspect of the knee. In addition, these lesions can be shown to be in continuity with the popliteal artery.[34] Frequently, popliteal aneurysms contain a large thrombus with a resultant high risk of distal embolization. Surgical repair is undertaken to prevent distal embolus, as these aneurysms rarely rupture. Sonography is useful for accurately determining the dimensions of a popliteal aneurysm; the frequent presence of thrombus may cause the aneurysm's size to be underestimated by arteriography.[33] Because blood flow through the lesion may be sluggish, flow may not always be detectable with Doppler studies. In such cases, documentation of continuity of the mass with the superficial femoral artery is necessary to confirm the diagnosis of aneurysm.

A thrombosed vein also may present as a popliteal mass. Thrombus is present if the vein cannot be completely obliterated by compression with the transducer.

Nerve tumors such as neurofibromas or schwannomas can present as popliteal masses. Normal nerves appear as linear, fibrillar, and hyperechoic structures in proximity to vessels. The diagnosis of nerve sheath tumor is suggested by a fusiform, hypoechoic enlargement of the nerve.[6,35]

Ganglia and synovial and meniscal cysts can all present as popliteal masses. Sonography is very accurate in the characterization of these masses and also can guide their aspiration.

REFERENCES

1. Fornage BD: Musculoskeletal evaluation. p. 1. In: Mittelstaedt CA (ed): General Ultrasound. Churchill Livingstone, New York, 1992
2. van Holsbeeck M, Introcaso JH: Musculoskeletal ultrasonography. Radiol Clin North Am 30:907, 1992
3. Resnick D, Niwayama G: Diagnosis of Bone and Joint Disorders. 2nd Ed. WB Saunders, Philadelphia, 1988
4. Marchal GJ, van Holsbeeck MT, Raes M et al: Transient synovitis of the hip in children: role of US. Radiology 162:825, 1987
5. Peck RJ: Ultrasound of the painful hip in children. Br J Radiol 59:205, 1986
6. van Holsbeeck M, Introcaso JH: Musculoskeletal Ultrasound. Mosby-Year Book, St. Louis, 1991
7. Mori Y: Debris observed by arthroscopy of the knee. Orthop Clin North Am 10:559, 1979
8. Gilsanz V, Bernstein BH: Joint calcification following intraarticular corticosteroid therapy. Radiology 151:647, 1984
9. Richardson ML, Selby B, Montana M, Mack LA: Ultrasonography of the knee. Radiol Clin North Am 26:63, 1988
10. van Holsbeeck M, van Holsbeeck K, Gevers G et al: Staging and follow-up of rheumatoid arthritis of the knee. Comparison of sonography, thermography, and clinical assessment. J Ultrasound Med 7:561, 1988
11. Aisen AM, McCune WJ, MacGuire A et al: Sonographic evaluation of the cartilage of the knee. Radiology 153:781, 1984
12. Teitz CC: Ultrasonography in the knee: clinical aspects. Radiol Clin North Am 26:55, 1988
13. Barrie HJ: The pathogenesis and significance of meniscal cysts. J Bone Joint Surg 61:184, 1979
14. Casser HR, Sohn C, Kieckenback A: Current evaluation of sonography of the meniscus. Results of a comparative study of sonographic and arthroscopic findings. Arch Orthop Trauma Surg 109:150, 1990
15. Fielding JR, Franklin PD, Kustan J: Popliteal cysts: a reassessment using magnetic resonance imaging. Skeletal Radiol 20:433, 1991
16. Peetrons P, Allaer D, Jeanmart L: Cysts of the semilunar cartilage of the knee: a new approach by ultrasound imaging. J Ultrasound Med 9:333, 1990
17. Burk DL, Dalinka MK, Kanal E et al: Meniscal and ganglion cysts of the knee: MR evaluation. AJR 150:331, 1988
18. Mathieson JR, Connell DG, Cooperberg PL et al: Sonography of the Achilles tendon and adjacent bursae. AJR 151:127, 1988
19. Lindgren PG, Willen R: Gastrocnemio-semimembranous bursa and its relationship to the knee joint. I. Anatomy and histology. Acta Radiol 18:497, 1977
20. Ferrari FS, Tirobocchi A, Monetti G et al: Ultrasonography in the study of synovial disorders and tendon lesions in rheumatoid arthritis. Radiol Med 80:631, 1990
21. Canoso JJ: Bursae tendons and ligaments. Clin Rheum Dis 7:189, 1981
22. Genovese GR, Jayson MI, Dixon AS: Protective value of synovial cysts in rheumatoid knees. Ann Rheum Dis 31:179, 1972
23. Fornage BD, Rifkin MD, Touche DH, Segal PM: Sonography of the patellar tendon: preliminary observations. AJR 143:179, 1984
24. Fornage BD, Rifkin MD: Ultrasound examination of tendons. Radiol Clin North Am 26:87, 1988
25. Roels J, Martens M, Mulier JC, Burssens A:

Patellar tendinitis (jumper's knee). Am J Sports Med 6:362, 1978
26. Merkel KH, Hess H, Kunz M: Insertion tendinopathy in athletes. A light microscopic, histochemical and electron microscopic examination. Pathol Res Pract 173:303, 1982
27. Clemente CD (ed): Gray's Anatomy. Lea and Febiger, Philadelphia, 1985
28. Hughston JC, Andrews JR, Cross MJ, Moschi A: Classification of knee ligament instabilities. I. The medial compartment and cruciate ligament. J Bone Joint Surg 58:159, 1976
29. Nicholas JA, Hershman EB: The Lower Extremity and Spine in Sports Medicine. CV Mosby, St. Louis, 1986
30. Mink JH, Reicher MA, Crues JV III: Magnetic Resonance Imaging of the Knee. Raven Press, New York, 1987
31. Gruber G, Harland U, Gruber GM: Ultrasound image of the Lachman test in lesions of the anterior cruciate ligament. Sportverletz Sportschaden 6:123, 1992
32. McDonald DG, Leopold GR: Ultrasound B-scanning in the differentiation of Baker's cyst and thrombophlebitis. Br J Radiol 45:729, 1972
33. Pathria MN, Zlotkin M, Sartoris DJ et al: Ultrasonography of the popliteal fossa and lower extremities. Radiol Clin North Am 26:77, 1988
34. Sarti DA, Louie JS, Lindstrom RR et al: Ultrasonic diagnosis of a popliteal artery aneurysm. Radiology 121:707, 1976
35. Fornage BD: Peripheral nerves of the extremities: imaging with US. Radiology 167:179, 1988

14
Ankle and Foot

Marnix van Holsbeeck
Alex Powell

Sonography can be used as a cost-effective method of evaluating muscles, tendons, ligaments, and synovium of joints and bursae.[1] It is therefore an important adjunct to clinical examination and radiographic evaluation of patients with musculoskeletal disorders of the foot and ankle.[2] The availability of ultrasound equipment throughout the world and its affordable price make this technology very accessible for patients with foot and ankle disease.[3] The dynamic character of ultrasound examination is an additional advantage. Sonography has the potential to become a screening method that allows one to distinguish between significant and trivial soft-tissue injuries, not unlike the way radiography is used to exclude the possibility of fractures.

Acute and chronic foot and ankle pain cause thousands of patients each year to visit orthopaedic surgeons, rheumatologists, physiotherapists, sports doctors, and podiatrists. Unfortunately, radiographs of painful feet usually are negative or show bone spurs that may or may not be the cause of the patient's symptoms. Many times, the true pathology within the foot is related to soft-tissue injuries, abnormal fluid collections, or soft-tissue masses. All of these can be visualized by high-frequency sonography. Unlike the often-varied opinions that come from interpretation of soft-tissue abnormalities seen on standard radiographs, more standardized diagnoses of soft-tissue diseases of the foot and ankle are possible with sonography. This more objective approach to foot and ankle pain can better identify surgical candidates and improve subsequent surgical outcomes.

LIGAMENTS AND TENDONS

Ankle injuries are common. Among the most frequent are inversion injuries resulting in a lateral ankle sprain. Following this type of injury, the lateral ankle often will be swollen and tender. Despite significant ligamentous injury, radiographs often will be negative. The most commonly injured ligament is the anterior talofibular ligament. Tears can also involve the calcaneofibular ligament and, rarely, the posterior talofibular ligament. Disruption of these ligaments and the associated hematoma can be detected sonographically.[4,5] However, the finding of ligament disruption does not change treatment

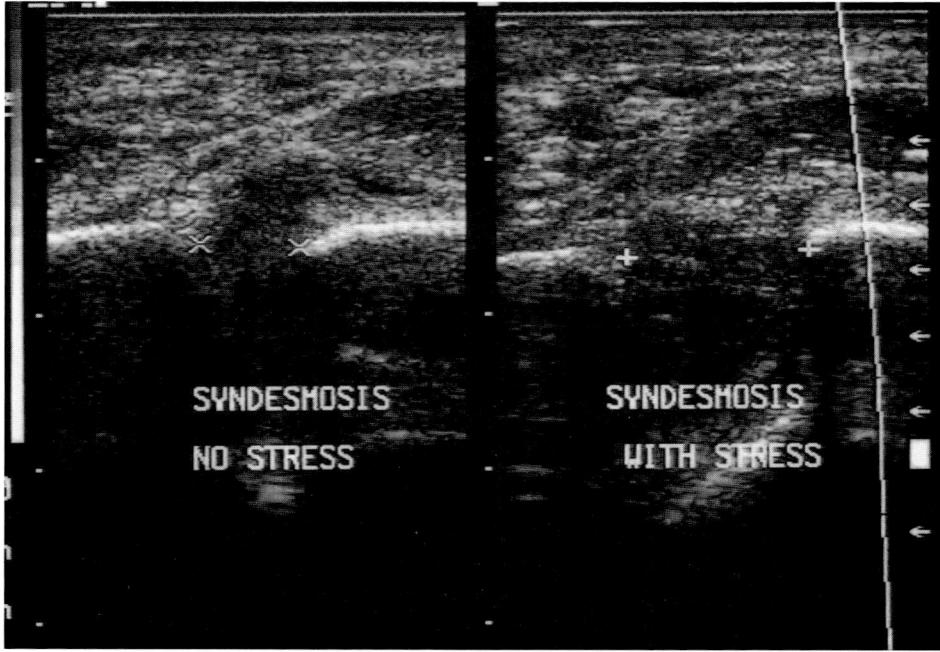

Fig. 14-1 Stress testing during a real-time sonographic examination following an external rotation injury with associated ligament rupture. Under stress, abnormal widening of the tibiofibular syndesmosis can be seen.

in most cases. The exception to this is an external rotation injury resulting in rupture of the anterior tibiofibular ligament with associated instability of the ankle syndesmosis; surgical treatment is indicated for an unstable and ruptured syndesmosis. The instability can be measured sonographically (Fig. 14-1).

Nearly all tendinous injuries of the foot and ankle can be evaluated by ultrasound. These injuries occur when excessive forces result in abnormal elongation of the tendons. The most common injuries involve the Achilles tendon, posterior tibial tendon, or peroneal tendons. Rare cases of inflamed or torn anterior tibial and flexor hallucis longus tendons have been observed. Soccer players and ballet dancers are two groups of patients in whom such injuries have been seen. It should be noted that not all patients with acute ankle injuries should be evaluated with ultrasound. Sonography should be used in patients suspected of having an Achilles tendon rupture or inflammation and in patients with persistent ankle pain and swelling lasting longer than 4 to 6 weeks after an acute injury. The incidence of tendinous injury is markedly increased in patients with persistent pain.

Achilles Tendon

An acute ankle dorsiflexion injury can cause Achilles tendon rupture. Although this more commonly occurs in patients with a history of Achilles tendinitis, rupture also can occur in patients with no such history. A significant risk factor for Achilles tendon rupture is tendinitis previously treated with steroid injection. The diagnosis of Achilles tendon rupture is easily made sonographically. The characteristic findings include a tear at the

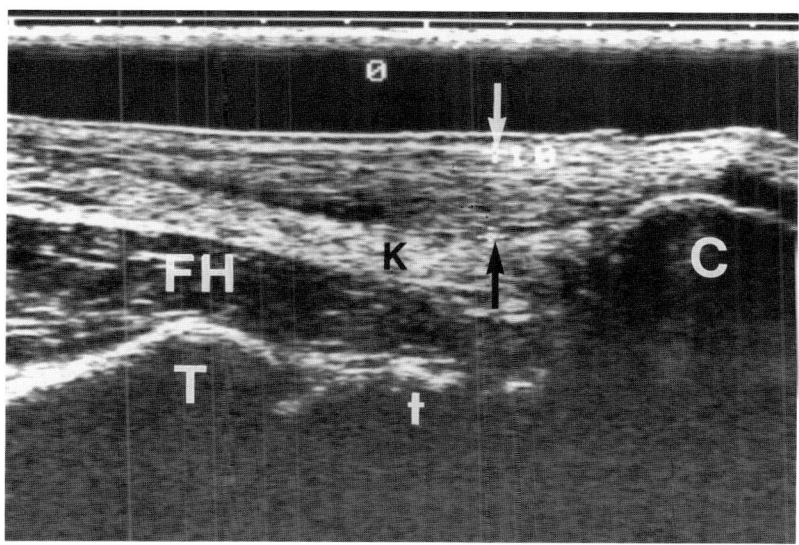

Fig. 14-2 Achilles tendinitis. Longitudinal sonogram of the Achilles tendon shows fusiform and hypoechoic tendon swelling (arrows). C, calcaneus; FH, flexor hallucis longus muscle; K, Kager's fat pad; T, tibia; t, talus.

level of the posterior malleolus,[1] anechoic fluid or a blood-filled defect in the tendon, and retracted and swollen proximal and distal ends of the tendon.[1,6] The irregular edges of the tendon will cause refraction and can cast acoustic shadows in the adjacent soft tissues.[1] Dynamic evaluation is an essential aspect of proper ultrasound examination of the Achilles tendon. Careful dorsiflexion of the ankle will cause maximum separation of the torn edges of the tendon; plantar flexion will approximate the split tendon fragments, and the minimum defect can then be measured.[3] It is likely that the information garnered from dynamic ultrasound examination will soon be used in deciding between conservative cast treatment and surgical intervention.[3]

Achilles tendon swelling and pain are more common in aging athletes than in younger ones. Although there may be a history of trauma, the pain most often develops after a strenuous training program or when athletic activity is increased. The tendon typically swells at the level of the posterior malleolus.[7] With swelling, the distance between tendon bundles increases. The tendon appears fusiformly swollen and hypoechoic (Fig. 14-2).[8,9] Diagnosis is made by right-left comparisons of longitudinal scans.[8] In fact, carefully performed right-left comparisons can be used to confirm tendinitis in any location (Fig. 14-3). Measurement is rarely necessary, but differences in the anteroposterior dimension of 2 mm or more between the right and left Achilles tendons are considered significant.[1] More distal swelling at the tendon insertion may be due to tendon disease, bursa swelling, or simultaneous tendon and bursa disease[3]; this kind of disorder is the most resistant to treatment.[10]

Posterior Tibial and Peroneal Tendons

The posterior tibial tendon is the second most common tendon affected in ankle disease (Figs. 14-4 and 14-5). Shear load can rupture the posterior tibial tendon around

224 MUSCULOSKELETAL ULTRASOUND

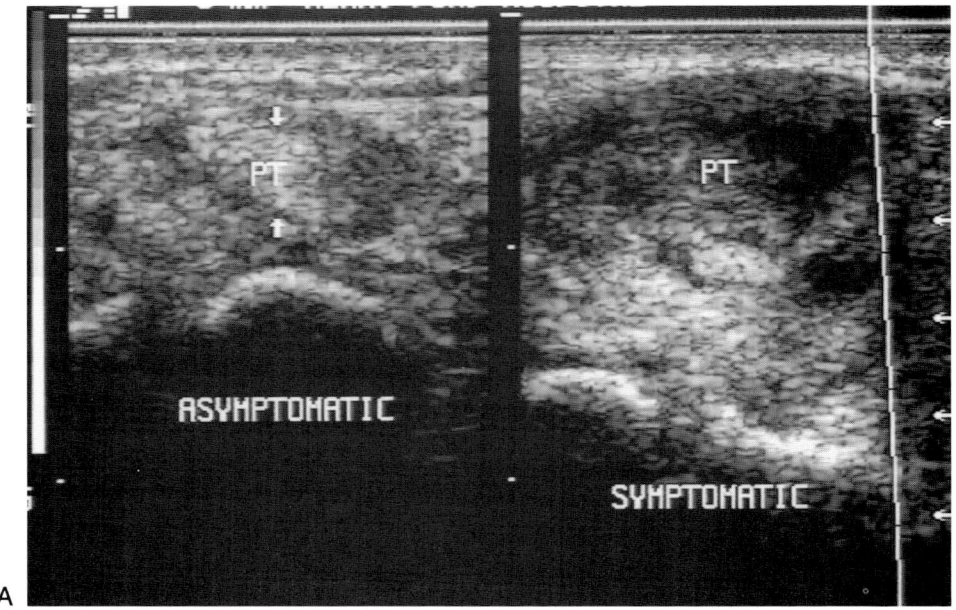

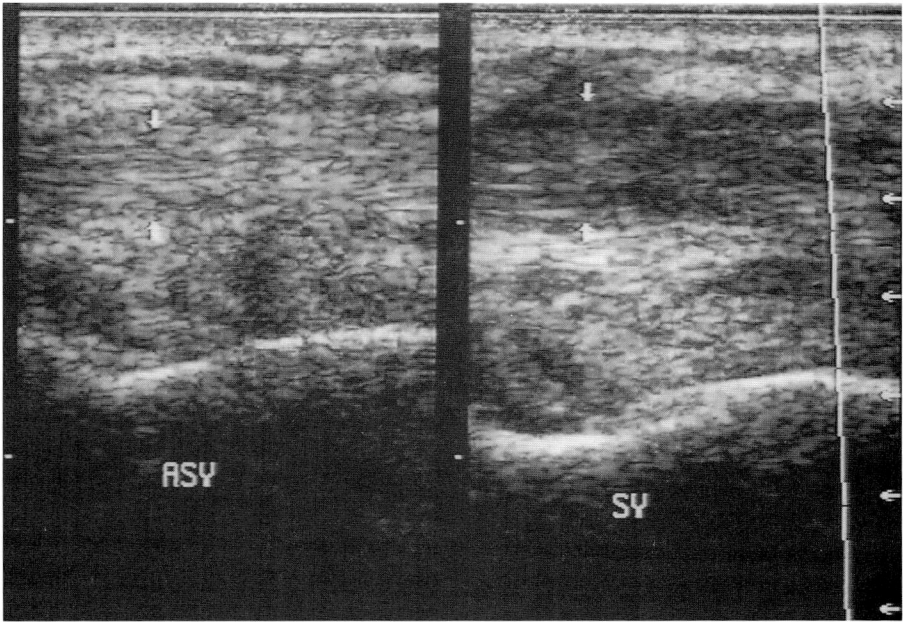

Fig. 14-3 Intrasubstance tear of the posterior tibial tendon. **(A)** Transverse sonograms. Left: Normal left posterior tibial tendon (PT and *arrows*) is oval and hyperechoic, with a surrounding hypoechoic halo. Right: The abnormal and symptomatic right posterior tibial tendon (PT) appears hypoechoic and is swollen to three times the normal diameter. **(B)** Longitudinal sonograms. Left: The asymptomatic (ASY) left inframalleolar posterior tibial tendon appears hyperechoic (*short arrows*). Right: The abnormal tendon (SY) is hypoechoic and swollen (*short arrows*).

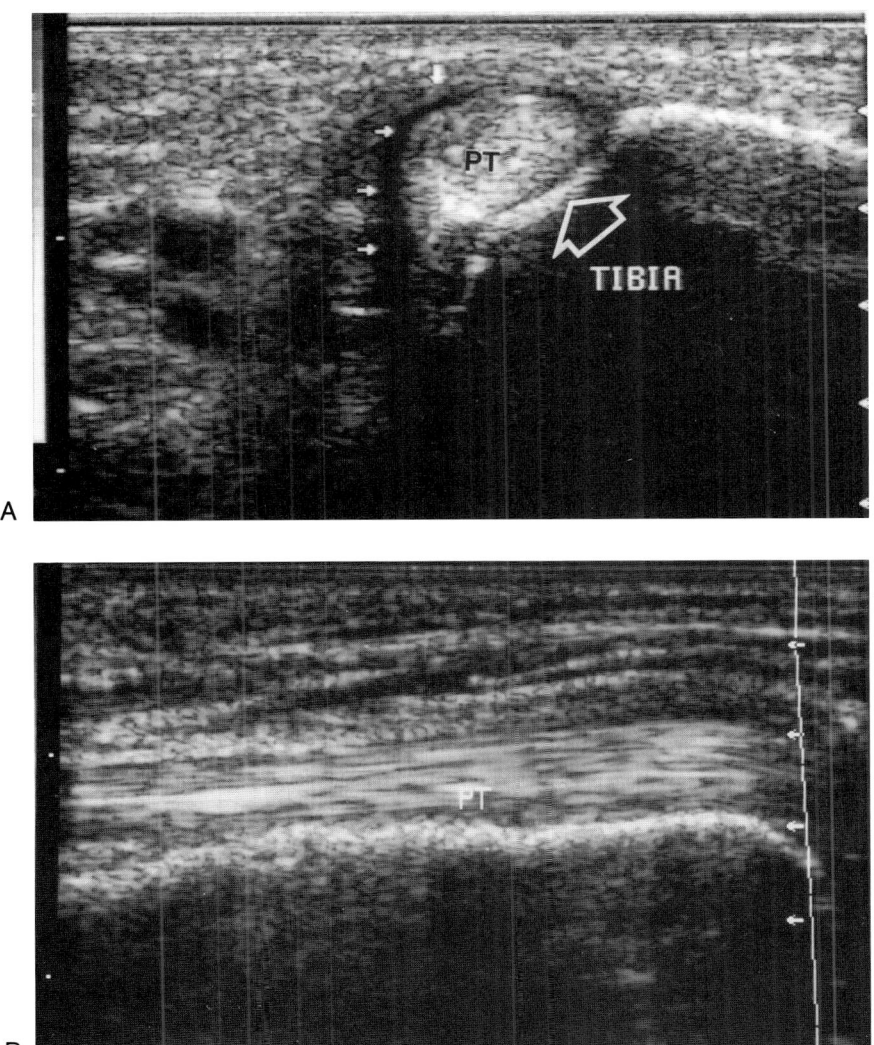

Fig. 14-4 Normal posterior tibial tendon. **(A)** Transverse sonogram through the normal posterior tibial tendon (PT). The supramalleolar tendon is hyperechoic and oval, with a diameter of from 4 to 6 mm. A thin hypoechoic tendon sheath (*short arrows*) surrounds the normal tendon. The shallow groove (*open arrow*) in the malleolus is a recognizable landmark. **(B)** Longitudinal sonogram through the normal posterior tibial tendon (PT). The transducer has been rotated 90 degrees from the position in Figure 14-4A. The fibrillar pattern of the tendon is now clearly visible.

the medial malleolus and the peroneal tendons at the lateral malleolus.[1,2] Such tears can occur in patients with malleolar fractures and in patients who have sustained pure soft-tissue injuries. Persistent perimalleolar pain is the most common symptom of such ruptures. The posterior tibial and peroneal tendons usually tear longitudinally (Figs. 14-6 and 14-7).[1,2] A longitudinally torn tendon appears elongated and swollen, with one or more splits cleaving the tendon in the long axis. Transverse tears of these tendons occur

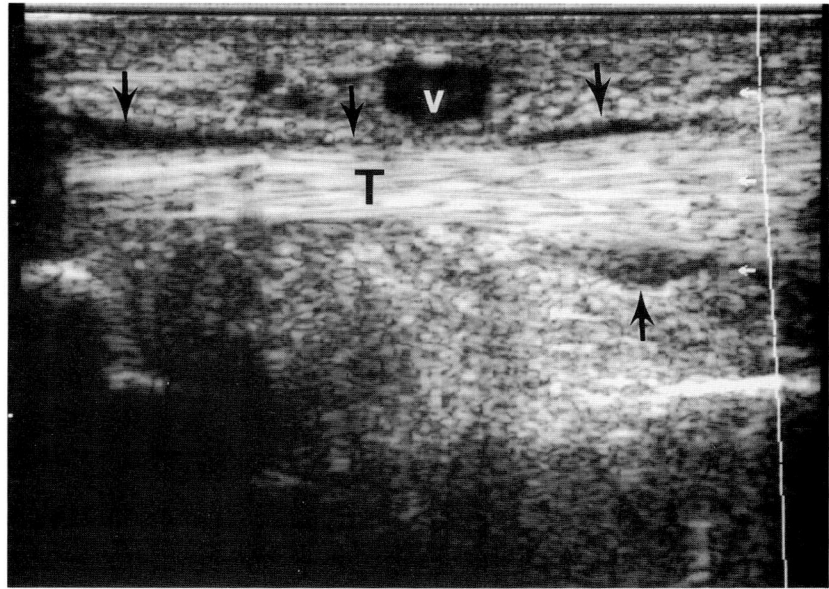

Fig. 14-5 Normal posterior tibial tendon. On a longitudinal sonogram, the normal inframalleolar tendon (T) appears hyperechoic and fibrillar. The tendon sheath is anechoic and contains a thin layer of fluid (*arrows*). A small vein (v) circumscribes the distal tendon.

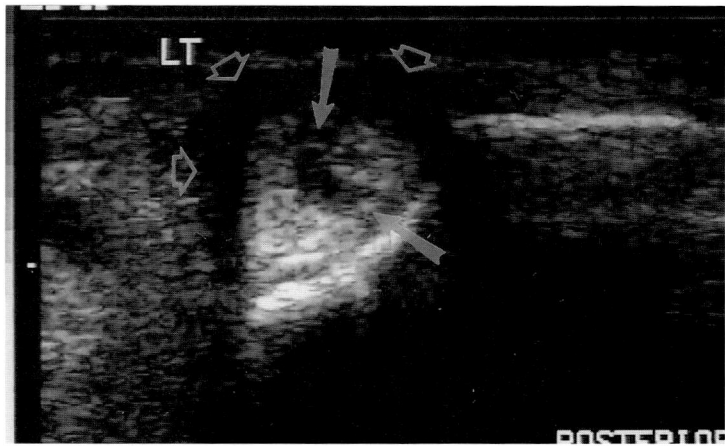

Fig. 14-6 Tear of the posterior tibial tendon. Transverse sonogram shows a cleft (*arrows*) in the posterior tibial tendon and a fluid-filled tendon sheath (*open arrows*) in a patient with a surgically proven longitudinal split in the posterior tibial tendon. Compare to the normal tendon in Figure 14-4A.

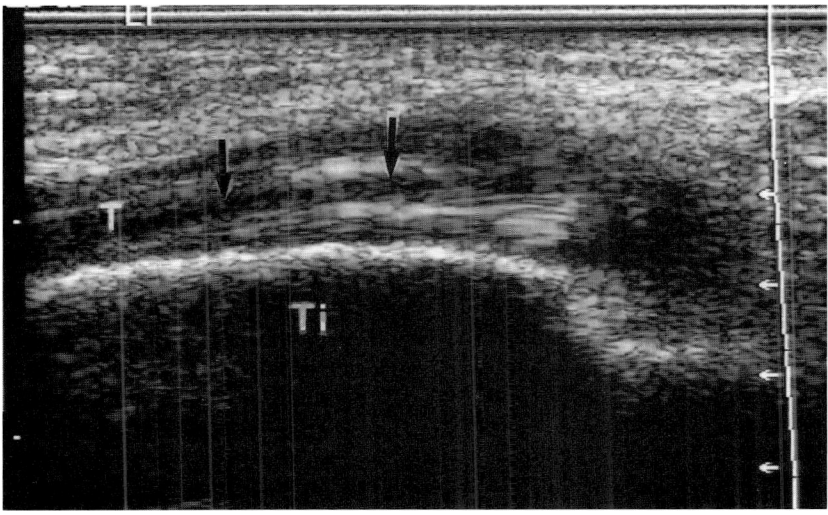

Fig. 14-7 Longitudinal tear in the posterior tibial tendon. Longitudinal sonogram shows a longitudinal cleft (*arrows*), which follows the long axis of the tendon (T). Ti, medial tibia.

less frequently. Patients with transverse tears of the posterior tibial tendon have a painful flat foot deformity and heel valgus.[11] Pain is present over both malleoli. The lateral pain is probably related to foot malposition. Characteristically, a transverse tear appears on sonograms as a hypoechoic gap within the tendon substance, with excessive fluid filling the tendon sheath (Fig. 14-8). There may be retraction of the interrupted tendon ends. In

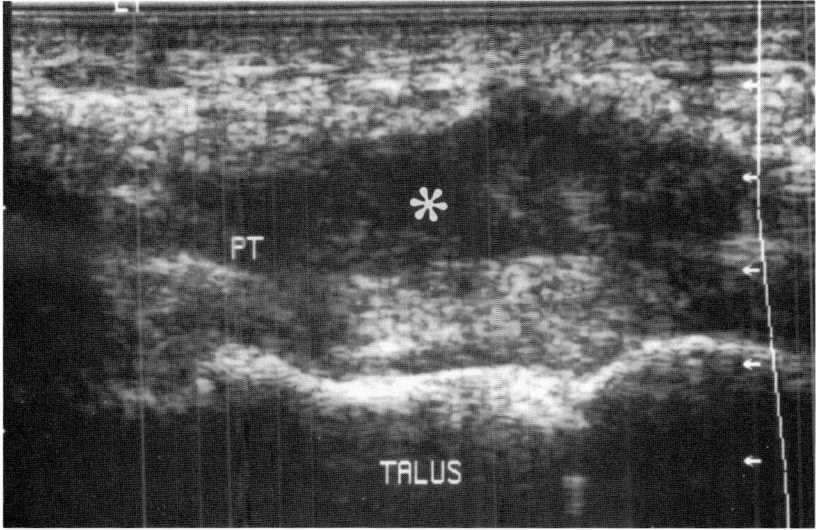

Fig. 14-8 A torn, retracted posterior tibial tendon (PT) with an empty tendon sheath (*asterisk*). Compare to the sonogram of a normal tendon shown in Figure 14-5.

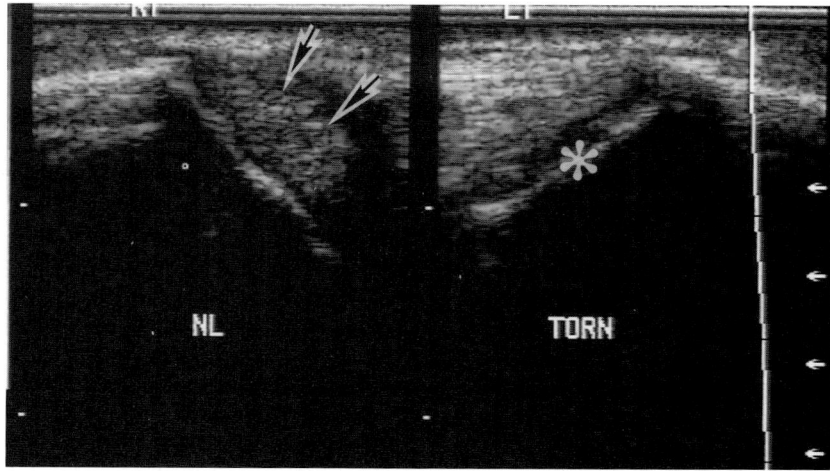

Fig. 14-9 Chronic complete transverse tear of the supramalleolar left posterior tibial tendon. *Left:* The normal (NL), oval right tendon (*arrows*) fills the bony groove. *Right:* An empty groove (*asterisk*) is seen in the left ankle.

chronic and late-stage injuries, fluid may be absent and the retracted tendon invisible. In these cases, the supramalleolar groove is often empty (Fig. 14-9). Sonography allows differentiation between patients with tendon edema or tendon sheath swelling and those with true tendon tears. Clinically, this is an important distinction because surgery is indicated for longitudinal or transverse tears but conservative therapy is used for intrasubstance tears or tenosynovitis.

A common sequela of trauma to the lateral ankle is rupture of the superior peroneal retinaculum. The peroneal tendons then subluxate out of the bony groove behind the lateral malleolus when the patient walks on uneven surfaces such as sand, gravel, or rock. Subluxation may be associated with an audible snap or click and invariably causes a sharp pain at the outer ankle. On sonography, most of these patients will show signs of peroneal tenosynovitis, with small fluid collections or thin layers of synovial thickening of the peroneal tendon sheath. The anechoic fluid moves within the tendon sheath when pressure is applied to the transducer.

Synovial thickening is hypoechoic and feels soft when pressure is applied. In patients with long-standing and severe problems, fusiform thickening of both the peroneus longus and brevis tendons will be noted at the tip of the malleolus (Fig. 14-10).[1] Longitudinal splits and fraying can be seen in advanced disease (Fig. 14-11).[1] Dislocation of the tendons over the lateral aspect of the ankle is rarely seen. In some cases, intermittent subluxation can be shown with both dorsiflexion and eversion of the foot.[1]

Chronic perimalleolar pain is not always due to tendon tears. Scar formation in the tendons (Fig. 14-12), capsules, and ligaments; tenosynovitis (Fig. 14-13); chronic malleolar bursitis (Fig. 14-14); or a ganglion also can cause swelling and pain. These structures can be easily distinguished from tendon tears sonographically.[1,12,13]

Although most tendon injuries of the ankle are closed, injuries caused by direct or penetrating trauma to the tendons do occur. Tendinous injuries have been seen in association with gunshot or stab wounds, motor vehicle

ANKLE AND FOOT 229

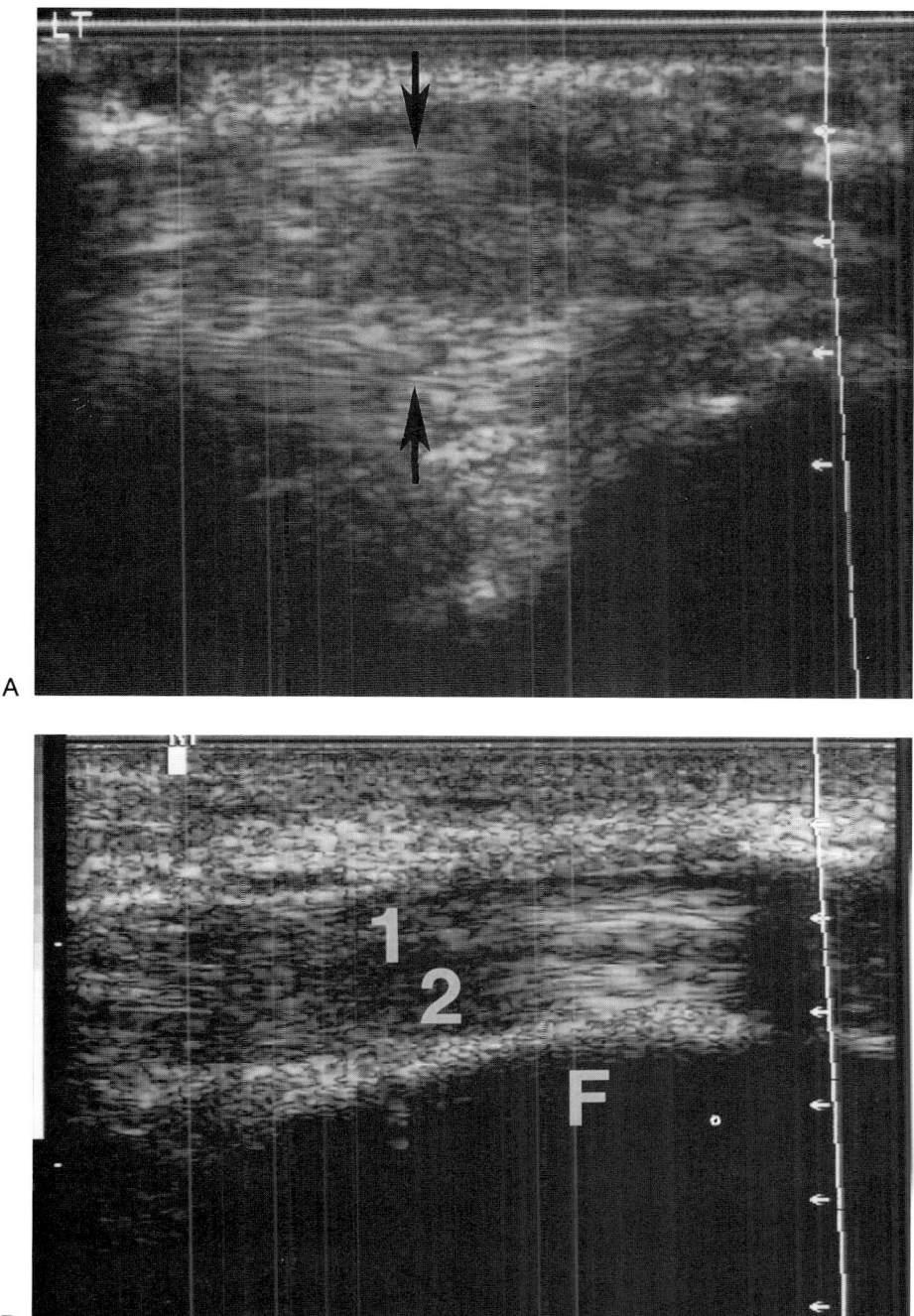

Fig. 14-10 Chronic instability of the peroneal tendons. **(A)** Unstable peroneal tendons are swollen (*black arrows*) at the level of the lateral malleolus. **(B)** Longitudinal sonogram of normal peroneus longus (1) and brevis (2) tendons for comparison. F, fibula.

230 MUSCULOSKELETAL ULTRASOUND

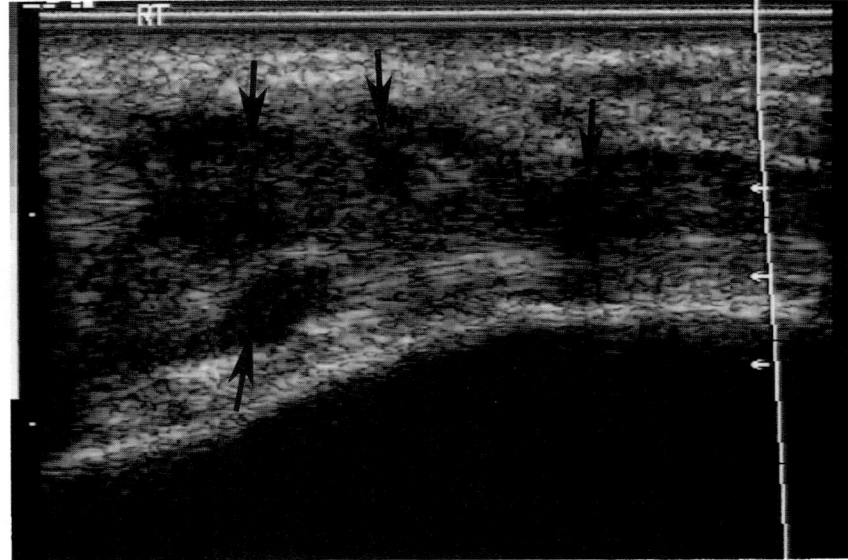

Fig. 14-11 Peroneal tendon degeneration. Longitudinal sonogram shows swelling and decreased echogenicity (*black arrows*) of the tendon substance.

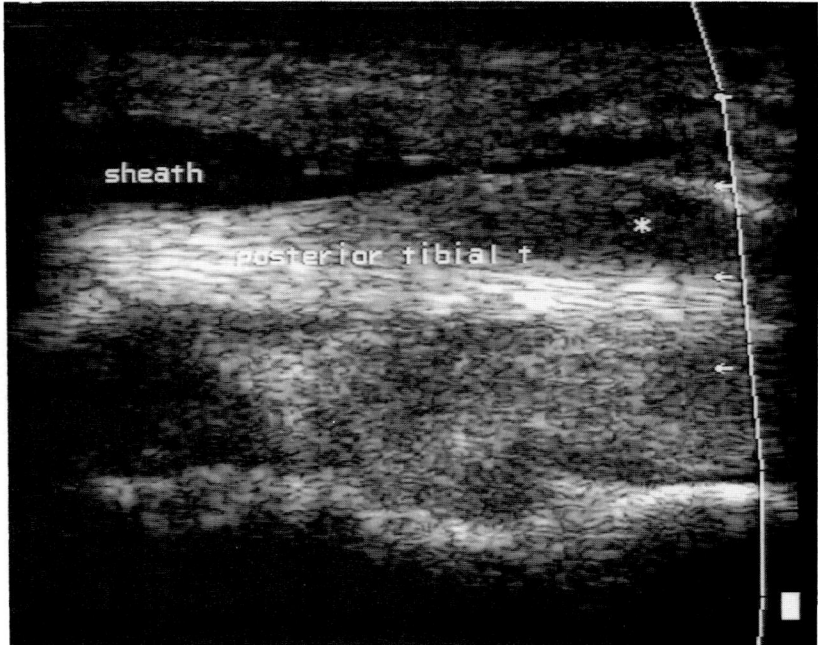

Fig. 14-12 Old scar (*asterisk*) in the distal posterior tibial tendon as a cause of continued pain. Flexor digitorum graft interposition was necessary to relieve the patient's symptoms.

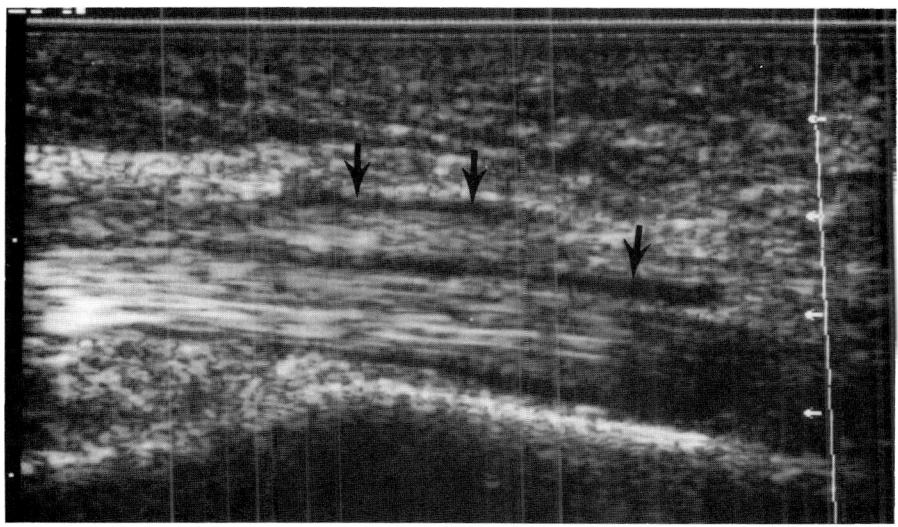

Fig. 14-13 Peroneal tenosynovitis. An increased amount of synovial fluid (*black arrows*) is noted. Tenosynovitis is a more benign process than tendinitis.

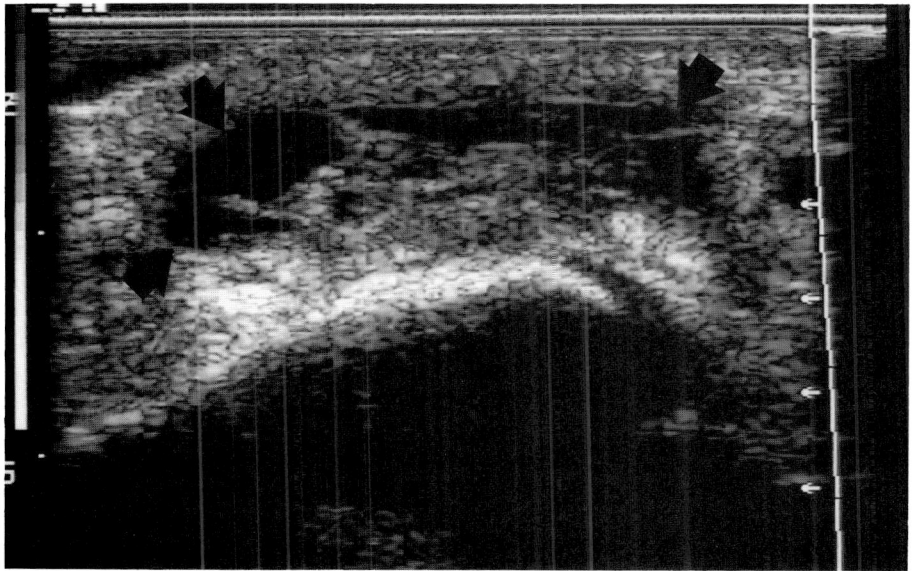

Fig. 14-14 Lateral malleolar bursitis. An inflamed lateral malleolar bursa (*arrowheads*) caused localized pain and tenderness.

accidents, and foreign body penetration. The clinical examination of these injuries is often limited by hematoma and tissue damage. Inspection of the wound often fails to demonstrate tendinous injury. Sonography is highly useful in cases in which direct examination is limited and tendon stress tests suggest injury. To prevent infection in these cases, the transducer is first wiped with alcohol, and sterile gel is used. Sonography can determine whether it is safe to suture the soft tissues without surgical exploration of the tendon. The sonographic appearance of tendon laceration does not differ from that of an elongation injury, although the location of the tear is more variable: penetrating injuries can occur all along the course of the tendon, whereas elongation injuries occur at predictable weak spots.

JOINTS

Both the ankle and foot consist of multiple joints in a complex anatomic arrangement. Clinically, it is often difficult to distinguish joint swelling from swelling of the extra-articular soft tissues. Sonography can detect capsular or synovial pathology. Initial evaluation of the tibiotalar joint should be performed with the foot in plantar flexion. Optimally, a linear-array transducer is used; however, a curved linear-array transducer may be used if plantar flexion is limited. The joint is studied by moving the transducer from malleolus to malleolus along the anterior aspect of the ankle.[1] The posterior aspect can be visualized with the patient in a ventral decubitus position, with the Achilles tendon serving as an acoustic window. Longitudinal orientation of the transducer is preferred throughout the study of the tibiotalar joint.

Effusions appearing as anechoic intra-articular fluid collections are a common finding following uncomplicated ankle injuries. The majority of these effusions will clear in the first couple of weeks following injury. Although their pathogenesis is unknown, the presence of such intra-articular fluid collections is a relatively good prognostic factor, as it seems to indicate that the ankle's collateral ligaments are intact. It has been speculated that the fluid is related to cartilaginous injury. Gout, rheumatoid arthritis, Reiter syndrome, and psoriatic arthritis also can cause such effusions and, in some cases, synovial proliferation in the hindfoot. Persistent monoarticular synovitis in young patients can be induced by an intra-articular foreign body or osteoid osteoma.

Arthrosonography has great potential in detecting loose bodies in the ankle joint (Fig. 14-15). Both CT arthrography and MRI have difficulty detecting small intra-articular loose bodies. Sonography has a distinct advantage because examination of the joint can be performed throughout the range of motion. Active or passive joint mobilization can show how the loose bodies move around the joint recess. Because of synovial irritation, a symptomatic joint containing loose bodies will have an increased amount of intra-articular fluid. This fluid serves as a natural contrast agent in the detection of intra-articular lesions. To evaluate a joint for loose bodies, the sonographer should compress the medial and lateral joint recesses with the free hand while scanning with the other. This allows the examiner to move any intra-articular debris. In cases in which there is little or no intra-articular fluid, injection of 10 ml of sterile saline and 0.5 ml of epinephrine may be useful. This technique can help to determine whether a calcification is located in the capsule or in the joint space.

SOFT-TISSUE MASSES

Benign soft-tissue masses are common in the foot. Abscesses, soft-tissue granulomas, and ganglia can be located and diagnosed with ultrasound. In addition to being cost-effective, sonography may allow for both di-

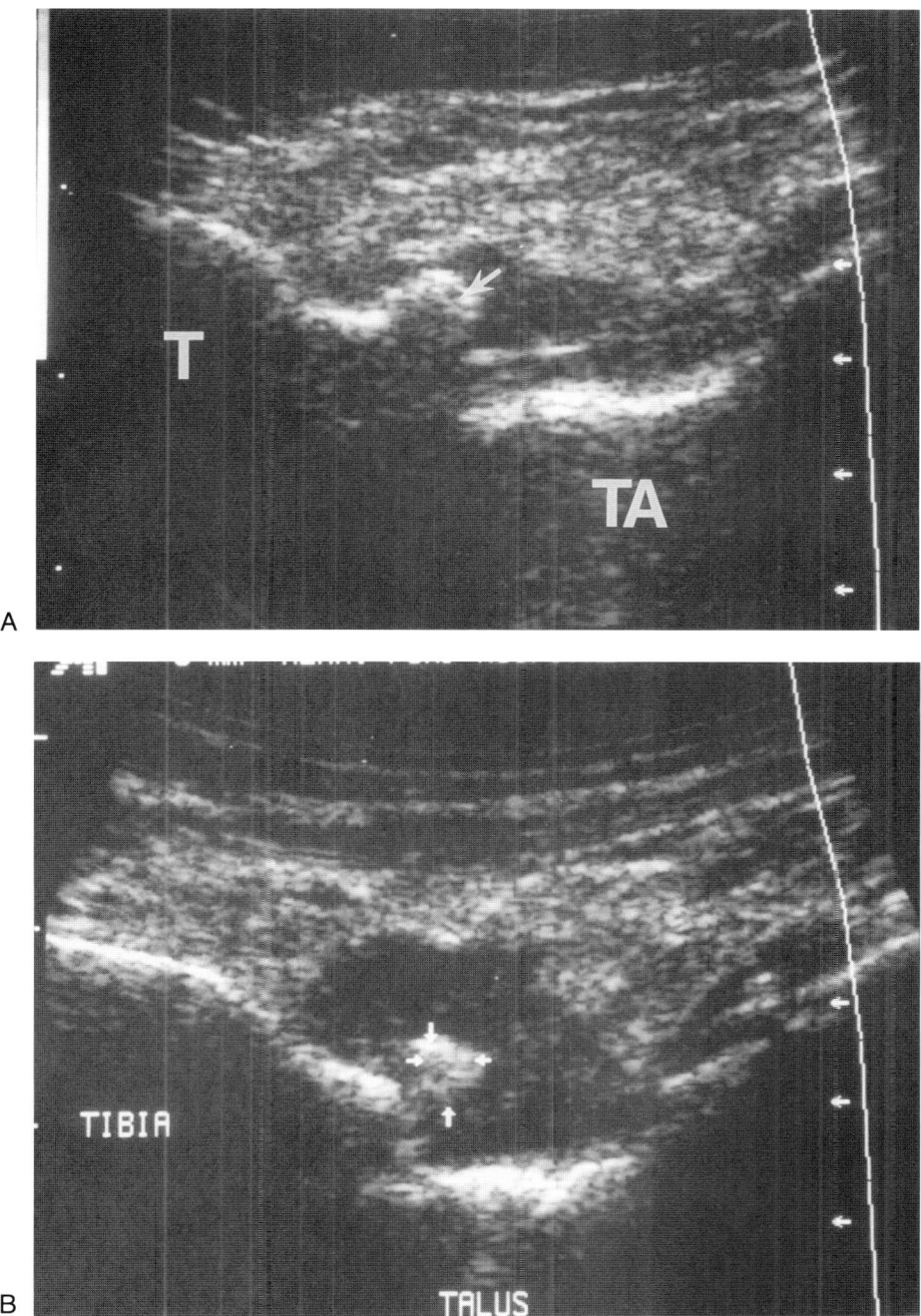

Fig. 14-15 Diagnosis of loose bodies in the ankle. **(A)** Longitudinal sonogram through the anterior tibiotalar joint recess shows nodular hyperechoic structure (*arrow*). It is unclear whether this structure is attached to the anterior distal tibia (T). TA, talus. **(B)** Following injection of sterile saline, the intra-articular location and free mobility of the loose body (*arrows*) are obvious.

234 MUSCULOSKELETAL ULTRASOUND

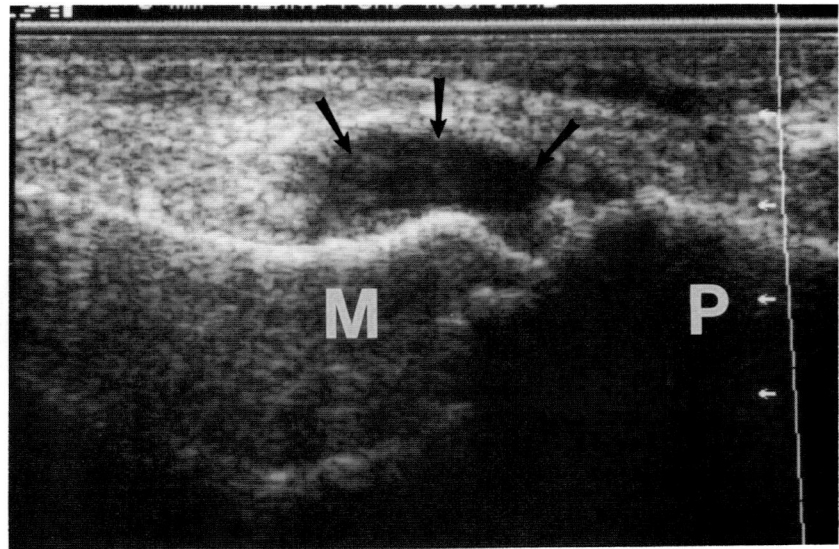

Fig. 14-16 Infection from penetrating injury. This patient presented with diffuse swelling of the entire foot after stepping on a rusty nail 2 weeks previously. Exploratory surgery failed to show the cause of the swelling, and no retained foreign body was found. Longitudinal sonogram shows arthritis (*black arrows*) of the first metatarsophalangeal joint. Joint fluid culture was positive for *Pseudomonas aeruginosa*. M, first metatarsal; P, first phalanx.

agnosis and treatment during the same imaging session. Ultrasound-guided aspiration is often used.

Ganglia are found around the ankle or over the dorsum of the foot.[1] Possible communication with a joint or a tendon sheath should be assessed. Typically, ganglia are anechoic or hypoechoic. Aspiration of the ganglia reveals clear fluid of variable viscosity.

Sonography is also highly useful in the evaluation of foot infections. Typically involving the plantar aspect of the foot, penetrating injuries or foreign bodies often precipitate infection. With sonography, the extent of small fistulous tracts can be followed. Extensions into abscesses, infected tendon sheaths, and joint cavities have all been identified. Infected masses (Fig. 14-16) often appear hypoechoic. This appearance in itself is nonspecific, and the diagnosis relies heavily on the clinical history and results of aspiration. On occasion, sonography can identify a radiolucent foreign body as the source of infection. In addition, sonography is particularly useful in assessing the foot postoperatively for possible infection. Artifacts caused by metallic implants do not limit ultrasound examination as they do other frequently employed modalities.

Preoperative staging of most solid foot masses should be done with MRI. The superior soft-tissue contrast resolution of MRI provides the most accurate preoperative staging of soft-tissue sarcomas or bone tumors with soft-tissue invasion. However, sonography does play a role in the diagnosis and evaluation of suspected plantar fibromas and Morton's neuromas. These benign solid soft-tissue tumors can be diagnosed sonographically by their typical location and distribution. Plantar fibromas are frequently bilateral and occur in patients with a family history of Dupuytren's disease. They are of-

ANKLE AND FOOT **235**

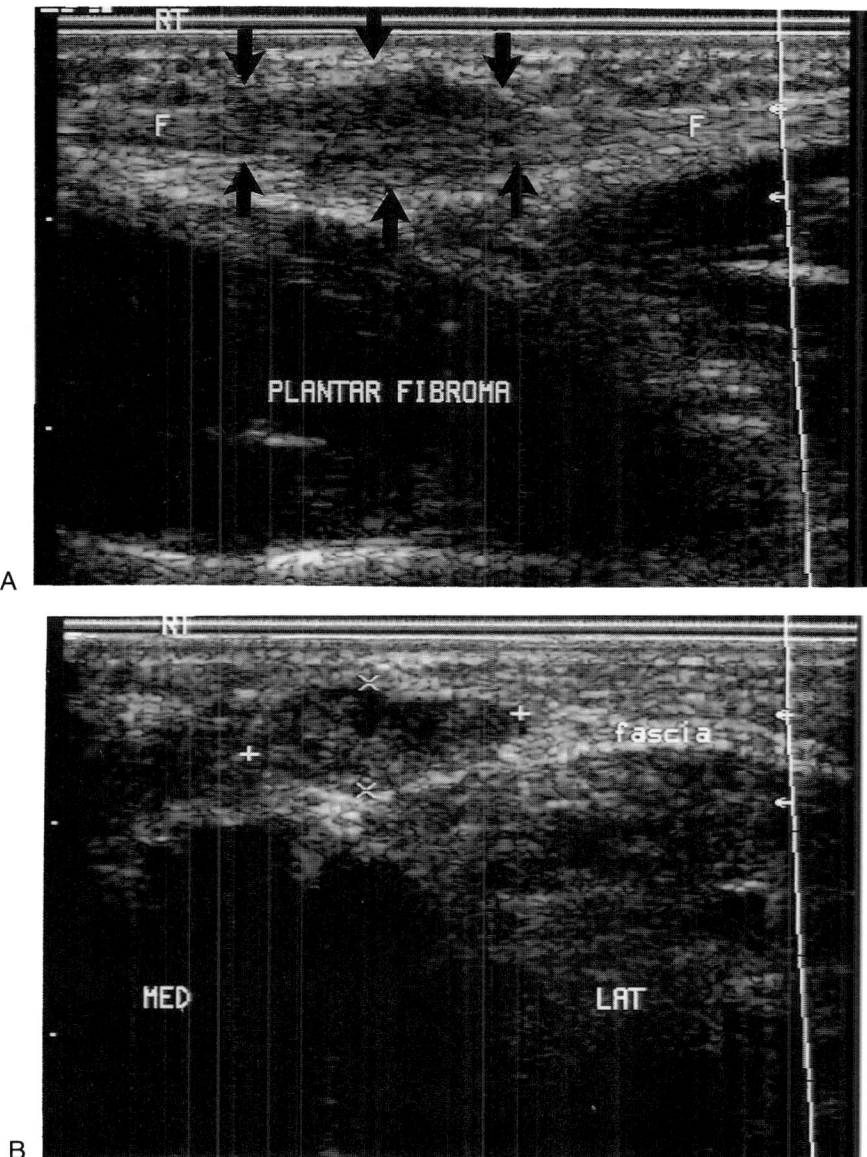

Fig. 14-17 Plantar fibroma. **(A)** Longitudinal sonogram shows a hypoechoic plantar fibroma (*black arrows*) invading the plantar fascia (F). Typically, such lesions have a fusiform shape, extending along the plantar fascia. **(B)** Transverse sonogram shows the relationship between the hypoechoic mass (*calipers*) and the echogenic fascia. It often is not possible to determine whether the lesion originates from the fascia or whether it erodes into the fascia.

236 MUSCULOSKELETAL ULTRASOUND

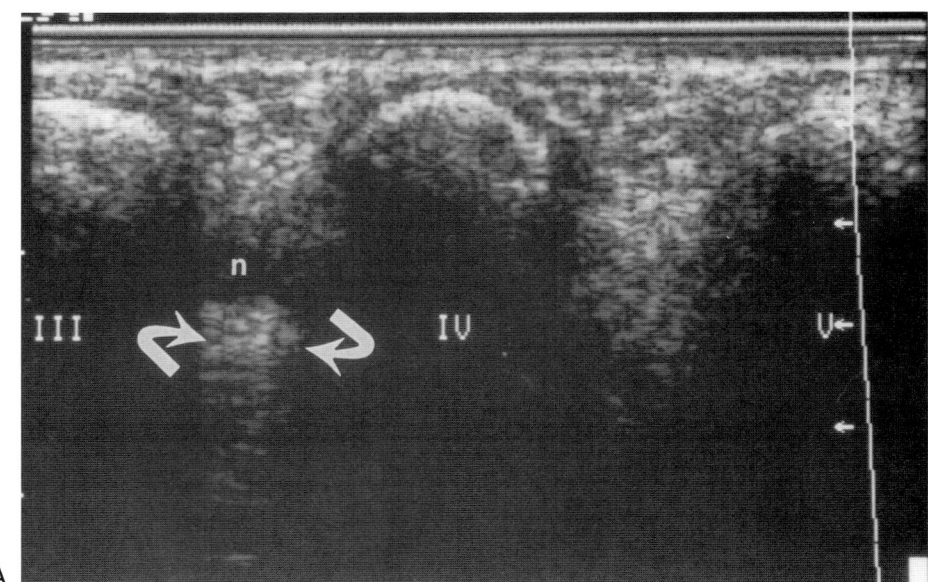

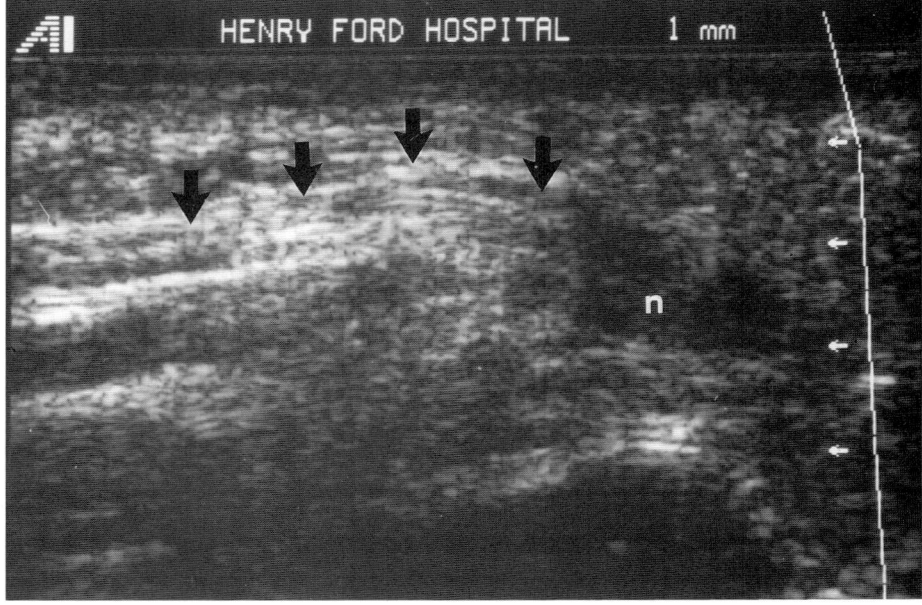

Fig. 14-18 Morton's neuroma. **(A)** Transverse sonogram with the transducer over the plantar aspect of the metatarsal heads. Careful screening for increased through-transmission (*curved arrows*) helps detect a small neuroma (n). The majority of Morton's neuromas are located in the third intermetatarsal space between metatarsals three (III) and four (IV). **(B)** Morton's neuroma (n) is best visualized when the transducer is then rotated 90 degrees in the intermetatarsal space. The lesion is hypoechoic and fusiform in shape. In some cases, the small interdigital nerve (*central arrows*) can be distinguished proximal to the mass.

ten associated with knuckle pads over the dorsal aspect of the fingers. These fibromas are hypoechoic relative to the subcutaneous layer in which they are located (Fig. 14-17). They are frequently oval and may adhere to the plantar fascia.

Morton's neuromas are typically located in the third intermetatarsal space. Occasionally, they are bilateral and occur in several interspaces. Women are more often affected than men. Ultrasound evaluation[12,14] for possible Morton's neuroma is first performed in the transverse plane at the level of the metatarsal heads (Fig. 14-18A). The transducer is placed over the dorsal and plantar surfaces of the foot. The index finger of the examiner's free hand should be run over the interspace in the sole while the foot is scanned from the dorsal aspect of the foot. This technique of sonographically guided palpation allows correlation with the patient's symptoms. Of further benefit, the palpating finger brings the lesion into better focus by compressing the soft tissues. Longitudinal scanning should also be performed as part of a complete examination (Fig. 14-18B). Longitudinal scanning may demonstrate continuity of the Morton's neuroma with the interdigital nerve. Not only is it important to confirm the diagnosis of a suspected Morton's neuroma, but exact localization of the damaged nerve sheath is important in preoperative planning.

REFERENCES

1. van Holsbeeck M, Introcaso JH: Musculoskeletal Ultrasound. Mosby-Year Book, St. Louis, 1991
2. van Holsbeeck M, Introcaso J, Kolowich P: Sonography of tendons: patterns of disease. Instructional Course Lectures 43:475, 1994
3. Thermann H, Hoffmann R, Zwipp H, Tscherne H: The use of ultrasonography in the foot and ankle. Foot Ankle 13:386, 1992
4. Perin B, Cremonini L, Davi L, Gardellin G: Ultrasonic assessment of the capsulo-ligamentous structures of the ankle. Normal features. Radiol Med 83:737, 1992
5. Friedrich JM, Schnarkowski P, Rubenacker S, Wallner B: Ultrasonography of capsular morphology in normal and traumatic ankle joints. J Clin Ultrasound 21:179, 1993
6. Neuhold A, Stiskal M, Kainberger F, Schwaighofer B: Degenerative Achilles tendon disease: assessment by magnetic resonance and ultrasonography. Eur J Radiol 14:213, 1992
7. Laine HR, Harjula ALS, Peltakallio P: Ultrasonography as a differential diagnostic aid in achillodynia. J Ultrasound Med 6:351, 1987
8. Fornage B: Achilles tendon: US examination. Radiology 159:759, 1986
9. Blei CL, Nirschl RP, Grant EG: Achilles tendon. US diagnosis of pathologic conditions. Radiology 159:765, 1986
10. Mathieson JR, Connell DG, Lloyd-Smith RD: Sonography of the Achilles tendon and adjacent bursae. AJR 151:127, 1988
11. van Holsbeeck M, Katcherian D, Wu KK, Introcaso JH: Patterns of posterior tibial tendon abnormality (abstract). Radiology 185(P):143, 1992
12. van Holsbeeck M, Introcaso JH: Musculoskeletal ultrasonography. Radiol Clin North Am 30:907, 1992
13. Peetrons P, Moerman C, Creteur V: Echographie ostéoarticulaire: présent et avenir. Radiologie (Journal du CEPUR) 12:14, 1992
14. Redd RA, Peters VJ, Emery SF et al: Morton neuroma: sonographic evaluation. Radiology 171:415, 1989

Index

Note: *Page numbers followed by f indicate figures; those followed by t indicate tables.*

A

Abductor pollicis longus tendon, *152*
Abscess, soft-tissue, 23–25
Acetabulum, 180, 191
Achilles tendon
    rupture of, 222–223
    tendinitis of, 223, 223f–224f
Acromioplasty, postoperative appearances of, 126–129, 128f–130f
Actinic keratosis, 90, 91f
Air bubbles, in subcutaneous tissue, 103, 105f
Anatomic snuffbox, 152
Aneurysmal bone cyst, 39f, 65
Aneurysm
    of popliteal artery, 29, 217
    sciatic nerve displacement by, 79, 80f
Angiolipoma, 31
Angioma, 30–31, 31f
Angiosarcoma, 33. See also Sarcomas, soft-tissue
Ankle
    Achilles tendon, 222–223, 223f–224f
    joint effusion, 44–45
        loose bodies in, 232, 233f
    ligamentous injuries of, 221–222, 222f
    peroneal tendon, 225, 227–228, 225f–226f
    posterior tibial tendon, 223, 225, 227–228, 225f–228f, 232
    soft-tissue masses of, 232, 234, 234f–236f, 237
Arthritis. See Rheumatoid arthritis, Septic Arthritis.
Arthrocentesis, 195, 197
Arthroscopy, patellar tendinitis after, 212, 212f
Articular cartilage
    assessment of, 69–70, 69f–71f
    of knee, 203, 205, 204f–205f
    normal sonographic appearance of, 60f, 61
    technique of sonographic examination for, 59, 61, 60f–61f
    thinning of, 69, 69f
    traumatic lesions of, 61, 63, 62f
Artifacts
    comet-tail, 95, 100, 107
    from hairy skin, 103, 105f
    from metallic implants, 234
    ring-down, 23
Atrophy, spinal muscular, 16, 16t, 17f, 18

B

Baker's cyst
    in children, 61f
    differential diagnosis of, 55–56, 216–217
    dynamic study of, 44
    meniscal tears and, 206
    ruptured, 208, 210f
    sonographic appearance of, 27, 27f, 56f, 208, 209f–211f
Barlow maneuver, 180
Basal cell carcinoma, 94
Baseball finger, 165
Biceps femoris tendon, 213
Biceps tendon, normal ultrasound anatomy of, 114–115, 115f–116f
Bicipital groove, normal ultrasound anatomy of, 114–115, 115f, 116f
Biopsy, sonographically guided needle, of soft-tissue masses, 22, 96, 96f
Biopsy gun, 22
Blood vessels, foreign bodies in, 108, 108f–109f
B-mode equipment, for skin sonography, 85, 85F
Bone
    cysts, aneurysmal, 39f
    mineral density, quantifying, 59, 69–70, 71f
    normal sonographic appearance of, 60f
    osteomyelitis, 63, 65, 66f, 141–142
    traumatic lesions of, 39, 61, 62f, 63, 129–130, 131f, 141–142, 141f

Bone *(Continued)*
  tumors of, 36–39, 65, 67f, 172, 174f
Brachioradialis muscle, 137f
Bursa(e)
  of knee, 206–207, 207f
  subacromial-subdeltoid, 44, 116, 121-123, 124f, 126, 127f, 128
Bursitis, 26, 47
  lateral malleolar, 228, 231f
  olecranon, 51f
  prepatellar, 50f, 207, 207f
  trochanteric, 51f
Buttonhole deformity, 164–165

C

Calcifications
  in bursae, 103, 106f
  in knee, 203
  in lipomas, 30
  in malignant soft-tissue tumors, 35
  in soft-tissue granulomas, 105, 106f
  in tendinitis, 63, 65f, 131
  in traumatic myositis ossificans, 8f, 9
Capitellotrochlear groove, 135
Caput ulnae syndrome, 175
Carotid clamp, 107, 107f
Carpal tunnel syndrome, 158, 160, 159f–162f
Cartilage. *See* Articular cartilage, Hyaline cartilage
Cartilaginous abnormalities, of elbow, 142
Cavalryman's osteoma, 7
Cellulitis, 25, 25f, 95
Chondroma, synovial, 28
Clapper-in-the-bell sign, 4, 4f, 6
Clark classification system, for malignant melanoma, 92, 92t
Coccidioidomycosis, 106f
Common forearm extensor muscles, 137f
Common forearm extensor tendon, partial avulsion of, 145, 147f
Common forearm flexor muscles (CFFM), 139f
Common forearm flexor tendon, 138, 139f
Compressive neuropathies, of wrist, 158, 160–161, 159f–162f, 163
Congenital myopathies, 14
Contusion, muscular, 3, 4f
Coronoid process, 139f
Cruciate ligaments, 215, 217f
Crystal-induced arthropathy, 45, 46f
Cystic hematoma, 7

Cysts, 26–28, 27f–28f
  aneurysmal bone, 39f, 65
  Baker's. *See* Baker's cyst
  epidermoid, 169
  ganglion, 28, 28f
  hydatid, 25, 25f
  inclusion, 88–89, 105, 107f, 169
  meniscal, 28, 206, 206f, 218
  sebaceous, 88–89, 89f
  synovial, 27–28, 27f, 55–56, 56f, 218

D

De Quervain's disease, 152, 156, 156f
Dermatofibroma, 90, 90f
Dermatofibrosarcoma protuberans, 90
Dermatomyositis, 14, 14t, 16
Dermis, 86–87, 87f
Dermoepidermal junction, 87, 88f
Desmoid tumor, 32–33, 34f
Distal interphalangeal joint, traumatic injuries of, 165
Doppler sonography
  of elbow soft-tissue abnormalities, 146–147
  of inflammatory subcutaneous tissue conditions, 95
  of intramuscular blood flow, 2–3
  of malignant melanomas, 93
  of malignant soft-tissue tumors, 35
  for muscle trauma, 1
  of neurofibroma, 76, 77f
  of skin, 86
  for soft-tissue masses, 21–22, 35, 95, 96
  for synovial proliferations, 28
Duchenne-type muscular dystrophy, sonographic characteristics of, 13, 14t, 14f
Dupuytren's contracture, 29, 172–173, 175f, 234
Dystrophies
  congenital muscular, 14
  Duchenne-type muscular, 13, 14f, 14t
  Emery-Dreyfuss, 13, 14t
  Fukuyama-type congenital muscular, 14
  limb-girdle muscular, 13, 14t
  myotonic, 13
  progressive muscular, 13, 14f, 14t

E

Eczematous dermatitis, 90, 91f
Edema, of knee, 203
Elbow
  anatomy of, 135–136

bone abnormalities of, 141–142, 141f
cartilaginous abnormalities of, 142
congenital anomalies of, 146
fibrous-tissue injuries of, 142, 145, 146f–147f
joint abnormalities of, 142, 144f–145f, 145t
joint capsule, thickness of, 142, 145t
neoplasms of, 146, 148f
skeletal maturation of, 135
soft-tissue abnormalities of, 145–148
technique of sonographic examination, 136, 136f–137f, 138, 141
Emery-Dreyfuss dystrophy, sonographic characteristics of, 13, 14t
Epicondylitis
lateral, 142, 145, 146f
medial, 142, 145
Epidermis, 86–87, 87f–88f
Epidermoid cyst, 88–89, 105, 107f
of hand, 169
Epithelioid sarcoma, 33
Erythema nodusum, 25
Ewing sarcoma, 172, 174f
Extensor carpi radialis brevis tendon, 152
Extensor carpi radialis longus tendon, 152
Extensor digiti minimi tendon, 152
Extensor pollicis brevis tendon, 152

F

Fat necrosis, 25, 25f, 95
Fibromatosis
palmar. See Dupuytren's contracture
plantar, 234, 235f
Fibrosarcoma
of hand/wrist, 171–172, 173f
of soft tissue, 33
Fibrous histiocytoma, 90, 90f
Fibrous scar, of muscle, 7, 7f
Fibrous xanthoma, 90, 90f
Finger
baseball, 165
jersey, 165
trigger, 157, 157f
Fingernail, 88, 89f
Finkelstein's test, 156
Flexor carpi radialis tendinitis, 157–158, 158f
Flexor digitorum profundus tendon avulsion, 165
Flexor retinaculum, 152
Floppy infant syndrome with cerebral palsy, 18
Focal neuropathies, 16, 16t, 17f

Foot. See also Ankle
foreign bodies in, 103, 103f–104f, 105, 106f–107f
soft-tissue masses of, 232, 234, 234f–235f, 237
technique of sonographic examination, 221
Foreign bodies
in blood vessels, 108, 108f–109f
in esophagus, 108
in foot, 103, 103f–104f, 105, 106f–107f
in hand/wrist, 165, 166f
localization of, 108–109
in neck, 105, 107–108, 107f–108f
radiographic appearance of, 99
in subcutaneous tissues, 95, 95f
technique of sonographic examination, 99, 109–110
experimental studies, 99–100, 100f
in vivo, 100, 103, 101f–105f
Fractures, 61, 63
of elbow, 141–142, 141f
of humerus, 39, 62f, 129–130, 131f
of radius, 62f
of rib, 38f
Fukuyama-type congenital muscular dystrophy, 14

G

Gamekeeper's thumb, 163
Ganglion cyst, 28, 28f, 76, 77f
of foot/ankle, 234
of hand/wrist, 160, 162f, 165–167, 167f
of popliteal fossa, 218
Gardner syndrome, 32, 34f
Gastrocnemiosemimembranous bursa, enlarged See Baker's cyst
Gastrocnemius medialis muscle, detachment of, 6, 6f
Giant cell tumor, of tendon sheaths, 29, 167, 168f
Glass fragments, 100, 101f–102f
Glenohumeral joint effusion, 44
Glomus tumors, 33, 167–168, 169f
Glycogen storage diseases, 14
Granular cell tumor, 33, 35f
Granuloma, soft-tissue, 105
Guyon's canal, 153
syndrome, 160–161, 162f, 163

H

Hair follicles, 87–88, 87f
Hand
compressive neuropathies of, 158, 160–161, 159f–162f, 163
Dupuytren's contracture, 172–173, 175f

Hand *(Continued)*
    foreign bodies in, 165, 166f
    ganglia, 165–167, 167f
    normal ultrasound anatomy of, 152–154
    rheumatoid disease of, 173, 175, 176f, 177
    tendon diseases of, 154, 156–158, 155f–158f
    trauma of, 163–165, 164f
    tumors of, 167–172, 168f–172f
Hemangioma
    cutaneous, 89–90, 90f
    of hand, 171, 172f
    of soft tissues, 30
    subcutaneous, 95
    synovial, 54
Hemangiopericytoma, 33
Hemarthrosis, 142
Hematoma
    cystic, 7
    intramuscular, 3, 4f
        with partial muscle rupture, 5, 5f
        volume of, 6
    subcutaneous, 95, 105f
Hemophilia, pseudotumors in, 29
Hereditary sensorimotor neuropathies, 16
Hernia, muscular, 9, 9f
Hip sonography, in infants and children
    developmental dysplasia, 179–195
        anterior views, 192–193
        coronal-flexion view, 182, 184f–185f, 186–187
        coronal-neutral view, 182, 183f
        dynamic minimum standard examination, 191–192, 192t
        four-step examination technique, 181
        transverse-flexion view, 187, 188f–189f
        transverse-neutral view, 187, 191, 190f–192f
    joint effusion, 195
        indications for sonography, 197–198
        technique of sonographic examination, 195, 197, 196f–197f
Histiocytoma
    fibrous, 90, 90f
    malignant fibrous, 33, 35, 35f
Humerus
    capitellum of, 135
    fractures of, 39f, 129–130, 131f
Hyaline cartilage, 60, 60f, 203, 204f, 205f
Hydatid cyst, 25, 25f

I

Ilizarov technique, 66–68, 68f
Inclusion cyst, 88–89, 105, 107f, 169
Infant, hip of. *See* Hip sonography
Infraspinatus tendon, normal ultrasound anatomy of, 118, 121f
Injection injuries, of peripheral nerves, 79–80, 81f
Intramuscular myxomas, 32, 33f

J

Jersey finger, 165
Joint
    effusions, 28, 44–47
        of ankle, 232, 233f
        of elbow, 142
        of glenohumeral joint, 44
        of hip, 195–198
        of knee, 201, 202f, 203
    traumatic injuries of
        distal interphalangeal, 165
        metacarpophalangeal, 163–164, 164f
        proximal interplanangeal, 164
Jumper's knee, 211–212, 211f–212f
Juvenile rheumatoid arthritis, pannus formation in, 142, 145f

K

Kaposi sarcoma, 94
Knee
    articular cartilage of, 203, 205, 204f–205f
    bursae, 206–207, 207f
    cruciate ligaments of, 215, 217f
    joint effusions of, 201, 202f, 203
    jumper's, 211–212, 211f–212f
    lateral collateral ligament of, 215, 216f
    medial collateral ligament of, 213–215, 213f–214f
    menisci, 205–206, 205f–206f
    popliteal fossa masses of, 216–218
    synovial abnormalities of, 203, 204f

L

Lateral collateral ligament, of knee, 215, 216f
Leiomyosarcoma, 33
Ligament(s)

cruciate, 215, 217f
lateral collateral, of knee, 215, 216f
medial collateral, of knee, 213–215, 213f–214f
radial collateral
  of elbow, 135
  of finger, 163
ulnar collateral
  of elbow, 135
  of finger, 163
Limb-girdle muscular dystrophy, 13, 14t
Lipoma arborescens, synovial, 54
Lipoma
  and carpal tunnel syndrome, 158, 160f–161f
  of hand/wrist, 169, 171f
  of soft tissues, 30, 30f
  subcutaneous, 95, 96f
Liposarcoma, 33
Lymphadenitis, epitrochlear, 145–146, 148f
Lymphangioma, 90, 148f
Lymphedema, subcutaneous, 96f, 97,

M

Malignant fibrous histiocytoma, 33, 35, 35f
Medial collateral ligament, of knee, 213–215, 213f–214f
Median nerve compression, in carpal tunnel, 158, 160, 159f–162f
Melanoma, malignant, 92–93, 92t, 93f
Meniscal cyst, 28, 206, 206f, 218
Menisci, of knee, 205–206, 205f–206f
Metabolic myopathies, 14
Metacarpophalangeal joint, 153, 163–164, 164f
Metallic objects, 100, 103f–104f
Metastasis
  to bone, 65, 67f
  to skin, 94, 94f
  to soft tissues, 36, 38f
  to subcutaneous tissues, 96
Morton's neuroma, 78, 105, 234, 236f, 237
Muscle(s)
  abscess of, 23–25
  brachioradialis, 137f
  common forearm extensor, 137f
  common forearm flexor, 139f
  gastrocnemius medialis, detachment of, 6, 6f
  needle biopsy of, 22–23
  normal sonographic appearance of, 2–3, 3f, 12, 12f
  trauma-related injuries, 9–10

acute lesions, 3–6, 4f–7f
chronic lesions, 7, 9, 7f–9f
contusion, 3, 4f
cystic hematoma, 7
fibrous scar, 7, 7f
hernia, 9, 9f
mechanisms of, 2
myositis ossificans, 7, 8f, 9
rhabdomyolysis, 6, 7f
rupture, 3–6, 4f–6f
sites of, 1
strain, 3
technical considerations for, 1
tumors of, 29–36
Muscular dystrophy. See Dystrophies
Mycosis fungoides, 94, 94f
Myopathies, 12–16
  congenital, 14
  inflammatory, 14, 14t, 15f, 16
  metabolic, 14
  muscular dystrophies, 13–14, 14f, 14t
Myositis ossificans
  nontraumatic, 23, 24f
  traumatic, 7, 8f, 9
Myxoma, intramuscular, 32, 33f

N

Neck, foreign bodies in, 105, 107–108, 107f–108f
Nerve
  graft, 82f, 83
  median, compression of, in carpal tunnel, 158, 160, 159f–162f
  peripheral. See Peripheral nerves
  reconstruction surgery, 82f, 83
  sciatic. See Sciatic nerve
Nerve sheath tumors, 31–33, 32f, 74, 76, 77f
  of hand/wrist, 169, 170f
  of popliteal fossa, 218
Neurilemmitis, 80
Neurofibromas. See Nerve sheath tumors
Neuroma
  Morton's, 78, 105, 234, 236f, 237
  traumatic, 78, 79f
Neuropathies, 16–18
  compressive, of wrist, 158, 160–161, 159f–162f, 163
  sonographic findings in, 13t
Nevi, 89, 89f
Nodules, rheumatoid, 175

O

Ortolani maneuver, 180
Ossification centers, 59, 61f, 180
Osteoarthritis, 45, 46f, 203
Osteochondroma, 37, 65
Osteochondromatosis, 27, 54, 55f
Osteoma, 7
Osteomyelitis, 63, 65, 66f, 141–142, 145
Osteosarcoma, 39f

P

Palmar fibromatosis. See Dupuytren's contracture
Panniculitis, 25, 95
Pannus
   in juvenile rheumatoid arthritis, 142, 145f
   in knee, 203, 204f
   in rheumatoid tenosynovitis, 26, 27f
   in tenosynovitis, 47, 49
Parakeratosis, 90
Parasitic infection, 25, 25f
Patellar tendon
   normal ultrasound anatomy of, 208, 211
   tears of, 211
   tendinitis of, 211–212, 211f–212f
Pavlik harness, 194
Peripheral nerve(s), 73
   compression injuries to, 78–79, 80f
   inflammatory changes of, 80, 82f, 83
   injection injuries to, 79–80, 81f
   normal anatomy and sonographic appearance of, 74, 75f–76f
   tumors of, 74, 76, 77f
Peroneal retinaculum, superior, rupture of, 228
Peroneal tendons
   chronic instability of, 228, 229f
   degeneration of, 228, 230f
   tears of, 225, 227–228, 225f–226f
Perthes disease, 195, 197, 198
Pes anserinus tendon, 213
Peyronie's disease, 173
Phalen's wrist flexion test, 160
Pheboliths, 31, 95
Pigmented villonodular synovitis (PVNS), 28, 29f, 51–52, 53f, 54
Plantar fibromatosis, 234, 235f
Polymyositis, 14, 14t, 15f, 16
Popliteal artery, aneurysms of, 29, 217
Popliteal cyst. See Baker's cyst

Popliteal fossa masses, 216–218
Posterior tibial tendon, tears of, 223, 225, 227–228, 225f–228f, 232
Prader-Willi syndrome, 18
Progressive muscular dystrophies, sonographic characteristics of, 13, 14f, 14t
Proximal interphalangeal joint, traumatic injury of, 164
Psoriasis, 92
Pyomyositis, 24–25
Pyrophosphate arthropathy, 45, 46f

Q

Quadriceps tendon
   normal ultrasound anatomy, 212
   tears of, 212–213, 213f

R

Radial collateral ligament
   of elbow, 135
   of finger, 163
Radial head, 135
   fracture of, 141–142, 141f
Radiohumeral articulation, 136, 136f–137f
Radioulnar articulation, 136, 138f
Retinaculum
   extensor, 152
   flexor, 152
   peroneal, superior, rupture of, 228
Rhabdomyolysis, 6, 7f
Rhabdomyosarcoma, 33
Rheumatoid arthritis
   Baker's cyst and, 56, 208
   of elbow, 142, 145f
   of hand/wrist, 162f, 173, 175, 176f, 177,
   of knee, 201, 203, 204f
   nodules, 29, 49
   of shoulder, 48f, 49, 49f
Rotator cuff, 113–133
   arthropathy, 129, 130f
   normal ultrasound anatomy
      of biceps tendon, 114–115, 115f–116f
      of bicipital groove, 114–115, 115f–116f
      of infraspinatus tendon, 118, 121f
      of subscapularis tendon, 115–116, 117f
      of supraspinatus tendon, 118, 119f–120f
      of teres minor tendon, 121

sonography of
  instrumentation for, 113–114
  pitfalls in, 130–131
  postoperative appearances, 126–129, 128f–130f
  tears, 121–129
  technique of examination, 114

S

Sarcoma, soft-tissue, 33–36
  biopsy of, 35, 37f
  color Doppler of, 35
  local recurrence of, 35–36, 37f
    magnetic resonance imaging of, 36, 37f
  sonographic appearances of, 33, 35, 35f
  subtypes of, 33
  treatment response, evaluation of, 35, 36f
Scar, fibrous, of muscle, 7, 7f
Schwannomas. See Nerve sheath tumors
Sciatic nerve
  displaced by aneurysm, 80f
  graft, 82f
  injection injury of, 79–80, 81f
  normal, 76f
  traumatic neuroma of, 79f
Sebaceous cyst, 88–89, 89f
Seborrheic keratosis, 90, 91f
Septic arthritis
  of elbow, 142, 144f
  of hip, 45f, 195, 198
Shoulder. See also Rotator cuff
  effusion, 44, 46f
  trauma, 129–130
Skin, 85–94
  color Doppler imaging of, 86, 91, 93, 94
  inflammation, 90–91
  normal ultrasound anatomy of, 86–88, 87f, 88f
  pathologic conditions of. See also specific pathologic conditions
    benign, 88–92, 89f–91f
    malignant, 92–94, 93f–94f
  technique of sonographic examination, 85–86, 86f
Snuffbox, anatomic, 152
Soft-tissue masses, 21–42
Spinal muscular atrophy, 16, 16t, 17f, 18
Squamous cell carcinoma, 94
Standoff pads, 1, 21, 22, 85, 109, 114, 151
Stener's lesion, 163
Stenosing tenosynovitis, of hand/wrist, 156–157, 156f

Strain, muscular, 3
Stress testing, during ankle sonography, 222, 222f
Subacromial-subdeltoid bursa, 44, 116, 121, 122, 123, 124f, 126, 127f, 128
Subcutaneous tissues
  abscess in, 95
  air bubbles in, 103, 105f
  benign tumors of, 95, 96f
  cellulitis in, 25, 25f, 95
  color Doppler, imaging, 25, 86, 95–97
  erythema nodosum, 25
  fat thickness, measurement of, 96–97
  foreign bodies, 95, 99–111, 165
  hematoma in, 95
  malignant tumors of, 96, 96f
  normal ultrasound anatomy of, 86–87, 87f
  panniculitis, 25, 95
  pathologic conditions of, 94–97, 95f–96f
  technique of sonographic examination 85–86, 97
  trauma to, 95, 95f
Subscapularis tendon. See Rotator cuff
Supraspinatus tendon. See Rotator cuff
Synovial cyst, 27–28, 27f, 55–56, 56f, 213
Synovial diseases, 43–57
  masses, 28, 29f, 51–55
  membrane, 43
  sarcoma, 33
  synovitis, 44–47, 48f–49f, 142
    pigmented villonodular, 28, 29f, 51–52, 53f, 54, 203
    rheumatoid, 175, 176f, 203

T

Tendinitis, 25–26, 26f
  of Achilles tendon, 223, 223f
  of flexor carpi radialis tendon, 157–158, 158f
  lateral epicondylitis, 142, 145, 146f
  medial epicondylitis, 142, 145
  of patellar tendon, 211–212
Tendon(s)
  abductor pollicis longus, 152
  Achilles
    rupture of, 222–223
    tendinitis of, 223, 223f–224f
  biceps femoris, 213
  biceps, normal ultrasound anatomy of, 114–115, 115f–116f
  common forearm extensor, partial avulsion of 145, 147f

Tendon(s) *(Continued)*
  common forearm flexor, 138, 139f
  extensor carpi radialis brevis, 152
  extensor carpi radialis longus, 152
  extensor digiti minimi, 152
  extensor pollicis brevis, 152
  flexor carpi radialis, tendinitis of, 157–158, 158f
  flexor digitorum profundus, avulsion of, 165
  infraspinatus, normal ultrasound anatomy of, 118, 121f
  patellar
    normal ultrasound anatomy of, 208, 211
    tears of, 211
    tendinitis of, 211–212, 211f–212f
  peroneal
    chronic instability of, 228, 229f
    degeneration of, 228, 230f
    tears of, 225, 225f–226f, 227–228
  pes anserinus, 213
  posterior tibial, tears of, 223, 225, 227–228, 225f–228f, 232
  quadriceps
    normal ultrasound anatomy of, 212
    tears of, 212–213, 213f
  subscapularis, 115–116, 117f
  supraspinatus, 118, 119f–120f
  teres minor, 121
Tenosynovitis, 26, 26f–27f, 47, 49, 51, 52f
  in hand/wrist, 154, 156, 155f, 175, 176f
  of peroneal tendons, 228
  septic, 49, 51
  stenosing, 156–157, 156f
Thumb, traumatic injuries of, 163, 164f
Tinel's sign, 160
Trauma
  to hand/wrist, 163–165, 164f
  muscular, 1–10
  to subcutaneous tissues, 95, 95f
Trigger finger, 157, 157f
Tumor(s). *See also specific tumors*
  benign soft-tissue, 30–33, 30f–34f, 95, 96f
    angioma, 30–31, 31f
    desmoid, 32–33, 34f
    giant cell, of tendon sheath, 29, 167, 168f

    glomus, 33
    granular cell, 33, 35f
    of hand/wrist, 167–169, 171, 168f–172f
    intramuscular myxoma, 32, 33f
    lipoma, 30, 30f
    lymphangioma, 33
    nerve sheath tumors, 31–32, 32f, 76–78
  bone, 37, 39f, 65, 67f, 172, 174f
  malignant soft-tissue, 33, 35–36, 35f–37f, 96, 96f
    of hand/wrist, 171–172, 173f
    metastases, 36, 37f
  sarcomas, 33, 35–36, 35f–37f

U

Ulnar collateral ligament
  of elbow, 135
  of finger, 163
Ulnar tunnel, 153
  syndrome, 160–161, 162f, 163

V

Volar plate injuries, 164, 165

W

Wood splinters, 100, 101f, 103f
Wrist
  compressive neuropathies of, 158–163
  ganglia, 165–167, 167f
  normal ultrasound anatomy, 152–154
  rheumatoid disease of, 173, 175, 176f, 177
  tendon diseases of, 154, 156–158, 155f–158f
  tumors of, 169, 170f, 171–172, 174f

X

Xanthoma, 29
  fibrous, 90, 90f

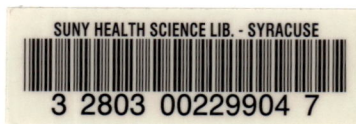

WE 141 M9859 1995
Musculoskeletal
ultrasound